DANCE AND SCIENCE IN THE LONG NINETEENTH CENTURY

UNIVERSITY PRESS OF FLORIDA

Florida A&M University, Tallahassee
Florida Atlantic University, Boca Raton
Florida Gulf Coast University, Ft. Myers
Florida International University, Miami
Florida State University, Tallahassee
New College of Florida, Sarasota
University of Central Florida, Orlando
University of Florida, Gainesville
University of North Florida, Jacksonville
University of South Florida, Tampa
University of West Florida, Pensacola

DANCE AND SCIENCE IN THE LONG NINETEENTH CENTURY

THE ARTICULATE BODY

Edited by Lynn Matluck Brooks,
Sariel Golomb, and Garth Grimball

UNIVERSITY PRESS OF FLORIDA

Gainesville/Tallahassee/Tampa/Boca Raton
Pensacola/Orlando/Miami/Jacksonville/Ft. Myers/Sarasota

This book will be made open access within three years of publication thanks to Path to Open, a program developed in partnership between JSTOR, the American Council of learned Societies (ACLS), University of Michigan Press, and The University of North Carolina Press to bring about equitable access and impact for the entire scholarly community, including authors, researchers, libraries, and university presses around the world. Learn more at https://about.jstor.org/path-to-open/

Cover: Plate 191 from *Animal Locomotion* by Eadweard Muybridge, 1887. From the Eadweard Muybridge Collection, 1870–1981. University of Pennsylvania Archives & Record Center (UPT 50 M993: OS 3, Folder 5).

Published in the United States of America

First cloth printing, 2025
First paperback printing, 2026

31 30 29 28 27 26 6 5 4 3 2 1

Library of Congress Cataloging-in-Publication Data
Names: Brooks, Lynn Matluck, editor. | Golomb, Sariel, editor. | Grimball, Garth, editor.
Title: Dance and science in the long nineteenth century : the articulate body / Lynn Matluck Brooks, Sariel Golomb, Garth Grimball.
Description: 1. | Gainesville : University Press of Florida, 2025. | Includes bibliographical references and index.
Identifiers: LCCN 2024025334 (print) | LCCN 2024025335 (ebook) | ISBN 9780813079264 (hardback) | ISBN 9780813081816 (paperback) | ISBN 9780813070889 (pdf) | ISBN 9780813073538 (ebook)
Subjects: LCSH: Dance—History. | Human body. | BISAC: PERFORMING ARTS / Dance / History & Criticism | HISTORY / Modern / 19th Century
Classification: LCC GV1595 .D27 2025 (print) | LCC GV1595 (ebook) | DDC 792.809—dc23/eng/20240814
LC record available at https://lccn.loc.gov/2024025334
LC ebook record available at https://lccn.loc.gov/2024025335

The University Press of Florida is the scholarly publishing agency for the State University System of Florida, comprising Florida A&M University, Florida Atlantic University, Florida Gulf Coast University, Florida International University, Florida State University, New College of Florida, University of Central Florida, University of Florida, University of North Florida, University of South Florida, and University of West Florida.

University Press of Florida
PO Box 140239
Gainesville, FL 32614
floridapress.org

GPSR EU Authorized Representative: Mare Nostrum Group B.V, Doelen 72, 4831 GR Breda, The Netherlands, gpsr@mare-nostrum.co.uk

CONTENTS

FIGURES

ACKNOWLEDGMENTS

This book would not have happened without the encouragement and advice of Mindy Aloff, and the early support of Cara Gargano and Olivia Sabee. We thank Stephanye Hunter, our editor at University Press of Florida, for her unflagging support and patience as this project came to fruition.

Introduction

Meeting Points, Overlaps, Escapes

Lynn Matluck Brooks, in conversation with Sariel Golomb and Garth Grimball

What links dance and the scientific disciplines? An initial response might focus on the human body, which serves as the medium of expression for dance and as the subject of scientific—particularly medical, anatomical, and kinesiological—inquiry. Connections between dance and science stretch back centuries to inquiries about the nature of bodies, society, and cosmos. *Dance and Science in the Long Nineteenth Century: The Articulate Body* reveals interplays between dance and a range of sciences as understood in the run-up to the nineteenth century, during that century, and in the turn to the twentieth; "long" in our title acknowledges the blurriness of time-period markers, and also earlier signs of cultural shift and endurance of practices and beliefs.

Why spotlight dance and science connections in this period? The nineteenth century was a time of marked transformation in dance performance, in tandem with forces of colonial expansion, industrialization, and urbanization that profoundly shaped cultures, knowledge systems, and political landscapes worldwide. The emergence of romanticism in Europe, appearance of close-embrace couple dances like the waltz throughout Europe and the Americas, circulation of "national" dances in theaters and ballrooms, and tensions between colonial and indigenous dancers and dances in imperial territories characterized this period. Change was also afoot in the scientific world: in European American research and practice, this period built on earlier developments to establish modern disciplines in general and to differentiate individual scientific fields.[1] Western science sought to consolidate its ascendancy over native practices and religious worldviews. Some emerging scientific fields focused on the human body and on those understood, by researchers engaged in those studies, as human "others." In some cases, such research fed the categorization and essentialization of human beings, serving as grounds for treatment of individuals or groups. In this context, dance might be understood as forming, controlling, liberating, or civilizing the body; as a threat to such control and categorization; or as a manifestation of bodily variation, failure, power, freedom, or excess. In some cases, the very groups targeted turned that research to their own purposes, undercutting arguments of European and American scientists overtly or covertly and developing alternative visions and practices in dance and the sciences of the body. Articulating the body to bring out its exceptional capacities, its likenesses to or differences from other bodies, and its expressivity became a critical project of the nineteenth century socially, politically, medically, and artistically.

Technology, too, influenced both theatrical and scientific developments. For example, optics had, from the seventeenth century onward, shaped analytical tools and introduced viewers to the focused gaze that both the microscope and the telescope offered; these tools became more widely available in the nineteenth century as amateur scientific societies proliferated in European, American, and colonial cities. Photography, a developing art and science throughout the nineteenth century, became subject and source of widespread interest, particularly as applied to understanding of animal and human motion. Eadweard Muybridge's remarkable *Animal Locomotion* series, an example of which graces our book cover, places dancing within the scope of technological research promoted by both educational and artistic institutions.[2] These and other emerging technologies allowed more precise measurement of distances, time, weather, and the organization of bodies themselves. Bodies whole or in parts, visible or invisible to the naked eye, now became articulately accessible to a wide swath of viewers through new instruments, means of illustration (including daguerreotypes and photographs), and dissemination techniques (including railroads) previously unavailable. Theatrical technologies contributed to innovative staging, such as the mysterious and magical effects enhanced by gas lighting.[3] Machinery also affected the ways the human body was shaped by industrial, mechanical labor as well as by refinements of dance pedagogy and practice.

This volume primarily attends to Western contexts or moments of interface between West and non-West; however, these essays also trouble "science" as—at least in part—an epistemological apparatus that introduces the "empirical" and "factual" as values wielded in the form of power. Some sciences presented here attempted to establish the rightness of researchers' endeavors, however violent, and to instill doubt in whatever stood as counterpoint—"belief," "religion," "culture," or "tradition"—making science in the long nineteenth century a colonial project and, subsequently, a world-altering phenomenon. Several publications address this and other intersections of theatrical performance and science, a list that the current volume expands. *The Cambridge Companion to Theatre and Science* (2020) offers perspectives on historical and contemporary exchanges among acting, performance, experimentation, corporeality, health, astronomy, meteorology, and ecology. Zooming into one historical moment and subject, Jane R. Goodall's *Performance and Evolution in the Age of Darwin: Out of the Natural Order* (2002) focuses on nineteenth-century "missing links" and "monkey-man" portrayals in museums and theaters. The ways that American Black thinkers, artists, and performers turned scientific views on their heads is revealed in Britt Rusert's *Fugitive Science: Empiricism and Freedom in Early African American Culture* (2017). *Performance and the Medical Body* (2016), coedited by Alex Mermikides and Gianna Bouchard, brings together a range of scholarly voices on the "biologization of theatre" and contemporary performance practices engaging the biomedical body. Felicia McCarren's *Dance Pathologies: Performance, Poetics, Narrative* (1998) explores interplays of madness, illness, dance, and their literary contexts in nineteenth-century Europe, while Kélina Gotman's *Choreomania: Dance and Disorder* (2018) follows the co-constitution of madness and movement in Europe alongside colonialism, exoticism, and medicine. The sciences of physics and chemistry inform Liz Heinecke's *Radiant: The Scientist, the Dancer, and a Friendship Forged in Light* (2021), which reveals how belle-époque scientific developments connected with performance technologies as practiced by Loïe Fuller. *Physics and Dance*, by dancer Emily Coates and physicist Sarah Demers (2018), investigates how these respective disciplines can enhance understanding of each field and of both everyday and cosmic questions. The essays in *Dance and Science in the Long Nineteenth Century: The Articulate Body* look at dance in these and other connections, in contexts from the social to the ritual, the theatrical to the educational.

At the core of this dancing, as noted above, is the body that dances, the body that draws attention *by dancing* to its physicality—shape, size, gender, skin tone, physiognomy, agility, movement patterns, social constructions, freedom, control, and other perceptible or implicit factors. In earlier centuries, Western natural philosophers and physicians undertook deep inquiry into the human body, a curiosity shared by painters and sculptors who sought knowledge of the physicality they depicted. This shared medical–artistic curiosity about bodily innards and workings led European and American scientific and artistic communities toward techniques of direct bodily observation: contested practices arose, including dissection of dead bodies (and the criminal acts often undertaken to procure them), human zoos displaying living people as odd or exotic physically, and museums where preserved examples of curious bodies were viewable. Such observational practices reached into the depths of the body while also reaching out for the disenfranchised, colonized, and powerless whose bodies served medical and visual-art dissection and display.[4] In Western sciences, this empirical turn connected to new understandings of the embodied nature of scientific practice, dependent not only on observation, but also on touch, material discovery, and invention of physical tools and machines.[5] Cultures less dualistic than the European West had engaged dance and aspects of what is now called science for millennia: shamans' healing powers might be sourced and expressed through dance movement;[6] cosmic motion might be expounded through human choreography;[7] spiritual worship and explanations of the material world might be expressed in dance.[8]

My (Lynn Brooks's) interest in the dance–science *pas de deux* arose from my research on antebellum theatrical dancing in Philadelphia, Pennsylvania; as playbills for the period 1820 to 1860 turned up dances purported to represent, or spring from, people of different national and racial types, I asked how audiences viewing such dances understood the bodies of those they saw "delineated"—a term then used for theatrical characterization—in performances. Beyond the stage, ballroom versions of perceived "national" or exotic dances—the "cracovienne," an "Arab Dance," "Ethiopian cotillions"—allowed social dancers to try movement patterns that created fads and fodder for dancing masters' teaching.[9] My question led me to medical and "ethnological" research burgeoning in that period, often promulgated by scientists connected to the same city, Philadelphia, that hosted the dances I investigated. These men of science joined an international cadre of researchers contributing to rankings and claims about human beings and their bodies. Their treatises and ideas were accessible not only to learned readers, but also to the literate public, eager to understand the ideas exchanged in the many scientific societies throughout the United States. Newspaper articles and pamphlets created accessible reportage on these findings, as did popular lectures on scientific subjects, which vied with theatrical performances as entertainment choices. Bodies of just the sort American scientists analyzed were on vibrant display, dancing on city stages and streets, where they were observed by the same audiences who read works asserting, as scientific fact, the rankings then determining US social order. The young nation's deepening turmoil, largely stirred up by the very questions that these scientific arguments sought to resolve, made such concerns incandescent.

What was understood by the term "science" in this period? As a discourse that promised to unlock truths linking all of nature, beyond distinctions of time and place, "science" appeared to many in the West as a methodology yielding powerful tools that could be deployed for ameliorating ills like disease and starvation, for dominating populations perceived as weak or barbaric, for penetrating the mysteries of the seas and stars, and, broadly, for subduing nature to serve human (or *some* human beings') economic, political, and social objectives. It was in 1834 that the term "scientist" emerged in English vocabulary, edging out prior descriptors like "naturalist," "natural historian," or "natural philosopher." Natural philosophy saw galloping developments in biology, medicine, entomology, psychology, marine studies, geology, physics, chemistry, astronomy, and engineering.[10] Branches of science and medicine were establishing professional boundaries and criteria that increasingly closed out the amateur, the citizen observer, and the apprentice—roles active early in the nineteenth century, and long maintained in folk practices as well as public performances of magic, mesmerism, and phrenology.

Observing and ordering all aspects of nature was key to this period's science, which focused on differences, rather than similarities, among subjects of study. Defending this classificatory drive, Philadelphia ethnologist Samuel George Morton, a leading figure in what has been termed "scientific racism," asserted in 1839 that "Natural Science . . . divested of arrangement, presents an uninviting chaos."[11] A century earlier, Swedish botanist and zoologist Carl Linnaeus had published his widely influential *Systema Naturae*.[12] The power of such systems appeared to sweep aside mythologies, religious beliefs, and social customs, challenging cultural assumptions and nearly destroying whole populations. The term "system" was, and is, applied to widely different phenomena—the solar system, the circulatory system, the oceanic system, the optic system, the periodic system, the nervous system, legal systems, educational systems, and more. In US theaters in the nineteenth century it applied to the "star system," whereby the most popular performers were featured and toured in established circuits—a practice financially and artistically detrimental, critics argued, to theater managers and local performers.[13] American dancing masters too advertised their "systems" of teaching based on "laws" and "principles"—essential foundations, they claimed, for effective education in dance and deportment critical to social rank.[14]

Stage dancing in the nineteenth-century European American sphere embraced the systematized art of ballet, which led to understandings of such dancing as "scientific"—a form of praise applied to individual dancers whose training and bodily mastery was thus highlighted.[15] The "ballet girl" herself became a taxonomical subject in a darkly humorous analysis of the mid-nineteenth century: Albert Smith's *Natural History of the Ballet-Girl* acknowledged "the unceasing labours of Cuvier, Linnaeus, Buffon, Shaw, and other animal-fanciers," who "had surmounted the apparently impossible task of marshalling all the earth's living curiosities into literary rank and file."[16] These taxonomers inspired Smith's putting "ballet girls," along with butterflies and fairies, under his literary microscope.

* * *

The essays in this volume grapple with interfaces between dance and science that span the long nineteenth century. Each contributor's research chapter calls up a response by another scholar, reflecting on ways that that chapter's themes open further pathways—perhaps developments older or recent, or in adjacent fields.

The book's first section explores modes of classification highlighting visuality and the meanings that spectating staged bodies brought to public awareness. Opening just before the turn to the nineteenth century, Chapter 1 brings to the fore then-current anatomical fascinations with the body, particularly the female body, in the inviting engagement with technology, medicine, display, and spectatorship of a wax model. Sariel Golomb's "Venus in Pieces: Choreographing the Anatomical Body with Clemente Susini's *Venus de' Medici*" explores the embodied nature of doing science, of representing findings, and of observing scientific subjects. The bare human body, even molded of wax or otherwise modeled, carries the weight of sexuality, gender, class, culture, and race, while stimulating imaginative possibilities of the life such a body might experience. Responses engage the viewer physically, emotionally, or socially, beyond the goal of objective knowledge-acquisition that was, by the nineteenth century, a desideratum of science. Reporting the history of such scientific investigation has been understood, by Lorraine Daston and Peter Galison, as "an account of kinds of sight,"[17] pointing to the sensual underpinnings of science acknowledged in Golomb's essay, which takes us choreographically through viewers' sensual encounters with Susini's enticing model. Science's rootedness in the senses stimulated "Choreographies of Knowledge: Touch and Vision in Anatomical Looking," Jane Desmond's response to "Venus in Pieces." Discussing "the confluence of 'art' and 'science,'" Desmond highlights the contemporary example of Gunther von Hagens's "Body Worlds" displays, popular with museumgoers worldwide, and featuring actual human bodies preserved by a patented technology. For different, culture-specific reasons, both the Anatomical Venus and the "Body Worlds" specimens were made inaccessible to visitors' touch, but their visual excitement remains, stimulating viewers' imaginative engagement of varied valences.

Choreographing our sensual encounter with this wax Venus, Golomb notes the figure's pearl necklace against its waxen flesh. This brought immediately to my mind Marie Taglioni's famously pearl-encircled neck, conjuring the atmosphere of early nineteenth-century ballet. Ballerinas like Taglioni were figures of obsession for some, as well as objects of public speculation and subjects of the burgeoning ballet print of the period—a kind of "pin-up" for the privacy of home, but also posted publicly in bookstores and barber shops.[18] The very real flesh-and-blood ballerina could transform herself on stage into ethereal figures—a sprite, a ghost, a fairy. Steven Ha's "Science under the Surface: Victorian Science in the Ballet *Ondine*" explores interplays of dance, marine science, stage technology, and audience engagement in a ballet taking as subject the water sprite Ondine. "Captivating and deadly," this naiad embodied fascinations with the monstrous minutiae then newly observed through technologies like the microscope, revealing a space once thought empty (water) to be rich in life, threats, and movement. Through Jules Perrot's choreography, *Ondine* allowed imaginative entry to this new world, enriching ballet repertory with novel movement, scenic effects, characterizations, and enduring fascination evident in later choreographic approaches to this theme. Ha brings Frederick Ashton's mid-twentieth-century work into conversation with the nineteenth-century treatment of the water sprite.

Like the watery world, botanical studies engaged scientists, planters, and public.[19] Flowers were understood as graceful, feminine, alluring life-forms—pairing perfectly with the image of the lithe, elegant, sometimes pluckable, and often short-lived ballet dancer. As Alexander Schwan reveals in "Phytology and Dance: The Impact of Plant Biology on Nineteenth-Century Flower Ballets," botanical study was in part classified, from Linnaeus onward, in light of then-current prescripts regarding human sexuality and romantic longings. Such meanings could enhance or suggest significance in ballet scenarios through floriographic "codes," understood by many, yet changing over time, locations, and textual descriptions. Choreographers including Henri Justament, Paul Taglioni, Marius Petipa, and Lev Ivanov employed such codes in works like the iconic "Waltz of the Flowers" that endure today, their original codes lost to current viewers. Exploring the sources of such works suggests possible inspirations, prejudices, and intentions of artists grappling with the knowledge, constructs, and curiosities of their times.

In responding to both "Science under the Surface" and "Phytology and Dance," Whitney Laemmli's essay, "New Sensations," highlights the popular grip of nineteenth-century scientific revelations on European publics of all classes. But, Laemmli notes, scholarly analysis has been slow to appreciate the role of *dance* in exploring and explicating the curiosities of the age's science. Both Ha and Schwan take on "the methodological hurdle of imagining how Victorian audiences saw" ballets like *Ondine* and *Thea* in light of the findings scientists shared through popular media. Laemmli also notes that both dance works and the mysteries of science these works grappled with were "entwined with private life, sensory experience, spectacle, and the supernatural." In an inverse example, Laemmli points to the performance of science as dance, current to our own age, in the sometimes charming and occasionally transgressive "Dance Your PhD" competition that draws submissions from budding scientists worldwide. These comments also point to our volume's "Outro," Emily Coates's "Observing the Observers," discussed below.

Like the marine world and plant biology, animal studies occupied nineteenth-century naturalists, organizing taxonomies of creatures encountered in their global reach. In these ranked systems, human beings stood atop the animal kingdom for their intelligence and capacity to understand and even change nature, as the genus and species name *homo sapiens* suggests. The category of human beings was further parsed as naturalists distinguished "varieties" broadly based on what is now known as "race," determined by nineteenth-century investigators' linking of geography, skin tone, hair type, gender, and other observable characteristics.[20] Euro-American naturalists ranked their own variety—the Caucasian (often specifically the Anglo-Saxon or Teutonic)—at humanity's pinnacle. Part II of this volume brings dance together with ideologies of nationalism, race, sex, and knowledge systems. Creating a bridge to this theme from the balletic focuses of Ha's and Schwan's essays in part I, we open this second section with Elizabeth Claire's chapter on developments shaping the Paris Opera's ballet training and troupe under Dr. Louis Véron's direction. A trained physician and man of wide-ranging intellect and connections, Véron's short tenure at the Paris Opera has loomed large as an influence on the history and the historiography of ballet, with the Opera as one of—if not *the*—center(s) of balletic emergence. Claire investigates connections among Véron's medical thinking, concepts of public hygiene, and French theatrical practices shaping understandings of dancers, their bodies, and their social-psychological influences on viewers. Responding to Claire's essay, Olivia Sabee turns to popular literature and theater, including *physiologies* and *vaudeville,* to explore the role that medical and public-health perceptions played in imaginaries of theatrical dance and dancers. Along with taxonomies of plants, animals, and humans, the ballet girl—the *rat de l'opéra*—became a classificatory subject of "scientific" study.

The French also shared the intense interest, displayed by Western scientists positioned within the period's colonial power structures, in peoples whose lands were claimed by European rule. This included the "Indostanic Family" of humankind, such as the "Hindoos"; the "Americans," referring to people native to what are now the Americas; and those belonging to the "Negro Family," arising in Africa and enslaved in Europe and the Americas. In the case of

India, as Pallabi Chakravorty explores in "The Paradox of the 'Subtle Body': Dance, Tantra, and Science," traditional practices of health and healing came into contest as well as dialogue with Western empirical systems brought by British rule to India. That encounter transformed some Indian yogic practice, which also came to influence Westerners seeking to import and adapt Eastern spiritual practices. Valuing the "subtle body" in this transformed yogic philosophy elevated certain classes of Indian practitioners while delegitimizing others; the mystic, refined body of this new vision appealed to European and American dance practitioners and to Western spiritual seekers, influencing early modern dance and other artistic forms in both India and the West, while stripping traditional practices of their powerful sensuality. We return to the French scientific gaze with Tiziana Leucci's response essay, "Viewing Indian Dance across Time and Space," demonstrating how Western fascination and condemnation fed one another over the course of centuries. Leucci identifies encounters reported by eighteenth-century French naturalists and missionaries with yogis, Sufis, and female dancers, emphasizing the latter's perceived fanaticism and moral failures—qualities that upper-class Indians reforming yogic practices altered or expunged at the expense of long-standing spiritual and dance conventions.

In the US, those practicing science had proximate access to significant representation of three of the racial groupings noted above—the "Caucasian," the "American," and the "Negro." In his influential book *Notes on the State of Virginia* (1785), Thomas Jefferson remarked, "To our reproach it must be said, that though for a century and a half we"—white American naturalists—"have had under our eyes the races of black and of red men, they have never yet been viewed by us as subjects of natural history."[21] That would soon change, as US scientists' investigations led to worldwide recognition of the "American School" of anthropology in the 1840s and 1850s.[22] Ephraim G. Squier, who turned attention to Native Americans, exemplified American naturalists' essentialist, hierarchical views.[23] His investigations convinced colleague Josiah Clark Nott, author and editor of a voluminous tome (1854) that exemplified what is now called "race science." Nott considered America's "aboriginal barbarous tribes" as a distinct and "savage" human type, whose failure to grasp modes of Europeanized life available to them in the Americas made it "as clear as the sun at noonday, that in a few generations more the last of these Red men will be numbered with the dead." Nott was wrong: despite removal, murder, and cultural suppression, Native Americans sustained their identities and worldviews through engagement in practices that included dance.

Educational methods and institutions played roles in both proscribing and facilitating cultural transmission and aspiration among African Americans.

In post–Civil War Atlanta, as Carrie Streeter illuminates in "Exhibiting (Scientific) Grace: American Delsartism and Black Citizenship in the New South," African American women adopted, practiced, and made their own techniques and promises of movement-based entertainments developed from investigations by French music and acting pedagogue François Delsarte. Black women in the US, long denied descriptors such as "graceful" or "elegant" by many scientific and social classifiers, grasped opportunities to engage in systematically organized movement practices often aligned with progress, modernism, and white superiority. At universities and colleges in the Jim Crow South, teachers Mae B. Peckham and Adrienne McNeil Herndon taught young Black women the Delsarte system, encouraging creative displays of "aesthetic gestures" and emotionally affecting choreographies. Such visible statements of artistry and movement refinement undermined persistent claims by scientists and social leaders of African Americans' failure to deserve equal political, aesthetic, or social treatment. Responding to Streeter's chapter, Susan Cook's "Interrupting Jim Crow" considers ways that "performative acts of speech and attendant movement could be taught, practiced, and drawn upon" in resisting the "monstrous" work of late nineteenth-century minstrelsy. Performers like Aida Overton and Jayna Brown, Cook explains, could offer "contradictory narratives of citizenship" that allowed African Americans to reclaim humanity and rights, including that of free expression.

Education was a goal of Émile Jaques-Dalcroze who, like Delsarte, emerged from the world of music, addressing concerns of self-care, self-improvement, and health, highlighted in Part III of this volume. In Andrea Harris's "The 'Muscular Sense' and Therapeutic Modernism in the Eurhythmics of Émile Jaques-Dalcroze," the roots of Dalcroze's thinking in nineteenth-century study of the nervous system reveal then-widespread concerns about "modern" social ills. Dalcroze's rhythmic movement system aimed to ameliorate what Harris identifies as "the chaotic experience of industrial society," perceived as resulting in "physical fatigue, moral confusion, and the erosion of individual autonomy," particularly in cities. To address the rhythmic failings that Dalcroze considered the root of such problems, movement was understood not as utilitarian or aesthetic, but rather as therapeutic, healing human neurasthenia by drawing on investigations of the "muscular sense" (kinesthesia, proprioception). Dalcroze's Eurhythmic system became an effective tool not only for therapy but also for creativity, influencing artists who became important exponents of twentieth-century modern dance—Hanya Holm, Rudolf Laban, Marie Rambert, and Mary Wigman, among others.[24] Dick McCaw's "Dualism in Jaques-Dalcroze's Theory of Movement," responding to Harris's essay, brings out connections of Dalcrozian theories to actor training, to developments in neuroscience, and to Laban movement studies. McCaw points to common concerns and models shared by these artistic, philosophical, and scientific fields. Might Dalcroze's work still be a source for physical, moral, intellectual, and artistic integration? Might it offer glimpses of new visions that can move us beyond his own systemization?

A highly personal program of healing and self-regulation is treated in Johanna Pitetti-Heil's "From Animal Magnetism to Materialist Transcendentalism: Margaret Fuller on Fanny Elssler." In the late eighteenth century, a new health treatment was popularized by German physician Franz Anton Mesmer, who proclaimed as the source of animal vitality a universal magnetic fluid, the free flow of which he considered essential to health. As controversies developed over Mesmer's work, esteemed men of science ultimately discredited his claims, but mesmerism remained popular well into the nineteenth century. This treatment offered a powerful component of self-practice for Margaret Fuller—influential Boston feminist, Transcendentalist, and writer-editor. Her engagement with animal magnetism as a treatment and means of self-knowledge deepened Fuller's appreciation of her body and helped her with physical ailments and with her gendered position in professional, medical, and personal contexts. Witnessing megastar Fanny Elssler during the ballerina's US tour (1840–1842), Fuller not only appreciated the performances aesthetically, but also as expressions of spiritual-corporeal harmony and female selfhood. Responding to Fuller's response to Elssler, Claudie Jeschke's "The Labor and Laboratory of Kinesic Interplay in Nineteenth-Century Dance Theory" investigates "praxeological considerations of inner perception, or proprioception" as evidenced in early nineteenth-century dance treatises which, I note, laid groundwork for the "muscular sense" that Dalcroze, discussed above, sought to harness. Focusing on an 1838 text by Paulo Bruno Bartholomay, Jeschke illuminates dancing masters' investigations of anatomy, movement, and balance that shaped dance training and appreciation as ballet technique became increasingly systemized and virtuosic, as demonstrated by Elssler.

Cross-influences between science and dance are evident as the nineteenth century turned into the twentieth. The interest stimulated by the "sleep dancing" of Madeleine Guipet is explored in Chantal Frankenbach's "Hypnotic Dancing and the Science of Sleep and Dreams: The Controversial Case of Madeleine G." Behind this apparently "cataleptic" dancer was a century of medical texts on sleep, nervous disease, and the emerging field of psychology, which Guipet's expressive dancing seemed to engage. Public and scientific fascination with hypnotism, as well as with questions of movement motivation and human creativity, made Guipet's performances widely attractive, for her dancing appeared to reveal deep states of consciousness rarely accessible to observers.[25] Scientists and stage artists, physicians and philosophers found her performances unsettling, informative, questionable, and often moving. Was Madeleine a hoax, a victim, a patient, or, perhaps, an artist? To some viewers, her expressive capacity surpassed that of Isadora Duncan—then a young contender for dance fame. In Kélina Gotman's response to this chapter, "Ecstatic Fervors: Of Trance, Dance, and Self-Possession," deep "corporeal states of noncontrol" are explored for their perceived revelation of movement inspiration. The longing of viewers—physicians, artists, and theater-attendees—to witness (somehow, to experience) the "primitive" body, freed of Western social constraints, was perhaps more telling about such viewers than was Madeleine's dancing informative about deep psychological states.

Mentions of "technique" and "technology"—artistic and scientific—have arisen throughout this introduction. Janice Ross's "The Depths from the Surface: Interlaced Histories of Technologies and Dance in The Nineteenth Century" treats such themes directly. Reviewing science's fascination with the human body's interior, Ross investigates how technology's revelation of organs, bones, and expression revealed not only information and new treatments to scientists, physicians, and the public, but also widespread anxieties, prejudices, and (dis)tastes. These technologies had ramifications in the dance world. The beating heart amplified by the stethoscope in the early nineteenth century magnified audience readings of Giselle's fragile heart as they watched her staged death by heartbreak. Later, as Thomas Edison attempted to infuse mechanical dolls with such human qualities as speech and locomotion, the balletic figure of Coppelia represented an alluring mechanical woman, comic and threatening. Edison also explored the frighteningly probing X-ray, imaging the body in motion in one of the first moving pictures of dance—a "Couchee" dancer at the 1893 World's Columbian Exhibition in Chicago, a theme highlighting colonialist connections. Responding to the mechanization of corporeal representation in Ross's essay, Claudia Jeschke's "Movement-Machines: Reflecting New Technologies in Doing and Scoring Dancing" discusses emerging nineteenth-century techniques of dance notation based on understanding bodily motion as a mechanical system. Jeschke elucidates the notation systems of Giovanni Léopold Adice and Arthur Saint-Léon as means of seeing the body in segments, differentiating body and movement elements in new methods of choreographic notation.

Like other figures mentioned in this volume, artists like Adice and Saint-Léon employed the scientific turn of their age for their own pedagogical and expressive purposes. Today, the engagement of dance and science remains lively and critical. The work of dancer-choreographer Emily Carson Coates closes this volume out (thus, the "Outro"), bringing to light a contemporary creative conversation between these domains. In "Observing the Observers," Coates takes the reader on her investigative journey into Dartmouth College's Shattuck Observatory, built in 1854, as impetus for her choreography. Moving between past and present, the doing of science and the making of art, the written word and the dancing body, colonizers and native populations, Coates reveals encounters with—and transformations of—the nineteenth-century drive to observe and grasp the natural world. A phenomenological unfolding both conceptual and physical, Coates reminds the reader of the layered, sensual histories that braid together science, art, and experience, as well as ancient cosmic visions, nineteenth-century views of the universe, and contemporary creation. Responding to Coates, Christian DuComb's "Querying the Cosmos" explores understandings of cosmic motion from Plato to Louis XIV, from the Haudenosaunee people of the Great Lakes to the Barasana of the Amazon, from Thomas Elyot to George Lakoff and Mark Johnson. Our answers have varied widely, but peoples across time and geographies have queried the cosmos and danced their place in it.

* * *

This volume was conceived and developed as the COVID-19 pandemic unfolded. As data, science, and technologies concerning the coronavirus updated daily, the scientific methods and approaches developed generations ago and newly unfolding captivated a shuttered world awaiting return to "normalcy." What role could/should dance articulate in grappling with this disease physically, emotionally, and socially?[26] With the passage of time and experience, we newly appreciate the reach and influence of the nineteenth century on contemporary discourses about anatomy, bodily autonomy, and bodies as means to social, artistic, and political ends. In an age of wellness culture, health anxieties, and body hacks, physical movement is subject to data collection and output, to scientific questions that slip into market-value research and sales, to artificial-intelligence manipulations that question human corporeality itself. The research in this volume offers an historical lens to the slippage of raw data among scientific advancement, motivations of control, and drives for creation and artistry.

No one volume can comprehensively cover the story of dance and science in the long nineteenth century; indeed, there is no center nor margin to history, no fixed bounds. We hope this volume stimulates further investigations that view "science" as an invitation to think through the ways that systematic inquiry, healing methods, processes of construction and observation, and explanation of events and phenomena shape and articulate relations between self and world in cultural contexts. In some such contexts, noted earlier in this introduction, relations between self and ecology have motivated sense-making that contrasted markedly with inquiries directed by distinguishing the familiar from the Other. Dance, too, engages these varied processes and understandings; in this sense, we look forward to ongoing investigations of historical and contemporary interplays between dance and science.

Notes

1 Shumway and Messer-Davidow, "Disciplinarity," esp. 204–11.

2 Miller, Hamsath, and McConaghy, "Biographical Note."

3 Foster, *Corporealities,* 5; Grimsted, *Melodrama Unveiled,* 76–81. An example of such "Scientific Effects" is a playbill for the Philadelphia Academy of Music's production of *Black Agate, or Old Foes with New Faces* (by Charles Kingsley), in Historical Society of Pennsylvania (HSP) Academy of Music Scrapbook, collection 3150.

4 Abbattista, *Moving Bodies, Displaying Nations;* Ghosh, "Human Cadaveric Dissection"; Impey and MacGregor, *The Origins of Museums.*

5 Wolfe and Gal, "Embodied Empiricism."

6 Cumes, *Africa in My Bones;* Zarcony and Hobart, *Shamanism and Islam.*

7 Gargano, "Bodies, Rest, and Motion"; Miller, *Measures of Wisdom.*

8 Holdrege and Pechilis, *Refiguring the Body;* Millones, *Taki Onqoy;* Schweinfurth, *Prayer on Top of the Earth.*

9 See example in Carpenter, *Amateur's Preceptor;* Ferrero, *The Art of Dancing;* Hillgrove, *Hillgrove's Ball Room Guide.*

10 An overview of Western science in this period is Williams, *Album of Science;* see the book's "Foreword," by L. Bernard Cohen, on William Whewell's proposing the term "scientist" (p. xi). The term "biology" came into use around 1800 to encompass "a comprehensive study of the living organism" (Russett, *Sexual Science*), 4. See also Herschthal, *The Science of Abolition,* 3–6; Rusert, "Introduction," 1–32 in *Fugitive Science.*

11 Morton, *Crania America,* 4. The focus on "differences" as opposed to "unity" is discussed in Russett, *Sexual Science,* 6–7; Schiebinger, *Nature's Body,* 144; West, "Race and Modernity," *The Cornel West Reader,* 55–86. On "scientific racism," see Bay, *The White Image in the Black Mind;* Ouellette, "American Golgotha"; Owens, *Medical Bondage.*

12 The enduring reach of Linnaeus's work is evident in the article "On the Species of Varieties of the Human Race," intended for popular consumption, published in a Philadelphia newspaper, the *National Gazette* (Feb. 28, 1826): 2. Schiebinger, *Nature's Body,* discusses Linnaeus's taxonomic influence on eighteenth- and nineteenth-century biology. On interplays of such views with established religious beliefs, see Bailyn, "Introduction"; Russett, *Sexual Science;* and Stanton, *The Leopard's Spots.*

13 Durang, *History,* 3 ser., ch. 8, p. 25 and v. 6, ch. 102, p. 323; Wilson, *A History of the Philadelphia Theatre,* 28; Wood, *Personal Recollections,* 355, 397.

14 Carpenter, *Amateur's Preceptor,* 7, 74; Durang, *Durang's Terpsichore,* 23; Gourdoux-Daux, *Elements and Principles of the Art of Dancing,* 14.

15 On ballet as "scientific," see *National Gazette* (Philadelphia, Feb. 2, 1838), on young US

ballerina Augusta Maywood; *Public Ledger* (Philadelphia, Nov. 21, 1838), reviewing French-born Philadelphia ballet master M. Hazard; HSP Academy of Music Scrapbook, collection 3150, clipping (Sept. 26, 1857) on Domenico Ronzani's *Faust.*

16 Smith, *The Natural History of the Ballet-Girl* (1847), 7. Smith, the son of a surgeon, had trained as a physician before becoming a writer.

17 Daston and Galison, *Objectivity,* 9.

18 Moore, "Prints on Pushcarts," 6. An example of a ballet-print posted in a barbershop is given in a *New-York Observer* article (Mar. 3, 1827), defending French ballerina Eugenie Lecompte despite a "rowdie" print highlighting her fleshy exposure.

19 See Herschthal, *The Science of Abolition.*

20 Morton, *Crania Americana,* 3–93. See also Dixon Gottschild, *The Black Dancing Body,* Schiebinger, *Nature's Body,* 117–20; West, "Race and Modernity," 77.

21 Jefferson, "Notes on the State of Virginia," 153.

22 Hall, *A Faithful Account of the Race,* 64; Patterson, "An Archaeology of the History of Nineteenth-Century U.S. Anthropology," 463; Wallis, *Black Bodies, White Science,* 41–42.

23 Stanton, *The Leopard's Spots,* 81–88, 98–99.

24 These artists are discussed in Hodgson, *Mastering Movement,* 68–72.

25 For an example of a contemporary choreographer exploring "preconscious movement" and therapeutic applications of dance, see Martin, "Dance Leads to Therapeutic Potential."

26 Among many commentaries on dance in COVID times, see Ding et al., "The Effect of Dance-Based Mind-Motor Activities"; Gingrasso, "Practical Resources for Dance Educators"; and Kim et al., "OTT Streaming Distribution Strategies for Dance Performances."

Bibliography

Abbattista, Guido, ed. *Moving Bodies, Displaying Nations: National Cultures, Race and Gender in World Expositions Nineteenth to Twenty-first Century.* Trieste: University Press Italiane, 2014.

Academy of Music Scrapbook, Historical Society of Pennsylvania, Collection 3150.

Bailyn, Bernard. "Introduction: Reflections on Some Major Themes." In *Soundings in Atlantic History,* edited by Bernard Bailyn and Patricia L. Denault, 1–43. Cambridge, MA: Harvard University Press, 2009.

Bay, Mia. *The White Image in the Black Mind: African-American Ideas About White People, 1830–1925.* New York: Oxford University Press, 2000.

Carpenter, D. L. *Amateur's Preceptor on Dancing Etiquette and Waltzes, Gallopes, Quadrilles, &c.* Philadelphia: Lee & Walker, 1854.

Cumes, David. *Africa in My Bones: A Surgeon's Odyssey into the Spirit World of African Healing.* Claremont, South Africa: New Africa Books, 2004.

Daston, Lorraine, and Peter Galison. *Objectivity.* Brooklyn: Zone, 2007 (5th ed., 2021).

Ding, Yi, Chenchen Guo, Shaohong Yu, Peng Zhang, Ziyun Feng, Jinglong Sun, Xiangxia Meng, Li Li, and He Zhuang. "The Effect of Dance-Based Mind-Motor Activities on the

Quality of Life in the Patients Recovering from COVID-19." *Medicine* 100, n. 11 (Mar. 19, 2021), https://www.ncbi.nlm.nih.gov/pmc/articles/PMC7982229/.

Dixon Gottschild, Brenda. *The Black Dancing Body: A Geography from Coon to Cool.* Basingstoke: Palgrave MacMillan, 2003.

Durang, Charles. *Durang's Terpsichore, or Ball-Room Guide.* Philadelphia: Turner and Fisher, 1848.

Durang, Charles. *History of the Philadelphia Stage, between the years 1749 and 1855, arranged and illustrated by Thompson Westcott.* Philadelphia: Thompson Westcott, 1868.

Ferrero, Edward. *The Art of Dancing, Historically Illustrated: To which is Added a Few Hints on Etiquette.* New York: The Author, 1859.

Foster, Susan. *Corporealities: Dancing, Knowledge, Culture, and Power.* New York: Routledge, 1996.

Gargano, Cara. "Bodies, Rest, and Motion: From Cosmic Dance to Biodance." *New Theatre Quarterly* 16, n. 3 (2000): 211–18.

Ghosh, Sanjib Kumar. "Human Cadaveric Dissection: A Historical Account from Ancient Greece to the Modern Era." *Anatomy and Cell Biology* 48, n. 3 (Sept. 2015): 153–69.

Gingrasso, Susan. "Practical Resources for Dance Educators! Choreographing our Way through COVID-19." *Dance Education in Practice,* 6, n. 3 (2020): 27–31.

Gourdoux-Daux, Jean-Henri. *Elements and Principles of the Art of Dancing, as used in the Polite and Fashionable Circles,* trans. Victor Guillot. Philadelphia: J. F. Hurtel, 1817.

Grimsted, David. *Melodrama Unveiled: American Theater and Culture, 1800–1850.* Chicago: University of Chicago Press, 1968.

Hall, Stephen G. *A Faithful Account of the Race: African American Historical Writing in Nineteenth-Century America.* Chapel Hill: University of North Carolina Press, 2009.

Herschthal, Eric. *The Science of Abolition: How Slaveholders Became the Enemies of Progress.* New Haven, CT: Yale University Press, 2021.

Hillgrove, Thomas. *Hillgrove's Ball Room Guide and Practical Dancer.* New York: Dick and Fitzgerald, 1863.

Hodgson, *Mastering Movement: The Life and Work of Rudolf Laban.* New York: Routledge, 2001.

Holdrege, Barbara A., and Karen Pechilis, eds. *Refiguring the Body: Embodiment in South Asian Religions.* Albany: SUNY Press, 2016.

Impey, Oliver, and Arthur MacGregor, eds. *The Origins of Museums: The Cabinet of Curiosities in Sixteenth- and Seventeenth-Century Europe.* Oxford: Clarendon Press, 1985.

Jefferson, Thomas. *Notes on the State of Virginia* (1785). Philadelphia: Prichard and Hall, 1788.

Kim, Jian, Eunhye Kim, and Aeryung Hong. "OTT Streaming Distribution Strategies for Dance Performances in the Post-COVID-19 Age: A Modified Importance-Performance Analysis." *International Journal of Environmental Research and Public Health* 19, n. 1 (2022): 327.

Martin, Lisa. "Dance Leads to Therapeutic Potential." *TCU Magazine* (Spring 2020). https://magazine.tcu.edu/spring-2020/dance-relief-cerebral-palsy-nina-martin/.

Miller, Amy, revised by DiAnna Hemsath and Mary D. McConaghy. "Biographical Note." Eadweard Muybridge Collection, UPT 50 M993. https://archives.upenn.edu/collections/finding-aid/upt50m993/. Accessed Mar. 3, 2024.

Miller, James. *Measures of Wisdom: The Cosmic Dance in Classical and Christian Antiquity.* Toronto: University of Toronto Press, 1986.

Millones, Luis. *Taki Onqoy: De la enfermedad del canto a la epidemia.* Santiago de Chile: El Centro de investigaciones Barros Arana de la Biblioteca Nacional, 2008.

Moore, Lillian. "Prints on Pushcarts." *Dance Perspectives* 15. Brooklyn: Dance Perspectives Foundation, 1962.

Morton, Samuel George. *Crania Americana.* Philadelphia: J. Dobson; 1839.

National Gazette, Philadelphia newspaper.

New-York Observer, New York City newspaper.

Nott, Josiah, and George Gliddon. *Types of Mankind: Or, Ethnological Researches, Based Upon the Ancient Monuments, Paintings, Sculptures, and Crania of Races, and Upon Their Natural, Geographical, Philological and Biblical History, Illustrated by Selections from the Inedited Papers of Samuel George Morton and by Additional Contributions from L. Agassiz, W. Usher, and H. S. Patterson.* Philadelphia: J. B. Lippencott, Grambo, & Co., 1854.

"On the Species of Varieties of the Human Race." *National Gazette* (Feb. 28, 1826): 2.

Ouellette, Jennifer. "American Golgotha." *Ars Technica* (Oct. 4, 2018). https://arstechnica.com/science/2018/10/theres-new-evidence-confirming-bias-of-the-father-of-scientific-racism/.

Owens, Deidre Cooper. *Medical Bondage: Race, Gender, and the Origins of American Gynecology.* Athens: University of Georgia Press, 2017.

Patterson, Thomas C. "An Archaeology of the History of Nineteenth-Century U.S. Anthropology." *Journal of Anthropological Research* 69, n. 4 (Winter 2013): 459–84.

Public Ledger, Philadelphia newspaper.

Rusert, Britt. *Fugitive Science: Empiricism and Freedom in Early African American Culture.* New York: New York University Press, 2017.

Russett, Cynthia E. *Sexual Science: The Victorian Construction of Womanhood.* Cambridge, MA: Harvard University Press, 1991.

Schiebinger, Londa. *Nature's Body: Gender in the Making of Modern Science.* New Brunswick, NJ: Rutgers University Press, 1993–2013.

Schweinfurth, Kay Parker. *Prayer on Top of the Earth: The Spiritual Universe of the Plains Apaches.* Boulder: University Press of Colorado, 2002.

Shumway, David R., and Ellen Messer-Davidow. "Disciplinarity: An Introduction." *Poetics Today* 12, n. 2 (Summer 1991): 201–25.

Smith, Albert. *The Natural History of the Ballet-Girl.* London: D. Bogue, 1847.

Stanton, William. *The Leopard's Spots: Scientific Attitudes toward Race in America, 1815–59.* Chicago: University of Chicago Press, 1960.

Wallis, Brian. "Black Bodies, White Science: Louis Agassiz's Slave Daguerreotypes." *American Art* 9, n. 2 (Summer 1995): 38–61.

West, Cornel. "Race and Modernity." In *The Cornel West Reader,* 55–86. New York: Basic Civitas Books, 1999.

Williams, L. Pearce. *Album of Science: The Nineteenth Century.* Foreword by L. Bernard Cohen. New York: Charles Scribner's Sons, 1978.

Wilson, Arthur H. *A History of the Philadelphia Theatre 1835–1855*. New York: Greenwood Press, 1968.

Wolfe, Charles T., and Ofer Gal, "Embodied Empiricism." In *The Body as Object and Instrument of Knowledge*, edited by Charles T. Wolfe and Ofer Gal, 9–22. Dordrecht: Springer, 2010.

Wood, William B. *Personal Recollections of the Stage*. Philadelphia: H. C. Baird, 1855.

Zarcony, Thierry, and Angela Hobart, eds. *Shamanism and Islam: Sufism, Healing Rituals and Spirits in the Muslim World*. London: I. B. Taurus, 2013.

I

Learning How to Look

Regimes of Classification

1

Venus in Pieces

Choreographing the Anatomical Body with Clemente Susini's *Venus de' Medici*

Sariel Golomb

Introduction: Approaching the Glass Cabinet

A waxen woman reclines on satin sheets, her grazing finger and arms tensed with impulse toward motion (figure 1.1). She is nude but for a strand of pearls on her neck. A sheen of sweat on her skin draws attention to the ampleness of a healthy body. Small expressive details—a parted mouth, flushed cheeks, heavy lids, an exposed neck, and lifted chest—all read legibly as sexual climax. But consider the unfamiliar seam along her hip and pubic bones, demarcating a border between human and doll, which beckons to be parted by careful hands to see what other mechanical potential lies within (figure 1.2). Lift up the lid of her torso to reveal her innards: the warm tones of intestines, lungs, and stomach—the underbelly of the beautiful woman. Remove each piece to hollow the space within: detach the playful curl of a kidney, break open the uterus, even, to discover a developing fetus. Surveying her disemboweled frame, her once-erotic arch now reads closer to a gasp of terrible death. What is the role of the handler in this procedural choreography of anatomical undress that renders our ideal Venus abject? What medical knowledge is scripted in the intestinal wall; in the peering-over and looking-through of her lungs; in the final reward of a fetus? How does an encounter with this silent wax surrogate dance the line between reality and fantasy?

In this essay I bring a dancer's attunement to physicality and interaction to bear upon an unlikely subject: a late eighteenth-century inanimate wax body. This body, known variously as *Venus de' Medici* and the *Demountable Venus,* was created between 1780 and 1782 for the Florentine Museo della Specola, the first public anatomy museum in Europe, and embodies what I argue is the fundamentally choreographic nature of this period's approach to engaging the public in body knowledge.[1] La Specola was patronized by the Grand Duke of Tuscany as part of a larger project of social reform: he aimed to turn Tuscan "subjects" into "citizens" through public educational opportunities in empirical observation and natural philosophy.[2] And yet both the purposes and the audience of wax Venuses were enigmatic, belonging equally to physicians, artists, and the public. Europeans considered the human form to be as important to art, philosophy, religion, and civic life as it was to medicine in the late eighteenth century—far different from how we parse these epistemological frameworks today. As the fascination for wax models spread across the Western world in the late eighteenth century and throughout the nineteenth, they would eventually be exhibited in fairgrounds and private erotic collections. Yet we view this Venus today in her birthplace, the Museo della Specola (now under the care of the University of Florence) encased in a Venetian glass and rosewood casket, surrounded by others containing severed wax limbs and sliced perspectives of body.

I argue that the Venus was noteworthy as a sculptural tool for public educational usage that invites a procedural, choreographic encounter with her own organs. The Venus's "invitation to dance" with her is predicated upon her novel charisma, a complex imaginatory practice, and the suspension of time-based consequences of organic matter—consequences of decay and deterioration. This object encounter demonstrates several historical processes in motion for public science. First, it suggests that the integration of sight- and touch-based learning for anatomy—which was first brought into usage in sixteenth-century anatomy theaters—was here newly imagined as relevant and useful to citizens across classes as part of an emerging economy of edificatory entertainment. Second, the resulting need to render anatomy's practices visually simple and self-explanatory and its objects durable resulted in a vision of the body as a "fragmented whole" made up of agential parts of individual interest—each with its own will for action upon the body.[3] Finally, the hyperrealistic artificiality of the Venus points to a role for imagination in anatomical learning, but also reveals the implicit role of imagination in all medical training and practice. Anatomical knowledge is predicated upon an ideal and hypothetical image of the body that stands outside of and thus is not subject to elements of time, gravity, and sociocultural context.

Existing scholarship contextualizes the Anatomical Venuses within intersecting perspectives of medical and popular science, wax history, religious iconography, and the eroticization of the female corpse.[4] Yet what has been surprisingly neglected about these medical models—the *Venus de' Medici* the most famous among them—is how their physical deconstruction and reconstruction through separable organs charts new regimes for imagining

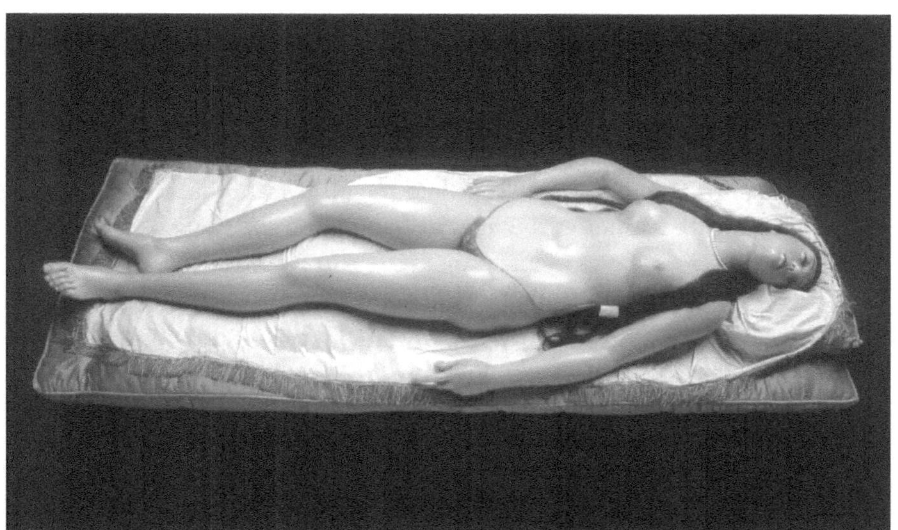

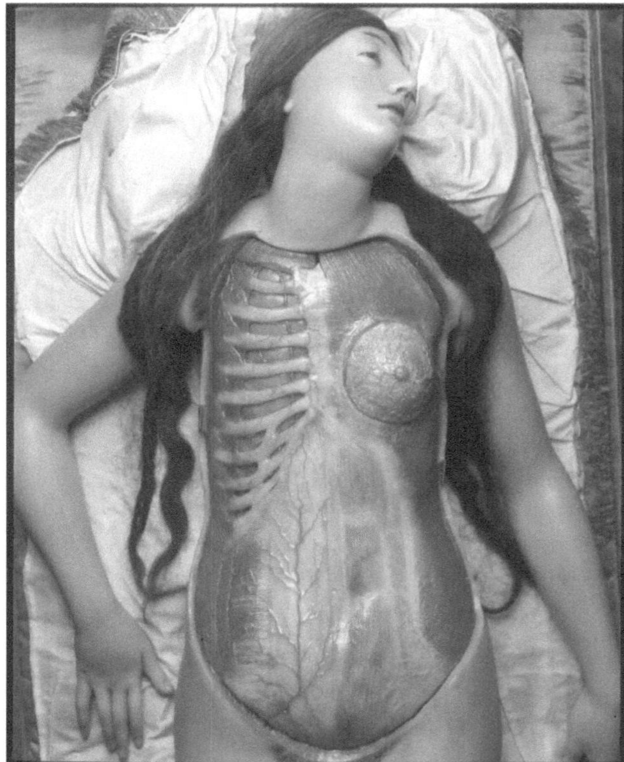

Figure 1.1 and 1.2. Workshop of Clemente Susini and Giuseppe Ferrini, Pregnant Female with Removable Skin and Abdominal Layers, also known as the Anatomical Venus, c. 1780. Wax with metal or wood skeleton, 180 × 80 cm. La Specola, Florence. Photos by Saulo Bambi, courtesy of Sistema Museale dell'Università degli Studi di Firenze, Sez. di Zoologia "La Specola" (fig. 1.1) and Museo di Storia Naturale dell'Università di Firenze, Sistema Museale di Ateneo, Firenze (fig. 1.2).

anatomy. The Venus's visual-haptic mechanism places her in conversation with what Rafael Mandressi, among others, has pointed to as the "sensorial program" of medical pedagogy practiced in anatomy theaters.[5] However, the fact that the wax model's instruction is repeatable, reversible, and mimetic—which a dissection of a real body is not—introduces a role for choreography to play in anatomical knowledge and the historical project of its public dissemination. The Venus encourages us to build on the study of the sensorial and dramatic qualities of anatomy theaters to consider how elements of dance—space, time, orientation, repetition, relations between simulation and the real, and transpositions between eye and hand—factor into the evolution of medical pedagogy. This exploration of choreography-as-hermeneutic offers a compelling bridge through which dance history and methodologies can inform the history of science and vice versa. Furthermore, the Venus provides a precedent for the forms of discerning gaze at the body in settings of public performance and display that would be elaborated upon across the nineteenth century.

I engage with the Venus through the framework of the choreographic (a poetics of choreography[6]), through which I refer to structures for understanding the physicality and motion—both intended and unintended—enacted by a body, regardless of sentience.[7] What I am concerned with is not her curiosity as static representation but what she conjures on a performative level by virtue of her details, and why it is significant that she was meant to be engaged with through repetitions of motion that were inscribed into her very construction. Mark Franko writes that choreography organizes "the physical potentials and limitations of the human body's movement."[8] Given that the Venus is a replica of the human body, we can view hers as an intertextual response-choreography: one that simulates the body's potentials while commenting upon the limitations of a choreography of dissection with an actual cadaver. Taking this choreographic approach offers an analytic for the peculiar charisma of this enigmatic object, which was invented in order to be handled physically, which traveled all over the Western world delighting audiences, and which also elicited forms of interaction that were off-script: a deviating choreography; a repetition with difference.[9] To fully appreciate the actors of a long nineteenth-century performance culture, we must first think about when, why, and how such discerning gazes onto bodies were honed—and consider that the gaze itself is not passive and distant but enfleshed, haptic, and agential. It touches and penetrates where it looks and grants an imprimatur of expertise about the body to the beholder.

The female body is often depicted in Western visual culture in fragments.[10] But to what extent can we consider the work of this demountable model as exceptional in encouraging an accessible, procedural encounter with organs: a spatial and temporal journey into the depths of the human body in which is scripted the handler's touch, fondle, and inquisitive multidimensional gaze? How might a dance studies perspective, attuned to the singularity of embodied knowledge, allow us to reconsider the development of medical hermeneutics and reassert that medical authority and diagnostic gazes are never not mediated, desiring, and socially situated? In light of this sensual body replica as a tool of medical purposes, I take up an assertion engrained in the very definition of scopophilia: that touch is integral to sight in that sight invariably involves the impactful touch of the gaze onto the body.[11]

My encounter with the Venus often veers into the hypothetical in that it responds to gaps in the archive. While La Specola (as just one of the Venus's exhibition sites) kept meticulous records of its visitors, the museum's idealistic initial belief that visitors should be entrusted to handle the models without oversight meant that much goes unaccounted for in terms of what the eighteenth- and nineteenth-century spectator thought of her body in pieces. Indeed, my preoccupation with the Venus is haunted by one of the few anecdotes about her public reception. Anna Maerker writes in *Model Experts: Wax Anatomies and Enlightenment in Florence and Vienna, 1775–1815* that La Specola's directors were "sorely disappointed when visitors . . . reacted to the models in a variety of idiosyncratic ways," including touching the models' genitalia—a trace borne out only in the forms of repair the models needed, and which eventually forced the museum to encase the figures in glass.[12] What could have gone so awry in the scripting of behavior toward these objects such that the directors envisioned a certain type of pedagogical touch, whereas users enacted an erotic touch? Maerker points to this as an important moment of reckoning between representational intention by State authorities and the assertion of agency and heterogeneity by its public.[13] I suggest that we consider what inspired such touch and also what the inability to dance with the Venus behind glass renders obsolete in her, even if she is still a visual marvel. This lingering anecdote, which corroborates how desire is implicitly written in the historical record, continues to motivate my attempt to think "idiosyncratically" alongside the Venus.

The Choreographic Encounter

The Anatomical Venus represents the birth of an Enlightenment ethos toward corporeality: a view that not only is the body a physical object discrete from and ruled by the mind, but that humans have a moral imperative to study the body to understand the divine principles of Nature. As Katharine

Park writes, depictions of the body in pre-Enlightenment visual culture were largely symbolic, narrativized, and interpersonal.[14] In considering what the Venus embodies of her era, we must acknowledge that the way we (as what could be called "lay practitioners" of anatomy) are commonly taught to conceive of "the" body—as functional, medical, consisting of layers and systems, and universal to all humans (regardless of vectors of differentiation like race or gender)—is fundamentally modern, departing markedly from pre-Enlightenment beliefs.[15]

Scholars remain uncertain as to the primary objective or intended audience of the Anatomical Venuses, but these figures evidently offered at least two functions for a newly secular-humanist Italy. First, they reduced anatomists' need for cadavers, the sourcing practices of which were coming under increasing public scrutiny.[16] Second, the wax Venuses served as the epicenter of a novel attraction: the public science museum. In 1775, Florence's Museo della Specola became the first natural history museum in Europe; the *Venus de'Medici* was created in the wax workshop at its center by its best-known ceroplastician, Clemente Susini.[17] La Specola and its wax anatomies became an instant sensation, providing a broad audience with what Joanna Ebenstein calls an "intuitive and pleasurable way" to understand the body's anatomical structures.[18] The wax anatomies demonstrate an effort to make rarefied scientific knowledge accessible to a public that included the high nobility, Grand Tour travelers, and even the "coarse mob."[19] But they also demonstrate ideation about the requirements for a new degree of legibility for scientific information. It would mean improving on existing technologies and knowledge forms to create a tool that could be scaled in its simplicity and functionality to accommodate different needs, contexts for display, and focal points in the body. This specific Venus was so highly regarded an anatomical tool that state leaders across Europe commissioned copies in the 1780s–1790s as they raced to create their own collections in pace with the Enlightenment ambition of social edification. By the end of the eighteenth century, notable collections included that of the Josephinum in Vienna and the Faculty of Medicine in Montpellier, commissioned by Napoleon; publicity circulated London for a "Florentine" Venus show in 1825, Signor Sarti's Anatomical Venus and Adonis in 1839, and Kahn's Anatomical Museum in 1857.[20] Wax anatomies "circulated" through museums, fairgrounds, medical teaching facilities, and even private collections.[21]

The Venus was designed not just to be seen but to be handled: sequentially uncovered and disemboweled by the hands of medical students and museumgoers alike. They would learn through her depths, looking from all perspectives to familiarize themselves with the organs and their fit. What is more, the viewer was made to understand that the body is made up of agential parts that serve distinct purposes and thus maintain some of their own charismatic power when regarded individually. Unlike the dissection of organic cadavers, one of the Venuses' great draws was what Margaret Carlyle notes is "their function as tactile objects—their removable parts were capable of withstanding the frequent handling typical of classroom setting usage."[22] Carlyle's use of "capable" is noteworthy: lying in her same glass casket at La Specola today as she did 240 years ago, the Venus's power is now largely imaginative, obstructed as it is from physical usage. Moreover, even in use, the purpose of an anatomical model is for touch that demonstrates capacity in anticipation of the real. But I want to push further on what the Venus was thought to be "capable" of that physical interaction with a cadaver could not offer: the hypothesis that there is an intuitive choreographics to a model replica—one that both newcomer and expert might find useful and that would turn any user into an anatomist. The Venus proposes that the act of holding, examining from new angles, rubbing, fondling, and reassembling the parts of the body are a means to understand their sequence and conceptualize them as one whole. Anatomy becomes a repeatable choreographics that a wide range of performers are capable of, predicated upon a mimetic yet hyperrealistic prop. And choreography becomes a hermeneutic, a means of embodying information and rehearsing in anticipation of a future encounter with a patient, or for the sake of simply knowing: gaining knowledge of one's own body through a surrogate's.

To be clear, the wax models' technology expanded upon wisdom about how to engage learners in anatomical knowledge that had circulated Europe for several centuries prior to the models' arrival. Rafael Mandressi writes that anatomy as a "knowledge based in practice" came into existence as a type of performance: "Space, action, time, instruments, and bodies made an ensemble of conditions for the anatomical enterprise to develop from its initial practice of dissection in Europe during the last decades of the 13th century."[23] As he argues, it was in the sixteenth century when French and Italian anatomists began to consolidate ideas about effective pedagogical practices, in particular through the successes of the public anatomy theater. Public anatomy was a highly curated and spectacular affair in which live dissections were performed for audiences comprised of doctors, medical students and the general public, "to expose publicly the anatomical apparatus, to render visible its new methods."[24] In these settings, where visual instruction on dissection was complemented by passing around body parts for audiences to individually handle, sensory experience motivated knowledge acquisition and became a "methodological leitmotif."[25] Oral instruction alone thus came to be considered insufficient for instruction on the body; as anatomist Jacopo Berengario da Carpi (d. 1530) wrote, "in this instance vision and touch are in fact indispensable."[26] The sixteenth century thus marks the birth of an oft-disputed convergence and elevation of the visual and haptic in medical knowledge, wherein the doctor's gaze is analogized as physical, even performative, traversing space and inciting changes in its environment through diagnosis followed by cure. Physicians were taught to treat illness in part by visualizing its location within the spatial layout of the body's organs, fibers, and vessels. As Foucault writes, as if describing a danced duet, "doctor and patient are caught up in an ever-greater proximity, bound together, the doctor by an ever-more attentive, more insistent, more penetrating gaze, the patient by all the silent, irreplaceable qualities that in him, betray—that is, reveal and conceal—the clearly ordered forms of the disease."[27]

Apropos the Venus, the procedure for public instruction in anatomy theaters was very much predicated on a choreographed order for unpacking the body. Here, however, the order was motivated by what parts of the body would rot first and need to be disposed of, starting with the entrails and ending with the extremities.[28] As the open cadaver became subject to time and gravity, instruments like bellows would be required to inflate the organs in order to demonstrate them in use.[29] This "linear sequence" would end with a displayed array of parts according to a "logic of fragmentation, of dismemberment, based on an analytical principle by which comprehending an object required its decomposition into segments."[30] In this way, public anatomy serves as a further precedent for the fragmentation of the Venus's wax body as a means of breaking down knowledge itself into intelligible parts.

The fame of a representational wax body as a "fragmented whole" shows us that in the Enlightenment era, a discourse of corporeality was disseminated in terms of the rehearsed physical knowledge of organization. Writing on bourgeois material culture, Susan Stewart theorizes a similar hermeneutic in the idea of the "collection." Ordering a collection requires the individual to "apprehen[d] . . . with eye and hand"—the visual and haptic combined—but it also depends on a movement between display and obscurity.[31] The sense of unpacking and the delight of the deeper finds that the Venus contains is caused not by any one piece of this "collection" but by the organizational structure of the body as a whole. As Stewart writes, the collection is about seriality, control, and containment: "One cannot know everything about the world, but one can at least approach closed knowledge through the collection."[32] This reflects the Enlightenment-era ambition to understand human anatomy as a means and metonym for understanding God. With the Cartesian theory of the mind as a power dominant over the body, we might understand the body as the ultimate collection for both physicians and the public to master.

In her rich account of the cultural and political context for the establishment of La Specola, Anna Maerker argues that the question of how to display the anatomical models embodied a conflict between two perspectives on teaching strategies during the Enlightenment. One position, held by La Specola's founding director, Felice Fontana, was that knowledge should be intuitive, enjoyable, and accessible to audiences of all walks of life, which motivated not only his advocacy for the creation of a wax workshop and museum in the first place, but also his continued insistence that the public be able to handle the models physically, despite his staff's protests.[33] The other position, held by Fontana's assistant, rival, and eventual successor, Giovanni Fabbroni, was that knowledge should be guided by the new Enlightenment persona of the "expert," a position he campaigned to bring to La Specola.[34] In the early years of the museum visitors were able to freely engage with the models, and Fontana is recorded as only offering demonstrations of the Venus to a few elite visiting groups.[35] But Fontana's confidence in the consummate authority of the Venus's implicit choreography over the viewer was thrown into question by those off-scripted touches and gropes. Maerker reads these subversive motions implicitly in the archive: through La Specola's frequent logs of necessary repairs and locks to the genitalia and monstrous fetus displays in particular, as well as the damages caused to those rooms by heightened traffic.[36] Eventually, in an effort to exercise more control over the public's interactions with the curiosities, La Specola's displays ended up behind glass.

Today, the *Venus de' Medici* is the only demountable Venus that still remains at La Specola, though she can be physically compared with other preserved anthropomorphized, demountable bodies that script the peeling back of layers to explore deeper realms of anatomy. The sixteenth-century German anatomical manikins—dissectible female figurines carved from ivory or wood, the size of a human hand—may have lacked the detail or accuracy of the Venus, but scholars believe they were used to teach well-to-do expectant mothers or new midwives about childbirth, or as decorative collectibles for *Wunderkammern,* or as a way for physicians to advertise their profession.[37] The sixteenth-century "fugitive sheets" or their nineteenth-century version, "flap anatomies," created by such anatomists as the Belgian Constant Crommelinck, were two-dimensional paper bodies in which flaps could be pulled back to show structures underneath.[38] Yet the Venus definitively departs in her hyperrealism, which indexes a transition from interaction with an object brought into relief as such (the moving of a flap) to interaction that simulates the haptics of the thing it represents—the body itself. Unlike the model in miniature or the paper anatomy, the wax model does not partially represent and then gesture toward further study; it is a self-contained realm of knowledge. The three-dimensional model ultimately allows the user to physicalize the "complex spatial puzzle" of the body, as Martin Kemp and Marina Wallace aptly put it.[39] Maintaining the necessity of the haptic for learning "intuitively . . . the relationship between the internal structures," even after the misuse of the anatomies led to their encasement, Felice Fontana turned from wax to wood as a preferred medium for anatomical teaching models. He spent his last years of life making a prototype of a wooden male model that could dissect into three thousand pieces. However, it was never realized, as humidity kept changing the size of the pieces.[40]

The Venus's spatial puzzle embodies the codification of the anatomy theater's pedagogy of sight and touch and its transposition to mimetic form; it also theorizes that anatomical choreography could be self-taught by the public to itself. Beyond the objective pursuit of knowledge, there is a notable erotics to how the gaze of a viewer is gratified in the Venus through invited touch. For this reason, the other theatrical dimensions of the Venus bear great importance in that they call upon the user's own embodiment. Reclining as she does, the Venus performs a certain passivity to the whims of the user. Yet her capability of breaking down into pieces that maintain their individual power symbolizes new modes of agency for the object: she invites her own obliteration with the safe promise of her return to wholeness.

Fantasizing the Scenario

Although she is a medical model, a scene and character materialize through the Venus's surface details. The very invocation of "Venus" primes one for a certain erotic spectatorship, even while the title was intended as a play-on-words: *medici* refers to both medics and the Medici family rule, while *Venus* is both the goddess and the sixteenth-century nickname for the beautiful city, *Venus-Fiorenza.*[41] With her surface's dewy gleam, the string of pearls, the landscape of smooth white skin, the active state of recline on a silken cushion, and the lengths of loose hair, she performs a woman who would not often be caught with her hair down. As modern viewers accustomed to medical tools that are removed from all cultural signifiers to perform their purposiveness, we have the advantage of appreciating the extremes of this representation. Still, even for her time, we can appreciate the ways in which the Venus's immanent dramaturgy distinguishes the experience from working with a real cadaver. What did an eighteenth-century ceroplastician think important to include in his wax model to be effective and spectacular, and how does this reflect upon the forms of movement done with her? If the Venus scripts a pleasurable fantasy on her surface, what fantasy plays out as she comes apart?

It is impossible to speak of the invitation to dance with the Venus without also speaking to the charisma of her surface details and to the irresistibility they must have conjured. Theater historian Joseph Roach describes an elusive form of cultural magic emerging in the eighteenth century that he terms the "It-Effect," whose fungibility enables historical icons who bear similar appearances and reputations to readily surrogate for each other in the "minds of fantasists."[42] This recursive fantasy is scaffolded by the bodily materia of the fetishized icon such as hair, skin, clothes, and accessories, which allow the ordinary person to be "in imaginative touch" with such a body.[43] The Venus invokes a palimpsest of such "role-icons": she not only shares a charisma with other Venuses like Florence's marble *Venus de' Medici* (c. 1598) (see note 1) or the *Venus de Milo,* but—with her long hair, satin bed, and recline—she could conjure Juliet Capulet, the Greek Psyche visited in the night by the god Eros, or one of the many iterations of a Sleeping Beauty tale that circulated Europe during this period. Maerker names a number of familiar reference points for viewers of the Venus that might have encouraged handling—among them the votives and relics they would have encountered in religious contexts, or even art sculptures, which some aesthetic theories of the day claimed required touch to be appreciated.[44] But by collecting a Venus or by joining in her physical assemblage, the public could also take part in a material connection with the diffuse desire for the mythological supine woman who does not protest being touched.

One significant aspect of the Venus's performance of charismatic femininity is the fusion of material accessories with her nakedness as if they were organs of her natural body. The Venus wears nothing but pearls and lies upon her bed, but the necklace and framing speak as loudly as her nudity and in fact transform the nudity by virtue of their statement. Given that before the creation of cultured pearls in the twentieth century, these gems were rare and expensive, the necklace situates the Venus as a woman of wealth and import and thus performs social status to the public.[45] The sheer surface area of white skin serves to consolidate an image of the respectable, white female body in European visual culture—what Roach calls a "synthetic experience of empire."[46] It is relevant here that eighteenth-century anatomists contended with a murky and unethical history of sourcing cadavers from the vulnerable poor, and in fact created wax models with the purpose of eliminating the need for real cadavers. As such, we might view the Venus as salvaging the project of anatomy from dishonorable associations. She suggests that the well-to-do might enjoy looking upon the divine project of science, or—as if anticipating that wax models would never officially do away with the need for cadavers—even donate their bodies to science. Substituting for the bodies of the poor while evoking the wealthy, the Venus thus occupies a liminal space between social classes.

This scenario ultimately contextualizes the choreographic choices that a physical interlocutor makes with the Venus. Like immersive theater, it allows them to take physical part in visual fantasy, coordinating a gaze of longing with touch. In the use of such accessories, the Venus suggests place and staging—the lush private quarters of an aristocrat—and her frozen, liminal expression of pleasure suggests timing: a momentary sequence that implies its own introduction and resolution. In this sense her body enacts a choreographics of a "little death" while invoking the specter of death itself.[47] While medicine and dramaturgy were already in conversation by way of the anatomy theater, the Venus innovates upon this by conjuring a fantasy of private encounter even as an object for public viewing.

Beyond her invitation to interact through her mechanical virtuosity, what distinguishes the Venus from other images of women in repose is that she is in the throes of waking experience—and pleasure, at that. Her expression of ecstasy alone is a contentious topic among historians, who disagree over whether it would have been read as explicitly sexual, given that the Venus was exhibited for a public that included women and children. Kemp, Wallace, and Ebenstein are quick to assert that the Venus's expression of ecstasy, which bears some similarity to Bernini's *Ecstasy of St. Teresa,* would have been historically regarded as chaste and sacred.[48] Yet this argument undermines the extraordinary work the Venus does in holding an enigmatic mixture of stated purposes and subconscious desires within her body. In keeping with both a Foucauldian genealogy of sexuality and extant perspectives within religious studies, I would contend that depictions of religious ecstasy—whether a female martyr or the Venus—function as a rhetorical channel for communicating sexual experience in a Catholic-dominated culture.[49] The off-script fondling of the museum's wax genitalia displays, which eventually forced our Venus behind glass, makes her implicit indexing of sexual desire further plausible.

With a tensed abdomen, trailing finger, and parted mouth, the vision of the Venus is liminal, captured almost photographically in a position between. An active state of recline, hers is a choreography of contradictions. Historians of science commonly distinguish the female wax models as showing repose to demonstrate the softness of musculature at rest, whereas their male counterparts depicted moments of muscular tension; Ludmilla Jordanova qualifies this by arguing that female models expressed experience where males showed action.[50] But the motion that the Venus depicts challenges this distinction—no matter the terms of her ecstasy, it clearly entails her own physical participation. Her expression of pleasure is perhaps the needed invitation for interaction with her mechanical capabilities.

The Venus's pleasure is mirrored in the pleasure of the user, which I argue is critical to her novelty as a technology of public medical pedagogy. To enter into the scenario the Venus proposes and then engage in her dissolution would at first glance seem to interpellate the user as a violator, which invokes the sordid history of anatomical practice and consequently equates the Venus to a corpse—a comparison on which Ebenstein and others premise their analyses.[51] But the violence of the deconstruction is curiously neutralized by her toy-like quality in a way that sanctions touch and suggests that an immanent vitality outweighs any deathliness. This quality frees the user's choreographic encounter from the specter of violation that haunts a true dissection in a way that renders anatomy approachable and also facilitates fantasy—whether erotic, or, as I will return to, medical.

More telling about the nature of such invited pleasure are those receptions of the Venus that went off-script. The anecdote that ghosts my writing—that La Specola's first audiences could not keep themselves from fondling the wax genitalia and "monstrous" anatomies—suggests that sex was not far from viewers' minds, and is why we cannot ultimately view the Venus's nudity or the ability to physically interact with her as unremarkable for the historical viewer.[52] But historical accounts suggest that the nonscientific, ornamental details that Susini enjoyed adding to his medical Venuses eventually caught the discerning eyes of private collectors, too.[53] This brought about an alternative industry and clientele for the Venuses as French courtiers began to collect "wax erotica" to store at home in private collections.[54] Some public wax exhibitions even included almanacs that contained pornography, and wealthy customers could buy custom models in sexually explicit poses, "perhaps to excite a novice partner," though none have survived to this day.[55] And when the famed Madame Tussaud moved to Britain in 1802 with the waxwork collection she had built alongside her uncle Philippe Curtius, she removed from her exhibits the grotesquerie and "'anatomical ladies' whose insides could be touched by gentlemen for an extra fee," restoring the wax museum to edifying family entertainment.[56] Thus, even beyond anecdotal evidence, we can clearly say that the Venus finds her place in a history of erotic entertainments.

Freezing time in the throes of pleasure, the Venus is both lovely and sinister. She was an anatomical innovation but also recognizable within a visual and affective realm of artistic depictions of women in her time. Given the repertoire of references and fantasies viewers could have drawn from when viewing her, we can appreciate the Venus's seemingly inherent approachability. What might be conveyed through the touch of a hand that follows such an appreciative gaze? Certainly there is a sense of ease and familiarity—and with it an element of entitlement—to enact such intimacy with a body replica; with those Venuses that were commissioned "to excite a novice partner," perhaps a more lustful touch that prepared to transfer from the replica to the real. But while I have argued for the importance of the Venus as an Enlightenment curiosity that demonstrates problem-solving toward the dissemination of lay medical knowledge, the Venus was also created in part as rehearsal for the future medical expert. This brings into consideration the role that fantasy-looking and -touching play even in medical practice. Indeed, the complex nature of the real is brought to the fore in danced duet with this wax form.

Dancing Outside of Time

Wax invites its partner to an intimate and sensuous dance. Awakened to moldable softness by the heat of the hand or a stoked fire, it follows the lead of the pouring vessel, the touch, pinching and pressing fingers, transitioning through space into new points of stasis. Wax is mobile, liminal, tractable. Suspended in the surface of our *Venus de' Medici* are traces of liveness: the pressings of cadavers whose bodies made the molds and the filmy imprints of centuries of hands that fondled her bodily parts, each time infinitesimally altering her wax topography through touch. Wax is not merely the Venus's visual medium—its materiality, too, contributes to her hermeneutic offering, by what I argue is a reconfiguration of the relation between body knowledge and time.

Wax is both the ideal and the abject material for the surrogation of a body in that its moist appearance, color depth, and what Ebenstein notes as its simultaneous transparency and opacity are uncannily flesh-like.[57] Wax is actually frequently equated with the uncanny, a sensation of not-quite-rightness that Freud theorized in 1919 as if describing the very experience of attraction and repulsion from the Venus's verisimilitude. The experience is frequently argued to be a product of the Enlightenment wherein stable binaries such as outside/inside, subject/object, and familiar/strange were first established.[58] Freud offers that the chilling uncertainty of the uncanny is evoked by encounters with dolls and automatons, which blur death with youth and immortality and suggest a likeness to the self. Most critical to the Venus's novel choreographics of deconstruction, the uncanny also unsettles time, and with it, order and sense. This unsettling is predicated upon a collapse of the real into unreal: here, the realities of medical skill and the fictions of a toy.

The way the Venus stands outside of time is perhaps her greatest technological capability and where the toy becomes a useful analytic in its material perpetuity. As Susan Stewart writes, toys are often fantasy objects meant to be puzzled over in solitude, even while they function as an other by stirring the imagination.[59] Yet according to Stewart, as an uncanny double, the toy contains an element of darkness tied to its temporality: it "threatens an infinite pleasure; it does not tire or feel, it simply works or doesn't work."[60] We might see this same uncanny interminability in the Venus whose body never decomposes and who remains frozen in pleasure as her dismemberment is infinitely repeated, never tiring in her mechanical abilities when we who "toy" eventually do. By cycling through a healthy, human life-likeness and a death-defying dismemberment, the Venus also obliterates the boundaries between reality and fantasy—what should happen when an anatomical form is disemboweled and what does not happen to hers. Going through the motions of deconstruction, the user is meant to experience something of what Foucault writes is the mythic "truth" of anatomy.[61] But a user could never repeatedly cycle through real dissection of the human cadaver: organic dissolution would get in one's way, integrity never regained.

In addition to their edification of the public, wax models held the potential to fundamentally clean up the act of medical research by replacing the rotting body with attractive frozen simulation; in so doing, I argue, they eradicate time as an element of dissection. But the further implication of this pedagogical simulation is the eradication of time from the episteme of the body altogether, in that knowledge and medical practice are predicated upon a theoretical body that is not subject to the decomposition or transfiguration that an actual encounter with organs would enact. The Venus implies a mode of medical thinking in the subjunctive: were we able to engage with the body outside of the ways that time and gravity changes it, this is how it would look and feel and this would be its structural integrity. As a body in the hypothetical, the Venus suggests that anatomy, like artifice, sculpture, and dance, is highly contingent upon imagination—holding the mimetic in mind even while dealing with the real.

The paradoxes of medical thinking in the congealed time of wax are epitomized in the Venus's hidden fetus. In *Unmarked*, Peggy Phelan writes of pregnancy as a subject par excellence of visibility: given that maternity is verifiable on the body by sight and touch while paternity (prior to modern technology) is mere hypothesis, "the authority of the patriarchy depends on a hierarchical relationship between the visible and the invisible, with the invisible (paternity and an invisible God) being the ascendant in the pair."[62] For an anatomical model molded with painstaking accuracy, the fetus is an oddity because it fails to reflect in the belly: it is visibly obscured as the final step of the choreography. The anonymous hands that uncover the Venus assume patriarchal authority through their command over visibility and, moreover, as what brings maternity to light by "discovering" the fetus. But without a swollen belly, the temporality of the Venus is jumbled. Her exterior shows a scenario in motion, in that she is tensed asymmetrically in a moment-between. Yet just one of her organs represents not its status in that moment, but its abstract functionality. In the end we might say that each piece of the Venus shows its "ideal" form: the uterus performs maternity because a uterus depicted empty would be pathological, while the exterior performs virginity as a body not yet claimed through pregnancy. Time collapses within her, and patriarchy into medical objectivity.

Indeed, in looking upon the Venus, can we rely on wax to be true or sincere—a term that only coincidentally contains the double entendre of "without wax"?[63] When and why could an anatomically correct simulacrum substitute for embodiment? Or could the deception performed by wax instead forebode a slippage, an inherent difference in how we sculpt, treat, and fantasize a life-like but not actually live or lived body? The Venus's emphasis on a visual-haptic continuum proves critical here. The Enlightenment medical exhibit centered around the Anatomical Venus illustrates an elevation of sight and touch by the idea that these specific senses can be taught and honed as skills of empirical observation that edify the citizen. The Venus did not teach the eager onlooker what a healthy body smelled or sounded like or ask what it feels like to be occupied. Moreover, she obscures from view certain intersections of the political and the anatomical, such as the poverty and neglect that brought cadavers to the operating table. These new models of medicine, after all, were fake wax bodies in pearls, politely frozen and lacking the sensations of organicism. The Anatomical Venus is a body that does not cry, stink, or suffer, and does not protest the viewer's sight, touch, and movement, but invites it.

Spectacular Neutrality

The Venus shows how an emergent need for awareness of a pluralistic European public radically shaped the way medicine was being interpreted in the Enlightenment. The project of cultivating enlightened citizens out of this public entailed teaching them how to look and marvel with discernment, and to utilize engagement with external subjects and objects to better understand themselves. I have considered the hermeneutical innovations of the Venus through its choreographic interactions, scenario, time, and repetition. I finally turn to display and gaze as domains of performance.

The Venus signals innovations in Enlightenment visual culture that would promote the consolidation of morphological differences as deviations from a "norm." Following the success of the Anatomical Venuses, cautionary wax exhibits emerged that showed the "undisciplined body, and . . . the dangers it posed for the subject if not properly controlled."[64] These wax models showed diseased and deformed body parts, the faces of criminals, as well as effects of pathologies and of "moral diseases" such as masturbation.[65] The European public were encouraged to attend exhibitions where they could enjoy discerning differences between themselves and the Other through a steady stream of visual lessons on how to look for details and gaze upon curiosities.[66]

Not only did anatomical museums develop to center on the medical abnormality; they also paved the way for the development of the nineteenth-century anthropological museum and its visual structuring of race. As metropoles brutally mined colonies for attractions, museums fed a fetishistic desire among European audiences to experience the Other from the comfort of the museum setting.[67] Popular science museums were often designed like a panopticon, where everything from animal curiosities, ventriloquists, freaks, and death masks, to "ethnic rarities" and ethnographic busts of "races of man" were exhibited in the same building.[68] This all-encompassing curation of anatomy, race, sexuality, fairground oddity, and disability violently conflated bodies into abject otherness. "Dr." Alsatian Spitzner's traveling Grand Musée d'Anatomie, which emerged in the mid-1800s but was still being publicized by 1908, featured demountable and diseased models alongside "Aztec and Hottentot" models.[69] Like the Venus, these visual troves offered aesthetic experiences where morbid curiosity gave way to the erotic, and where the symbology of physiological details are rewards in themselves.

This entertainment culture, centered on the practice of looking, would usher in the balletomania of the nineteenth century, which implicates proscenium dance in an educative spectrum of corporeality on display. By the first decade of the nineteenth century, anatomical collections were being built across European cities; Saartjie Baartman, the "Hottentot Venus," was forcibly exhibited in Europe between 1810 and 1815, and Marie Taglioni had her Vienna debut in 1822.[70] While this era was accompanied by an increased professionalization of dance and the spread of public theaters—arguably a distancing, even demotion of touch—the physical rift created between the audience and the admired body could be said to echo and magnify the Venus's encasement in glass. This distance does not ultimately prevent touch but showcases touch's complex means of consummation. This is evident in the material culture of memorabilia surrounding ballerinas like Taglioni following her rise to fame with *La Sylphide* in 1832, which, as Roach notes, allows the admirer to take physical part in the charisma of the dancer's body. It is also evident in the reviews of Théophile Gautier and contemporaries that dissect the ballerina's body into the granularities of anatomy—describing eyes, noses, necks, arms, feet, and toes to convey the magnitude of the visual event. Indeed, wax anatomical exhibits and the "cult of the ballerina" both reached their peak by the mid-nineteenth century.[71]

Between popular science museums and the Romantic ballet, a dialectical politics of visual display in the long nineteenth century brings the body itself into relief and renders it legible as site of the ideal and the abject. Public museum culture, concurrent with the Venus, opened a new realm of looking for pleasure, including its interplays with racial denigration, imperialism, and exploitation of the poor and disabled.[72] Yet these infinite ways that culture encroaches upon (and is the stuff of) medicine are obfuscated by the Venus, who represents the body as it is ostensibly meant to be: functioning, beautiful, unchanging, untouched by the world yet inviting the user's touch. Medicine's intersections with the political were increasingly withheld from sight in centuries to come as the Venus de Medici's pearls gave way to the positivist scientific body that we know today.

The Venus's "dancing" body offers compelling precedents to the dancing bodies of the nineteenth century: she contains a marvelous mechanical virtuosity, and like the *rats du ballet,* her visual pleasure is simultaneously sexual labor.[73] Notably, the Venus played a role as a critical marker in the genealogy of a "neutral body" within Western visual culture, a specter with which dance studies continues to contend and one that indelibly links dance to science. The "neutral body" obfuscated how bodies are situated in the world and conjured a vision of neutrality as white, young, slim, and able. Both Western medicine and dance have ideated this fantastical neutral body as a scaffold that supports their knowledge. But this body that stands outside of time, space, and culture is never attainable, is always a vanishing point.

Conclusion: "A Whole That Is an Aberration"

In this essay I have argued that what makes the *Venus de' Medici* exceptional is the accessible "choreographic" potential alive in her capability to demount and what this represents for the dissemination of medical knowledge. Through analysis of its choreographics, we can consider what this Enlightenment-era tool inherited and disavowed from anatomy theater practices in its repeatability, mimesis with hyperrealism, and internal dramaturgy. The Venus heralds new analytics for the intricacies of anatomical thinking at the turn of the eighteenth century into the nineteenth. This included sharing with the public a view of the body as spatially organized and made up of agential parts that constitute a whole, as well as the idea that the lay subject could learn anatomy through kinetic encounters with its interior. Moreover, in considering the cultivation of the lay citizen as body-expert, it is critical that contemporary scholars hold in mind how a medical gaze engenders forms of looking and discerning visual difference for pleasure.

The Venus also implies historical transformations through the role of fantasy in an interaction with her body. She conjures invitational fantasy that would have cited a familiar erotic visual repertoire for the eighteenth-century viewer as a means of sanctioning touch. But fantasy is also a tool for teaching the medical student to conceptualize a hypothetical body that stands outside of time and organicism, and this reveals a contingency in the Venus. In collapsing time and divorcing the study of anatomy from other sensorial and social elements, she inaugurates new dimensions of the body as a positivist, objective, and neutral thing, obscuring the pressing cultural issues surrounding anatomy and the visual of her era. This positivist body grounds much of today's dance studies discourse and yet is increasingly questioned as scholars and artists attempt to hold in mind the social contingencies and sordid histories of scientific thinking.

"The choreographic" becomes a performative mode of entry into the various aspects of the Venus as a way of thinking history in embodied and material terms—from touch with a representational body to spectacle, character, time, and repetition. I regard the Venus's body as a technology for the study of dance in the long nineteenth century because she suggests how choreographic thinking circulated far beyond the proscenium stage and even informed the medical sciences. And the spectacle of this female body on display, which was one of the harbingers of a visual entertainment culture, can be viewed as what Roach would call a "role icon" preceding the celebrated ballerinas of the century to come.[74]

The body, writes Stewart, is both contained and container, leaving us as viewing bodies perpetually concerned with its limits in space and as object.[75] But when taken apart and put back together, the individual pieces of the body are brought into focus, their gestalt cohesion becomes background, and a not-quite-rightness to the body emerges. It then becomes what Stewart calls "a whole that is an aberration."[76] In the spectacle of disintegration and reintegration, the body loses its power as either functional or aesthetic object.[77] There is something irreconcilable and yet uncannily liminal about the Venus's ideal exterior and abject interior, precluding the viewer from fully engaging on the terms of either one. The aberrant whole suggests that when one looks, one also thinks about this body's kinetic capability. As if spectating a dance, this gaze is anticipatory, awake to other possibilities that form can take and already designing their movement. Ultimately, this may answer for those anecdotal touches in the Venus's archive. Viewers are made aware that she invites their physical interaction, but how they respond to that invitation is theirs to choreograph.

Notes

1 This *Venus de' Medici* is preceded by an undated marble one of the same name, which came into the Medici family collection between 1598 and 1638, and currently stands in the Uffizi Gallery's *Tribuna,* its home since the 1680s ("Venus de' Medici," *Boston Athenæum*). Strikingly, this sculpture is not mentioned in the literature I have reviewed on the wax *Venus de' Medici,* but we can surmise that the ceroplasticians of La Specola would have encountered it, as the Uffizi opened to the public in 1769 ("Uffizi Gallery Tickets," *Florence Museum*).

2 For more on this history see Ebenstein, *The Anatomical Venus;* Stephens, *Anatomy as Spectacle,* 6; and "University of Cagliari," 161.

3 By "agential," I refer to the authority of the individual organ when considered both apart from and within the body. By the end of the eighteenth century the organ was being recognized semi-autonomously as a discrete contributor to internal systems (Foucault, *The Birth of the Clinic,* 5).

4 Ebenstein, *The Anatomical Venus;* Jordanova, *Sexual Visions;* Maerker, *Model Experts;* Stephens, *Anatomy as Spectacle.*

5 Mandressi, "Of the Eye and of the Hand," 61.

6 By this phrase I aim to identify an approach to choreography as an organized but non-prescriptive system of movement that often exists outside of formal aesthetic frames like "dance," but that nonetheless produces pedagogical, intellectual, erotic, or aesthetic effects upon the performer. An example is the way one moves through a museum with purpose in order to take in art physically, which may be habituated, but can also be deviated from.

7 My approach is modeled after Robin Bernstein's essay "Dances with Things," which proposes a method for unlocking past encounters between people and material culture by questioning the behaviors objects may have invited and encouraged beyond what their makers intended. This "agency [of material culture] emerges through constant engagement with the stuff of our lives," Bernstein writes (*Racial Innocence,* 12). I take up her call to engage with the interactions, desires, and ideas embedded in the historical artifact, because I believe the Venus's value was in large part her capacity to be dismembered: an embodied potential known of far beyond specific instances of her dismemberment's performance.

8 Franko, *The Work of Dance,* 1–12.
9 Auslander, "On Repetition," 88.
10 See Nochlin, *The Body in Pieces;* Davis, *Enforcing Normalcy;* Roach, *It.* Examples of fragmented depictions are *The Acts of the Christian Martyrs* (1954), *Torso of a Woman* (100 BC–400 AD), *the Venus de Milo* (150–125 BC), Caravaggio's *Medusa* (1596), Gustave Courbet's *l'Origine du Monde* (1866), *L'Inconnue de la Seine* (c. 1880), Pablo Picasso's and Hans Bellmer's women, Réné Magritte's *L'Evidence éternelle* (1930), Salvador Dali's *Portrait of a Passionate Woman* (1945), Marcel Duchamp's *Étant donnés: 1° la chute d'eau, 2° le gaz d'éclairage . . .* (1946–1966), Louise Bourgeois's *Sainte Sébastienne* (1992), and Cindy Sherman's *Untitled #250* (1992).
11 Groth, "Scopophilia," 321.
12 Maerker, *Model Experts,* 10.
13 Maerker, *Model Experts,* 130.
14 Park, *Secrets of Women,* 23.
15 Federici, *Caliban and the Witch,* 148.
16 Ebenstein, *The Anatomical Venus,* 42–43.
17 "University of Cagliari," 161; and Stephens, *Anatomy as Spectacle,* 6.
18 Ebenstein, *The Anatomical Venus,* 18, 32–33.
19 Maerker, *Model Experts,* 129.
20 "University of Cagliari," 161; *Madame Tussaud and the History of Waxworks,* 136; Ebenstein, *The Anatomical Venus,* 33, 148–49.
21 Wils, de Bont, and Au, "Introduction," *Bodies Beyond Borders,* 10–11.
22 Carlyle, "Artisans, Patrons, and Enlightenment," 33.
23 Mandressi, "Of the Eye and Of the Hand," 60, 74.
24 Mandressi, "Of the Eye and Of the Hand," 63.
25 Mandressi, "Of the Eye and Of the Hand," 62.
26 Mandressi, "Of the Eye and Of the Hand," 61–62.
27 Foucault, *The Birth of the Clinic,* 15–16.
28 Mandressi, "Of the Eye and Of the Hand," 69.
29 Mandressi, "Of the Eye and Of the Hand," 70.
30 Mandressi, "Of the Eye and Of the Hand," 74.
31 Stewart, *On Longing,* 155.
32 Stewart, *On Longing,* 161.
33 Maerker, *Model Experts,* 123, 128, 133–36.
34 Maerker, *Model Experts,* 135.
35 Maerker, *Model Experts,* 123.
36 Maerker, *Model Experts,* 132.
37 Ebenstein, *The Anatomical Venus,* 36–37; Kemp and Wallace, *Spectacular Bodies,* 6.
38 Deblon, "Imitating Anatomy," 120.
39 Kemp and Wallace, *Spectacular Bodies,* 63.
40 Kemp and Wallace, *Spectacular Bodies,* 63; Ebenstein, *The Anatomical Venus,* 36.
41 Ebenstein, *The Anatomical Venus,* 32.
42 Roach, *It,* 74.
43 Roach, *It,* 55.
44 Maerker, *Model Experts,* 132.

45 Ward, "The History of Pearls."

46 Roach, *It,* 165.

47 What it would mean conceptually to sculpt "death" (versus deadness) is a conundrum embodied in the Venus and further points to the inaccuracy of comparing her to a corpse. According to Tristanne Connolly, the contemplation of the spectacle of dying was a preoccupation of the Enlightenment ("Introduction," *Spectacular Death,* 3). For Connolly, this explains why the moments leading up to death are endlessly depicted, while the instance of death is in reality arbitrary and unspectacular.

48 Kemp and Wallace, *Spectacular Bodies,* 61; Ebenstein, *The Anatomical Venus,* 180.

49 See Burrus, *The Sex Lives of Saints;* Castelli, *Visions and Voyeurism;* Frankfurter, "Martyrology and the Prurient Gaze."

50 Jordanova, *Sexual Visions,* 55, 58.

51 See also Bronfen, *Over Her Dead Body;* and Pilbeam, *Madame Tussaud and the History of Waxworks.*

52 Maerker, *Model Experts,* 132–33.

53 While, like other eighteenth-century Venuses, Susini's went to a medical clientele, Susini consistently added ornamental details such as the bedsheets, arched back, open mouth, and pearls (Stephens, *Anatomy as Spectacle,* 37).

54 Pilbeam, *Madame Tussaud and the History of Waxworks,* 5.

55 Pilbeam, *Madame Tussaud and the History of Waxworks,* 16.

56 Pilbeam, *Madame Tussaud and the History of Waxworks,* 131.

57 Ebenstein, *The Anatomical Venus,* 70.

58 Freud, "The Uncanny," in Ebenstein, *The Anatomical Venus,* 203, 208; Kutze, "The Three Aspects of the Uncanny."

59 Stewart, *On Longing,* 56.

60 Stewart, *On Longing,* 57.

61 Foucault, *The Birth of the Clinic,* 48, 120

62 Phelan, *Unmarked,* 137.

63 This erroneous etymology, which claims *sin-cere,* meaning "without wax," stems from a Roman practice of filling cracked ceramics with wax, is common in scholarship on waxworks. The actual root of *sincere* is the Latin *sincerus,* meaning clean or pure ("sincere [adj.]" *Etymology Online*).

64 Stephens, *Anatomy as Spectacle,* 280.

65 Ebenstein, *The Anatomical Venus,* 140–41.

66 Stephens, *Anatomy as Spectacle,* 31.

67 Tyburczy, *Sex Museums,* 20; and Macdonald, "Corpse Stories," 73.

68 Ebenstein, *The Anatomical Venus,* 131, 140.

69 Stephens, *Anatomy as Spectacle,* 26; Pilbeam, *Madame Tussaud and the History of Waxworks,* 136.

70 Tyburczy, *Sex Museums,* 14, 66.

71 Stephens, *Anatomy as Spectacle,* 26; "Romantic Ballet," *Victoria and Albert Museum.*

72 Tyburczy, *Sex Museums,* 21.

73 McCarren, *Dance Pathologies,* 75

74 Roach, *It,* 39.

75 Stewart, *On Longing,* 104.

76 Stewart, *On Longing*, 108.
77 Stewart, *On Longing*, 105.

Bibliography

Auslander, Philip. "On Repetition." *Performance Research* 23, n. 4/5 (2018): 88–90.

Bernstein, Robin. *Racial Innocence: Performing American Childhood from Slavery to Civil Rights.* New York: New York University Press, 2011.

Bronfen, Elizabeth. *Over Her Dead Body: Death, Femininity, and the Aesthetic.* Manchester, UK: Manchester University Press, 1992.

Burrus, Virginia. *The Sex Lives of Saints: An Erotics of Ancient Hagiography.* Philadelphia: University of Pennsylvania Press, 2010.

Carlyle, Margaret. "Artisans, Patrons, and Enlightenment: The Circulation of Anatomical Knowledge in Paris, St. Petersburg, and London." In *Bodies Beyond Borders: Moving Anatomies 1750–1950,* edited by Kaat Wils, Raf de Bont, and Sokhieng Au, 23–50. Leuven, Belgium: Leuven University Press, 2017.

Castelli, Elizabeth A. *Visions and Voyeurism: Holy Women and the Politics of Sight in Early Christianity.* Berkeley, CA: Center for Hermeneutical Studies, 1995.

Connolly, Tristanne. "Introduction." In *Spectacular Death: Interdisciplinary Perspectives on Mortality and (Un)representability,* edited by Tristanne Connolly, 1–20. Bristol: Intellect, 2011.

Davis, Lennard. *Enforcing Normalcy: Disability, Deafness, and the Body.* London: Verso Books, 1995.

Deblon, Veronique. "Imitating Anatomy: Recycling Anatomical Illustrations in Nineteenth-Century Atlases." In *Bodies Beyond Borders: Moving Anatomies 1750–1950,* edited by Kaat Wils, Raf de Bont, and Sokhieng Au, 115–38. Leuven, Belgium: Leuven University Press, 2017.

Ebenstein, Joanna. *The Anatomical Venus: Wax, God, Death and the Ecstatic.* New York: D.A.P./Distributed Art Publishers, 2016.

Federici, Sylvia. *Caliban and the Witch: Women, the Body, and Primitive Accumulation.* New York: Autonomedia, 2004.

Foucault, Michel. *The Birth of the Clinic: An Archaeology of Medical Perception.* New York: Vintage Books, 1994.

Franko, Mark. *The Work of Dance: Labor, Movement, and Identity in the 1930s.* Middletown, CT: Wesleyan University Press, 2002.

Frankfurter, David. "Martyrology and the Prurient Gaze." *Journal of Early Christian Studies* 17, n. 2 (2009): 215–45.

Freud, Sigmund. "The Uncanny." 1919. https://web.mit.edu/allanmc/www/freud1.pdf. Accessed Apr. 23, 2022.

Groth, Miles. "Scopophilia." In *Encyclopedia of Gender in Media,* edited by Mary Kosut, 321–23. Thousand Oaks, CA: SAGE Publications, 2012.

Jordanova, Ludmilla J. *Sexual Visions: Images of Gender in Science and Medicine between the Eighteenth and Twentieth Centuries.* Science and Literature. Madison: University of Wisconsin Press, 1989.

Kemp, Martin, and Marina Wallace. *Spectacular Bodies: The Art and Science of the Human Body from Leonardo to Now.* London: Hayward Gallery, 2000.

Kutze, Donald. "The Three Aspects of the Uncanny." http://art3idea.psu.edu/locus/three_uncannies.pdf. Accessed Sept. 13, 2019.

MacDonald, Helen. "Corpse Stories: Anatomy, Bodies and a Colonial World." In *Bodies Beyond Borders,* edited by Kaat Wils, Raf de Bont, and Sokhieng Au, 73–90. Leuven, Belgium: Leuven University Press, 2017.

Maerker, Anna. *Model Experts: Wax Anatomies and Enlightenment in Florence and Vienna, 1775–1815.* Manchester, UK: Manchester University Press, 2001.

Mandressi, Rafael. "Of the Eye and of the Hand: Performance in Early Modern Anatomy." *TDR: The Drama Review* 59, n. 3 (2015): 60–76.

McCarren, Felicia. *Dance Pathologies: Performance, Poetics, Medicine.* Stanford, CA: Stanford University Press, 1998.

Nochlin, Linda. *The Body in Pieces: The Fragment as a Metaphor of Modernity.* London: Thames & Hudson, 1994.

Park, Katharine. *Secrets of Women: Gender, Generation, and the Origins of Human Dissection.* New York: Zone Books, 2006.

Phelan, Peggy. *Unmarked: The Politics of Performance.* New York: Routledge, 1996.

Pilbeam, Pamela. *Madame Tussaud and the History of Waxworks.* New York: Hambledon and London, 2003.

Roach, Joseph. *It.* Ann Arbor: University of Michigan Press 2007.

"Romantic Ballet." *Victoria and Albert Museum.* http://www.vam.ac.uk/content/articles/r/romantic-ballet. Accessed Jan. 26, 2024.

"Sincere (adj.)." *Etymology Online.* https://www.etymonline.com/word/sincere. Accessed Nov. 1, 2023.

Stephens, Elizabeth. *Anatomy as Spectacle: Public Exhibitions of the Body from 1700 to the Present.* Liverpool: Liverpool University Press, 2011.

Stewart, Susan. *On Longing: Narratives of the Miniature, the Gigantic, the Souvenir, the Collection.* Durham, NC: Duke University Press, 1993.

Tyburczy, Jennifer. *Sex Museums: The Politics and Performance of Display.* Chicago: University of Chicago Press, 2016.

"Uffizi Gallery Tickets." *Florence Museum.* https://www.florence-museum.com/uffizi-gallery-tickets.php. Accessed June 13, 2022.

"University of Cagliari: The Evolution of Anatomical Illustration and Wax Modelling in Italy from the 16th to Early 19th Centuries." In *Issues in Anatomy, Physiology, Metabolism, Morphology, and Human Biology,* edited by Q. Ashton Acton, 160–61. Atlanta: Scholarly Editions, 2011.

"Venus de' Medici." *Boston Athenæum.* https://www.bostonathenaeum.org/about/publications/selections-acquired-tastes/venus-de-medici. Accessed June 13, 2022.

Ward, Fred. "The History of Pearls." *Nova | PBS* Dec. 28, 1998. https://www.pbs.org/wgbh/nova/article/history-pearls/. Accessed May 17, 2022.

Wils, Kaat, Raf de Bont, and Sokhieng Au. "Introduction." In *Bodies Beyond Borders: Moving Anatomies 1750–1950,* edited by Kaat Wils, Raf de Bont, and Sokhieng Au, 9–20. Leuven, Belgium: Leuven University Press, 2017.

2

Choreographies of Knowledge

Touch and Vision in Anatomical Looking

Response to "Venus in Pieces: Choreographing the Anatomical Body with Clemente Susini's *Venus de' Medici*" by Sariel Golomb

Jane Desmond

Soft and smooth, warming under the hand, and infinitely shapable, wax is beguiling in its invitation to touch and its capacity to hold a form that lies on the edge of human-made-object and organic matter, coming as it does from the labor of bees. In her eloquent consideration of a specific anatomical model, the "Venus de' Medici" from late eighteenth-century Italy, Sariel Golomb reveals the probative value of bringing a choreographic analysis from the stage to the realm of objects. In this case, these are aesthetic and practical objects like wax "Venuses" meant to delight and inspire and, above all, to teach the public, as well as male students of anatomy, what might then have been regarded as the "mysteries" of female interior anatomy.

By bringing "a dancer's attunement to physicality and interaction" to bear on her project, Golomb expands the notion of choreographic study from the stage and the ballroom into the realm of everyday action and instruction through the matrix of vision, and, depending on the context of encounter, through actual or imagined touch, movement, assembly, and disassembly. Interbodily relationality, a key to analyzing dance movement, is here recast as a dance between bodies—one human, active, and living, the other an object representing a female human, formed and colored in wax as if formerly living. Historically and culturally specific conventions of artistic representation and gender haunt the interplay between historical actors (the students of anatomy, the museumgoers) and their pliable "partner" in this specific choreography of knowledge. That interplay choreographs what film scholar Jane Gaines in a different context has usefully termed "a right to look," both through its situation in the emergent European Enlightenment realm of the scientific and through the substitution of a wax body for a human one.[1] These three-dimensional forms, as Golomb argues, foreground the politics and erotics of touch or imagined touch.

Created between 1780 and 1782 for the first public anatomy museum in Europe, the wax "Venuses" can be analyzed in the wider context of shifting notions about the body, and the role of museums, which were moving from elite private collections to public institutions, instructing citizens in the values of the upper classes and the emerging secular notions of bodies as knowable collections of working parts. And, in noting the wider archival record that shows that such looking (and handling) of wax models also existed in a more explicitly erotic realm, in the private collections of wealthy men, Golomb also tracks the undercurrent of eroticism that coexisted with the emerging "scientific gaze." Such a finding reminds us that meaning-making through objects is never limited to their intended purpose, publics, or dominant discourses.

Golomb anchors her interpretive considerations in the historical record, noting its necessary gaps, while sketching a broader architecture of cultural histories that engages with larger concepts of cultural difference. These included notions of "madness," "race," and "normal" versus "freakish," categories that were played out in a variety of settings from museums to exhibits at fairs, and which, in different guises, had, and continue to have, long, lingering legacies intersecting with notions of "science."

One of the legacies that resonates with my own work on the contemporary display of dead bodies at the confluence of "art" and "science" is seen in the wildly popular "Body Worlds" shows, organized by Gunther von Hagens.[2] These shows (and their imitators) have drawn tens of millions of viewers to both art and science museums in many countries since premiering in Japan in 1995. Featuring exhibits of preserved human bodies displayed in everyday activities like playing chess or running, cadavers, donated by these individuals to support von Hagens's work, are prepared in a process of dry, odorless preservation called "plastination," often with their skin removed, and organs opened for display of interior structures.

While subject to some complaints based on religious objections, the exhibits have been among the most popular museum shows of all time, attracting tens of millions of visitors, from children on school field trips, to adults and families of all ages and demographic backgrounds. I explored the cultural work that these exhibits do in detail in an essay on "Postmortem Exhibitions" in my book *Displaying Death and Animating Life: Human-Animal Relation-*

ships in Art, Science and Everyday Life,[3] but here I want to contrast those to the wax Venus.

Unlike the wax "Venus" sculptures, which enabled a realism that spoke to the aesthetic and representational conventions of the time in Italy, von Hagens's contemporary exhibits showcase the "real" human body, and this is both the claim to uniqueness and the cause of their occasional censure and public protest. While the extensive exhibits reach museumgoers, von Hagens's patented "plastination" process for preserving human bodies is also used for anatomical teaching models. The price of public access to this exhibition of the "real," though, is an absolute prohibition of touch in the museum. The prohibition is intended not only to protect the physical integrity of the preserved tissues, but also to protect the individual after death from what in life would be transgressive touch—touch without permission.

The von Hagens pieces (which he terms "specimens") also invite the type of imaginary animation that Golomb ascribes to the reclining Venus figure, when she mentions the slight tensing of muscles. But with Body Worlds, the imagined animation of the figures for the viewer comes from their active poses—"The Runner" with one knee in the air, or "The Ballerina" posed midturn with her de-skinned foot balanced on a pink toe shoe. These animating postures choreograph a moment in a life of lively action, announcing both a subjectivity and a narration of experience in time. The von Hagens exhibits are built on signifiers of specificity of lives lived, even if, with most of the skin removed, markers of social differentiation like age and race are muted. The bodies thus oscillate between intense individuation ("this" human) and their status as universalized signifier of "the" human.

Notably, one of the controversial figures in the Body Worlds' exhibit was that of a pregnant woman (none of the individuals are named) who is presented with her Venus-like body opened to reveal the fetus in her womb. The fetus, unlike most of the other bodies on display, does not have the skin removed, resulting in a disturbing oscillation between the categories of the dead and the unborn. The woman herself has skin remaining in the areas of her lips and aureoles, sensitive and sensual parts of the body, echoing the potential erotic dimension that Golomb notes in her Venus discussion.

In the von Hagens plastinate, the reclining figure was disturbing in part because the languid pose and sexualized lips and nipples contradicted the "scientific gaze," intruding on it in too direct a way. Today's viewers bring a context to this piece that includes not simply anatomical questions, but medical ones, such as, why couldn't they save the fetus? Or, who made the decision to donate this body? These thrust us into the life of the individual before us in such an urgent way that "specimen" becomes once again "person."

Unlike the shift in status described in Golomb's reading of the wax Venus, as sensualized surface gives way through *sequential* disassembly to an interiority that becomes a "disemboweled frame," the von Hagens exhibit invokes a *simultaneity* of vision of inside and outside. We see both the mother-to-have-been and her fetus at the same time. We may look with wonder at the interiority we cannot normally see, but we cannot help but simultaneously confront the personhood of this individual before us. This is a type of psychic intimacy. And imagined touch is not really invited, or choreographed here—in fact it can be repelled. It is only the eyes that make contact, and in this the sensuality accorded the wax Venus figure is different, precisely because that model *is* a model, not a "real" human body. This Body Worlds "specimen" cannot invite us, as Golomb notes about the wax Venus, to "dance the line between reality and fantasy."

Golomb has extended an invitation to use choreographic analyses to understand the intersection of representation, imagination, and bodily sensations in the realm of scientific knowledge production in the late eighteenth century in Europe. But, beyond the specifics of the case she details, I suggest that, building on Golomb's analysis, a choreographic analysis can be widely useful when applied with historical and cultural specificity. This productive and powerful approach can deepen our understanding of other types of images, models, and exhibits in other specific times and places. These could range from manikins to paper dolls, to Virtual Reality training for surgeons, to 3-D models of cows for palpation training for veterinarians, and to the use of future technologies as yet unimagined to probe the body's depths and touch its surfaces. How, I suggest we should ask, is knowledge choreographed through vision and touch and imagination? How do bodily relationalities—depicted, imagined, speculated upon, rehearsed, or performed in material or de-materialized mediums, always enact a politics of knowing? And if what I am dubbing here as such "choreographies of knowledge" are found not only on the stage, in the studio, the dance club, or music video, but also in museums like Golomb notes, and beyond that too—in sculptures and video games and surgeon-training VR platforms—then developing the "dancer's attunement to physicality and interaction," which Golomb so powerfully articulates regarding anatomical models, might be a skill that scholars far beyond the worlds of dance should develop.

Notes

1 Gaines, "White Privilege and Looking Relations," 76.
2 von Hagens, "Body Worlds."
3 Desmond, "Postmortem Exhibitions," Chapter 2 in *Displaying Death and Animating Life*.

Bibliography

Desmond, Jane. *Displaying Death and Animating Life: Human-Animal Relationships in Art, Science and Everyday Life*. Chicago: University of Chicago Press, 2016.

Gaines, Jane. "White Privilege and Looking Relations: Race and Gender in Feminist Film Theory." *Cultural Critique* 4 (Autumn 1986): 59–79.

von Hagens, Gunther. "Body Worlds." https://bodyworlds.com/exhibitions/human/. Accessed June 3, 2022.

3

Science under the Surface

Victorian Science in the Ballet *Ondine*

Steven Ha

Ondine, ou La naïade (1843) is a Romantic ballet choreographed by Jules Perrot (1810–1892) and starring the ballerina Fanny Cerrito (1817–1909). In the course of this essay, I will argue that this work resonated with its audiences in unique ways, in part owing to interpretations viewers could make that connected *Ondine* with emerging scientific developments of the time.

Although inspired by Friedrich de la Motte Fouqué's novella *Undine* (1811)[1] about a water fairy who falls in love with a mortal man, Perrot's ballet diverged from its source material. The events of the ballet revolve around a festival of the Madonna in a Sicilian fishing village and preparations for a wedding between a humble fisherman, Matteo, and an orphan named Giannina. In a moment of solitude at sea, Matteo suddenly encounters the naiad Ondine. Captivating and deadly, she is a supernatural creature in the Romantic tradition, and nearly lures Matteo to death by drowning before he is fortuitously rescued by other villagers. Ondine, however, remains infatuated with Matteo and kidnaps Giannina to take her form and marry Matteo in her place. Ondine briefly succeeds at a life of earth-bound domesticity, until it becomes evident that her aquatic origins preclude her long-term survival on land. The weakening Ondine is eventually recovered by her mother, the Queen Hydrola, who also restores Giannina to her rightful place. Ondine revives once she is returned to the water and the ballet concludes with a spectacular finale of fountains, naiads, and dazzling stage effects.[2]

Ondine premiered against the backdrop of nineteenth-century London's deep engagement with science. At first glance, Perrot's ballet has little to do with the empirical investigations of the Victorians, investigations that spanned class lines and included professional and amateur scientists.[3] Scholars have elucidated ways in which science was entangled with the supernatural fairy: during this period, Victorians tried using scientific methods to prove the existence of fairies; they looked to the fairy as an escape from industrial modernity, and embraced the fairy in literature to embody and even teach the processes behind the phenomena of the natural world.[4] Supernatural characters—including fairies—were integral to some scientific literature, narrating stories, appearing in images, and explaining the invisible processes in nature.[5] Although such arguments establish the links between fairies and science, and the Romantic ballet is included in discussions of fairy art of the nineteenth century, the triangulation with the Romantic ballet and Victorian science warrants further investigation.

The impacts of science and technology on the Romantic ballet are frequently discussed in relation to innovations in stagecraft and the development of the pointe shoe, and predominantly focus on the sylph, or fairy of the air. Less attention has been given to the cultural impacts of science on the dancing fairies of the water, like Ondine, who offers some parallels to her sister sylphs, as well as a different view into Victorian fascinations with science. The increasing availability of microscopes and the proliferation of scientific literature on aquatic life-forms made water a particular subject of interest for the Victorians.[6] The espousal of studies in natural history for health and moral benefit also made marine flora and fauna popular objects for observation and collection.[7] These widespread interests in science created ripe conditions for *Ondine* to appeal favorably to the Victorian public.

Nina Auerbach argues that the theater was "the prism through which *all* Victorian artists and audiences . . . saw their world," identifying the theater as a vital component of Victorian culture.[8] Thus, I suggest that Victorian obsession with the underwater world meant that *Ondine* held unique appeal for its audiences. I will examine this ballet and speculate about how this dance circulated within discourses of fairies and science in its time: I argue that in *Ondine,* the Victorians saw a version of themselves and their relationship to contemporaneous scientific progress. I explore this notion through Victorians' enthusiasm for the supernatural in the visual arts and in popular theater; evolving technologies; the study of natural sciences among the general population; and scientific advances in the ocular, aerial, and aquatic realms.

The Fairy in the Visual Arts and Popular Theater

While the ethereality of dancing on pointe became a hallmark of the sylph in Romantic ballet, Ondine also used this technique to express her otherworldliness. Katherine Newey notes how developments in the pointe shoe contributed to producing the supernatural qualities of the sylph, who in practicing the most recent additions to ballet technique, embodied the illusion of flight.[9] The interconnection between fairies and flight was so pervasive that the Victorians apparently took no issue with a water fairy who could fly; Ondine is even described in an 1844 summary as a naiad "to whom the power of floating in the air is given equally with that of gliding upon, through, or beneath the waters."[10] That author does not explicitly state why both movement through air and water are ascribed to Ondine, perhaps simply making a straightforward comment about the setting of *Ondine* without reading anything deeper into the work. The nineteenth-century French poet/critic Théophile Gautier once dismissively remarked on ballet that "the literature of the legs is hardly a subject for contention," essentially lacking profundity and unworthy of literary analysis.[11] However, some audiences might have associated Ondine with a water bug. The proliferation of print materials such as books and magazines, of grand exhibitions, lectures, and scientific demonstrations indicate the extent to which science and recreation were integrated for the middle class and educated members of the lower class.[12] In this light, entomologically inclined Victorians who knew about orthopteran insects with amphibious life cycles or insects with adaptations for both flight and aquatic mobility might have understood Ondine as echoing observations in the natural world.

When *La Sylphide* (1832) debuted in Covent Garden shortly after its inaugural performances in Paris, Marie Taglioni created a sensation with London audiences.[13] Lithographs immortalized Taglioni as a winged fairy, corresponding with a visual precedent established by Thomas Stothard's illustrations in 1797 for Alexander Pope's *The Rape of the Lock* (1717). Stothard attached more accurate depictions of butterfly wings to the fairy body, "align[ing] the miniature fairy with the burgeoning disciplines of scientific naturalism in general and entomology in particular."[14] The insect/fairy metaphor persists in Albert Smith's *The Natural History of the Ballet-Girl* (1847), which describes the ballet dancer's growth from young student to professional in terms that echo insect metamorphosis: "This is the Ballet-girl in her first stage of transition from the flying fairy—the stages being Fairy, Extra, Corps de Ballet, and Coryphée."[15] While the original costume for Fanny Cerrito in *Ondine* did not have wings, an American-made lithograph from 1846 of the ballerina Fanny Elssler in the same role shows her with insect wings.[16] The conflation of Ondine with a winged fairy suggests that an availability of scientific knowledge and an inability to resist the ubiquity of fairies and flight all converge in this image, to produce an Ondine that resembles both fairy and insect.

Precision with morphological features of animals appears elsewhere in relation to *Ondine;* in its setting, marine fauna are depicted in detail. William

Figure 3.1. *Mademoiselle Fanny Cerito/in the Grand Ballet of Ondine, produced at Her Majesty's Theatre, June 22nd 1843.* Hand-colored lithograph by C. Graf, after drawing by Numa Blanc. Victoria & Albert Museum Collections. Image ©Victoria & Albert Museum, London. Given by Dame Marie Rambert.

Grieve designed the scenery, likely in collaboration with his family enterprise that produced scenes for London's most prominent theaters throughout the nineteenth century.[17] The Grieve family had a reputation for a Romantic style and landscapes that incorporated realism, as "audiences expected a reasonable degree of verisimilitude, but they wanted enchanting stage pictures as well."[18] It remains unclear if Perrot requested that Grieve create an accurate representation of nature. Perrot's ballet does draw elements from *Ondine, ou la Nymphe des eaux,* a *féerie* (a genre of French spectacular theater) by the playwright René-Charles Guilbert de Pixérecourt, including the Sicilian setting and the use of a shell as Ondine's mode of transportation.[19] Much to the delight of Victorian audiences, Ondine's first appearance has her rising through the stage on an enormous shell, which Susan Au attributes to numerous sources including Botticelli's *The Birth of Venus.*[20] Like Botticelli's rendition, Ondine's shell is identifiably a scallop, which, despite the issue of scale,[21] echoes a trend in the visual arts during the mid-nineteenth century when fairy painters like John Anster Fitzgerald (aptly nicknamed "Fairy Fitzgerald") incorporated more realism into their works by drawing foliage, animals, and other objects of nature with considerable accuracy.[22]

Supplementary artwork depicting Ondine's entrance in Charles Heath's *Beauties of the Opera and Ballet* (1844) follows suit; the illustration depicts Matteo's encounter with Ondine on her scallop shell, an image lifted from the ballet but placed in a natural setting. The foregrounded shoreline is littered with detailed sketches of a sea urchin skeleton, cowrie and abalone shells, et cetera.[23] The grotto that constitutes the background is also drawn with striking accuracy, such that the image blurs the boundaries between the natural world and the ballet. Similar juxtapositions of recognizable marine fauna with the fantastical image of the fairy also appear in fiction and in science literature, where the fairy is employed to cross the boundary between the supernatural and natural worlds.[24] The image of Ondine on a scallop shell symbolizes a melding of fantasy and science. Although mounting empiricism seemingly opposed the frivolity of superstition, the fairy persisted as a symbol of wonder and cultivated curiosity in nature.[25] Ondine can then be interpreted, in part, as an embodiment of Victorian interests in science and, perhaps, it reified the enthusiasm for studying water and marine fauna.

Stagecraft and the Supernatural

Ondine's grand entrance on the scallop relied on machinery and a trapdoor, exemplifying the dancing fairy as both a product and an embodiment of the industrial era and modernity.[26] Newey writes that the transformation scenes in British popular theater that transported audiences to fairyland depended on the technological innovations and skill of the craftsmen who built and operated the various apparatuses.[27] Perrot's ballet fully embraced such spectacle, enhanced by Grieve's designs, which satisfied the audience's expectations for realism with enchantment. One critic for *The Morning Post* wrote in 1843 that the scenes harmonized fantasy with reality in grand fashion, "at one moment representing nature with surpassing truth; at another embodying the poet's imaginings of fairy land. . . . The concluding *tableau,* with its rainbows, its gushing moving fountains, its transparent walls of rippling water, are only a few of the remarkable views. There is a *coup d'oeil* of most startling effect when *Matteo* rows *Ondine* across the water, and the Naiades are seen under the waves pursuing the boat."[28] Whereas Ondine's aerial qualities are largely attributed to the technology of her dancing, her supernatural qualities associated with water rely on the technologies of machinists.

Although the specifications for the technology in *Ondine* are not described in detail, they can be inferred from their use in contemporaneous theatrical traditions. A sketch of Ondine's appearance in the finale of the ballet depicts the ballerina standing knee-deep in water, surrounded by cattails and lily pads, underneath a two-tiered fountain.[29] She raises her arms overhead to create an oval shape, simultaneously framing her face and using her hands to part the curtain of water. By this time, fountains had been installed in several London theaters, fitted with water tanks and pumps.[30] They were marketed as a selling point for some productions, such as W. T. Moncrieff's *Giselle; or, The Phantom Night Dancers* (1841), a play that subsequently drew from the same libretto as the ballet *Giselle, ou les Wilis* (1841).[31] Moncrieff's *Giselle* contained ballet as a part of its spectacle but differs from the work choreographed by Jean Coralli and Jules Perrot. The finale of Moncrieff's interpretation of the story takes place in "The Palace of the Hundred Fountains" and stage directions for Myrtha's entrance indicate that "she is surrounded by playing fountains of real water, brilliantly illuminated by variegated fires."[32] Colored fire was produced by burning chemical powders in metal boxes that stagehands then held aloft on poles; a frequent device in transformation scenes representing fairyland.[33] References to rainbows and "coloured fountains"[34] in the finale of *Ondine* imply that the same effects were used for Perrot's ballet.

Innovations in lighting also played an important role. The installation of gas lights in 1818 in the King's Theatre (renamed Her Majesty's Theatre in 1837) enabled the darkening of the auditorium.[35] The shadowy atmosphere of dimmed gas lights enhanced the supernatural effects of the Romantic ballet and its numerous iterations of fairyland. In 1843, a critic for the *Examiner* noted *Ondine*'s capacity to indicate differences in location and the passage of time: "We have fantastic marine pictures glowing with cerulean splendour and fountains that rise up in abundance, each inhabited by its peculiar naiad. We have beautiful dioramic effects: and the same scene is made successively to pass through daylight, the tints of evening, and a brilliant moonlight."[36] Lighting effects were crucial to the famous *pas de l'ombre,* in which Ondine, having been transformed into a human in order to join her beloved Matteo, dances with her shadow, thrown into sharp relief by the theatrical lighting.[37] The impact of *Ondine* would have been greatly diminished without these developments in stage technologies to enhance the supernatural aspects of the ballet.

Victorians and Natural History

Science contributed to the shaping of Victorian attitudes toward women, a factor in understanding the impact of science on the ballet *Ondine*. Ann B. Shteir notes that with the European Enlightenment, science was incorporated into culture at large, reaching out to women with publications that espoused studies in various scientific fields as beneficial for morality, spirituality, conversation, and motherhood.[38] This outlook carried into the nineteenth century, exemplified by prescriptive texts such as *The Young Lady's Book: A Manual of Elegant Recreations, Exercises, and Pursuits* (1829), which included chapters on physical activities, morality, fashion, and natural history.[39] Although the natural history chapters on botany and conchology include instruction on the ornamental uses of flowers and shells, the majority of the text is dedicated to taxonomic classification of species with in-depth descriptions and diagrams of morphological characteristics. The duality of aesthetic concerns and science reflects the difficulties faced by Victorian women, who had to negotiate a societal encouragement of natural history study along with concerns about transgressing gender norms that discouraged the female scientist.[40] Nevertheless, many women contributed to science literature, especially in the latter half of the nineteenth century in such fields as botany, evolutionary biology, marine biology, conchology, ornithology, and entomology.[41]

The proliferation of science literature and activities for women and children also coincided with efforts to bring the distant habitats of the natural world into the Victorian home. Already, Victorians harbored a fondness for collecting objects of nature, which Thad Logan observes in distinctive categories such as: "(1) objects taken directly from nature—shells, minerals, fossils, feathers, and flowers; (2) objects found in nature but 'processed'—stuffed birds and animals, seaweed in albums, dried flowers, and skeletonized leaves; (3) objects containing live specimens—birdcages, Wardian cases, and aquaria."[42] Collections were maintained in the libraries of wealthy households, but Logan notes that "the parlour was frequently the repository for the small collection or the interesting single item, especially in more modest households," displayed with cultural artifacts on "whatnot shelves and overmantels."[43]

Aspiring to maintain live specimens then became a pursuit of the middle class, and both professional and amateur scientists endeavored to sustain such contained environments. The English doctor Nathaniel Bagshaw Ward's 1829 invention of the Wardian case, an early terrarium that became commonplace in Victorian parlors from the 1850s to 1870s, also stimulated the possibility of the marine version.[44] While various "proto-aquarists" had made progress in keeping contained aquatic ecosystems prior to the nineteenth century, developments in understanding the gas-exchange cycle of carbon dioxide and oxygen in relation to aquatic plants and animals spurred the momentum of the functional aquarium in Britain in the nineteenth century.[45] By the early 1850s aquariums were established as a hobby for the wealthy and a form of entertainment for all social classes at public venues, such as the first aquarium that opened at the Zoological Gardens at Regent's Park in 1853.[46] Publications espoused the ornamental values of the aquarium, and both freshwater and saltwater aquariums became prominent fixtures in the parlor, the latter type being more difficult and expensive to maintain, though far more popular.[47]

The primary challenge of keeping live specimens generally meant abiding by the balance principle of oxygen and carbon dioxide as the Victorians understood it, with growing knowledge of salinity and filtration, as well as of the necessary biodiversity to support a contained ecosystem severed from the natural environment.[48] Coincidentally, sustaining aquatic life outside of its natural environment is a theme that dovetails with the ballet *Ondine*. In a dream sequence, Matteo envisions Ondine in the underwater palace of Queen Hydrola, where Ondine renounces her immortality to pursue Matteo. Hydrola compares mortal life with the transience of a flower, to which Ondine (evidently crossing from dream into reality) responds by picking a rose near the sleeping Matteo, and according to one account from 1844, "she declares she would willingly perish *with* and *as* it, could she but have Matteo's love exclusively her own."[49] In this scene, Ondine equates herself with a living specimen, subject to decay when removed from her (super)natural environment. The editorial intent of Smith's *The Natural History of the Ballet-Girl,* as a light satire that details the life of female dancers with a documentarian approach, emphasizes such a comparison. The text treats dancers as biological specimens for observation, as a parody of natural history books.

As the ballet's plot progresses, Ondine assumes the *doppelganger* form of Matteo's actual betrothed and briefly lives her heart's desire. However, as the rose begins to wilt, Ondine's health too deteriorates, expressed in the *pas de la rose flétrié* ("the withered rose" dance). In this tarantella between Ondine and Matteo, "she pines and withers, and her youth and fondness cannot compensate to her exhausting frame for the surrender of its immortality."[50] This can, in part, be understood as the Victorian longing to bring the living marine environment into the home and the challenges of sustaining that life. When the household aquarium was finally realized, beginning in the 1850s, the Victorians anthropomorphized tank creatures and cast them, as Silvia Granata writes, "as actors on a stage," attributing human emotions and personalities to the unfolding spectacles of the tank.[51] Naturalist Shirley Hibberd even wrote in 1856 of his preference for the swimmerets of shrimps and prawns, over the dancing of French ballerinas.[52] While *Ondine* preceded this parlor trend, it is possible that some who attended its performances in the 1840s could see in the ballet the conundrum of sustaining aquatic life on land.

Through the Looking Glass

Industrialization in Victorian England enabled the mass production of materials, such as glass, and the widespread use of products containing glass elements. Isobel Armstrong describes the emergent "scopic culture" of what the author terms the Victorian "glassworld" as "a place where related and complex Victorian modernities played out their concerns."[53] Accordingly, I suggest that Victorian apprehensions about scientific discovery—in particular, microscopic organisms—can be mapped onto the ballet *Ondine*, where the beauty of the fairy alleviates some of those anxieties. I consider *Ondine* as related to the Victorian glassworld in two ways: first, by means of the audience's viewing experience through opera glasses, and second, in the nascent popularity of the microscope. The water fairy becomes an embodiment of the wonders of the microscopic world by drawing parallels between viewers' ocular experiences of *Ondine* through opera glasses and aquatic microorganisms through microscopes.

When *Ondine* debuted in 1843, opera glasses were already a part of mainstream Victorian culture. Monocular opera glasses were common by the late eighteenth century and in the early nineteenth century, binocular designs took precedence, made available by optician J. T. Hudson at London's most prominent theaters.[54] While opera glasses were not designed for scientific use, they utilized the same technology as their scientific cousins: Dr. William Kitchiner's 1818 treatise *Practical Observations on Telescopes, Opera-Glasses, and Spectacles* describes the specifications of various telescope models, explaining how they work in layman's terms, and proceeds to detail opera glasses in a similar manner.[55] The technology of opera glasses and telescopes converged later in the Victorian era as opera glasses adopted the dual-lens design of the Galilean telescope.[56] In addition to this similitude, the acquisition and display of opera glasses and of scientific devices was essentially the same. Opera glasses were available for purchase alongside "magnifying glasses, telescopes, microscopes and prepared slides" at shops stocked and maintained by opticians.[57] Jonathan Shears writes that in the Great Exhibition of 1851 at Hyde Park, T. W. Richardson's reflecting telescope was "one of many telescopes which jostled for space in a section that included new designs for opera glasses, microscopes, medical instruments for inspecting eye and ear."[58] The likeness among optical devices meant that the viewing experiences were largely analogous.

Aside from opera glasses, the Victorian microscope is especially salient to *Ondine* because the naiad's element of water was a popular subject underneath its lens. Throughout the early nineteenth century, advances in microscope technology fueled a preoccupation with magnifying the previously unseen in nature.[59] Victorians were engrossed with drops of water as the microscope revealed a nearly incomprehensible world in a substance once thought transparent.[60] As Armstrong writes, "water, and its invisible contents, polyps, infusoria, rotifers, water-bears, polyzoa, drawn from the sea or fresh water but particularly from the meeting point of land and sea, spawned a popular literature of microscopic investigation and, indeed, helped to bring into being the category of 'popular' science in an avid print culture."[61] The circulation of these texts and images of the mid-nineteenth century meant that "the 'wonders' of the microscope became a common trope."[62]

Consequently, I suggest that viewing *Ondine* through opera glasses emulated the experience of looking at aquatic life-forms through the microscope. In Lewis Carroll's *Through the Looking-Glass, and What Alice Found There* (1871), "when Carroll's Guard tries out different forms of prosthetic optical instrument, the farcical illusion is to the manifold types of lens available at the time, the monocular lens of the microscope, telescope, kaleidoscope, the binocular lens of the opera glass and stereoscope, all of which created different ways of seeing."[63] Furthermore, the act of seeing through a microscope engendered an otherworldly experience because the device cut off the observer from the reality of their surroundings, brought bizarre creatures into view, and distorted sensations of space and size.[64] Looking through opera glasses enacts a disconnect from one's immediate surroundings similar to the microscope's field of view, encircled by darkness. In both cases, the detachment from reality contributes to the focus on the observed object, shaping the overall experience of observation.

Ondine, as an object of study through opera glasses, thus bears a striking resemblance to the Victorians' observations of aquatic organisms through the microscope. Microscopy was certainly in conversation with the visual arts, as

Ursula Seibold-Bultmann notes that British fairy painting reached its zenith during the Victorian interest in microscopy.[65] Several artists exhibited a "predilection for circular fairy compositions," which Seibold-Bultmann links to the view through the microscope and the plethora of print media depicting those images.[66] If the microscopic view influenced fairy painters, I believe that audiences who had familiarity with both microscopes and opera glasses might have embodied parallel optical experiences; the fairy, thus, aided in reconciling tensions between science and the supernatural through a sense of wonder. Forsberg writes, "The conjunction of microscopes and fairies highlights Victorians' longing for scientific enchantment, which would allow them to see fairies through their microscopes and to espy natural wonders through their fairy lenses."[67] Even the devices themselves seemed to possess both scientific and magical properties.

The dancing fairy also becomes a pivotal figure as the subject of the lens. Irish-born American author Fitz-James O'Brien's short story *The Diamond Lens* (1858) indicates such a perspective, with a protagonist who repeatedly refers to a creature that he views in a drop of water through a microscope as a sylph, whom he even comes to prefer after comparing it to a fictional ballerina he calls Signora Caradolce, who dances as a fairy.[68] It remains unclear whether Signora Caradolce is inspired by a real person, though the Italian prefix and styling of the surname might draw speculation toward any one of the famous Italian ballerinas of the time. O'Brien did live in London before emigrating to America in November of 1851, and perhaps he even attended one of Cerrito's performances of *Ondine* a few months earlier at Her Majesty's Theatre.[69] Regardless, while O'Brien's protagonist clearly preferred the watery sylph over the ballerina, some Victorians may have preferred the dancer instead, for the shocking imagery of microorganisms in water induced a new set of anxieties, as viewers found themselves bewildered by incomprehensible creatures characterized as the "horrors lurking in a mere drop of water."[70] An 1828 etching by William Heath of a microscope slide featuring such organisms is titled *Monster Soup*, a play on such fears. The unease contained in the *Monster Soup* image suggests that some Victorians sought a remedy that could broker consternation with curiosity. Perhaps Ondine, as a beautiful fairy, offered just that possibility.

Into the Air

The most iconic moment of Perrot's ballet is the *pas de l'ombre* in which Ondine emerges onto a seashore and dances with her own shadow. This celebrated image of Cerrito, painted by G. A. Turner in 1843, shows the water sprite

Figure 3.2. *Fanny Cerrito as Ondine in "Ondine, ou La Naïade" by Jules Perrot.* Oil on canvas by G A Turner, 1843. Victoria & Albert Museum Collections. Image ©Victoria & Albert Museum, London. Cyril W. Beaumont Bequest.

in a moonlit clearing, torso pitched forward with her arms reaching toward the ground; one leg straightened beneath her while the other is bent at the knee and tucked behind.[71] Her gaze is cast downward, eyes transfixed on her shadow, which she sees for the first time. As she hovers just above the grass, gravity appears nonexistent, accentuated by the lofty volumes of her viridescent tulle skirt. At first glance, Cerrito seems to be mid-flight. The pointed toe of her straight leg does not even touch her shadow on the ground. However, by "kinesthetically seeing" this pose, the researcher might understand Cerrito as balanced atop her pointe shoe, as it would have been awkward to jump with the torso so far off-kilter.[72] Turner was just as likely to be captivated by the illusion of flight created by the pointe shoe, whether the moment was jumped or balanced. As Newey observes, "the critical encomia of these ballerinas . . . is marked with flights of fancy that seem to exceed the materiality of the actual performances reviewed."[73] I interpret Turner's depiction of Ondine as a sensationalized vision of what actually transpired on stage.

This perception of flight can also be associated with the Victorian fascination with advancements in aeronautical technology, concomitant with the recreational and experimental hot air balloons that enabled the Victorians to explore the sky. As Nicola Bown notes, "the mid-1830s saw a resurgence of 'balloon-fever' associated with two notable events: the first cross-channel balloon flight in 1837, and the highly popular Vauxhall balloon, installed in 1836. The latter was a fixed balloon, tethered at Vauxhall, in which, for a small fee, the public could make an ascent and view the panorama of London."[74] Written accounts by these aeronauts often described feelings of euphoria, or a "dream-like sensation of leaving the body behind," and Bown suggests that "to convey this feeling, balloonists turned to images of wings, butterflies, phantoms and fairies."[75] In sum, the Victorians were eager for and receptive to art works that could capture the emotions of this new experience.

Bown notes that the fascination with human flight influenced Victorian fairy painters to depict their fairies in mid-air and with wings. These allegories were paralleled and materialized by the Romantic ballerinas dancing on pointe. It is perhaps not a coincidence that one critic likened Cerrito's lofty style of dancing to a balloon in 1843: "By her lateral *jettés battus* she seemed to set at defiance the laws of gravitation. She bounded and flew about the stage (in a style that might rather be called ballooning than dancing), and executed such *voltes* and *pirouettes* high in the air, that the audience's plaudits at the end of each *pas* appeared to express relief at the dangers they imagined she had escaped, as well as delight at the daring feats she had realised beyond all expectation."[76] Although the French term *ballon* (in English, "balloon") refers to the easy, airy, and lightweight quality that ballet dancers aim for in jumps, the critic's use of the term as a verb links the statement to the recreational activity. The juxtaposition of danger and delight is not unusual; Elaine Freedgood observes how ballooning memoirs by Victorian writers engender a paradox of security and tranquility, despite the extreme risks.[77] Cerrito's Ondine embodies a similar interplay of safety and peril in relation to human flight, offering a mollifying yet thrilling image of a dancer who skirts the line between flight and precarity.

Into the Sea

Ondine's coastal setting likely appealed to the Victorians' changing relationship to the ocean. Granata writes that the seashore, "formerly seen mostly as a site of disaster, abandonment, or exile, began to be enjoyed—from both an aesthetic and conceptual point of view"—and, crucially, it was materially enjoyed as exemplified by the proliferation of the seaside holiday.[78] Amy M. King notes that in the earlier part of the nineteenth century, seaside holidays were seen as restorative and therapeutic, since bathing in and even drinking sea water purportedly had curative effects.[79] A belief in water's healing properties also appeared in urban settings: the increase in hydrotherapy centers, which developed water-based treatments to remedy various ailments, contributed to a larger culture of health tourism and domestic care that made water a ubiquitous subject.[80] The resolution of *Ondine* echoes this preoccupation; when Hydrola rescues Ondine, revealing the naiad as a simulacrum of Giannina and returning the human bride to Matteo, Ondine is restored to immortality after rejoining her aquatic kin. Critical response supports drawing connections to the ongoing discourse of water, as a writer for the *Morning Post* in 1843 states: "We have lived in a deluge of water for months, and there is a perfect fanaticism prevalent for its use to cure our moral as well as our physical maladies. We confess that, amidst the several hydraulic manias, *Ondine* and *Hydrola* are the only forms in which we accept the aqueous system—'*toto caelo*' [by the whole extent of the heavens]."[81] Even the music by composer Cesare Pugni played into water mania by echoing the rhythm of waves and including other atmospheric suggestions of the sea through sound.[82]

Seaside tourism grew with the expansion of the railways, the increasing leisure time of the middle class, and the widespread popularity of science, marketing its appeal to natural history enthusiasts with opportunities to see marine life and collect shells, seaweed, and fossils in the exposed cliffs of the coastline.[83] *Ondine* then would entice Victorian sensibilities by offering a fanciful vision of the underwater world. By the mid-nineteenth century, Victorians still had little knowledge of what lay beneath the ocean's surface.[84] They imagined the undersea realm as filled with shipwrecks, terrifying monsters, and bewitching but fatal mermaids.[85] Yet a wealth of new information became available in the 1830s, as technological developments in more industrialized countries enabled increasing exploration of the underwater environment.[86] Among these innovations were the early dredging expeditions, with exciting discoveries of new species, that continually made national news and captivated the public imagination.[87] These advances in the nascent marine sciences then inspired the arts, as people began to grapple with material evidence of a dramatically different world.[88]

Victorians were thus eager for submarine exploration. Improvements to the diving bell, especially throughout the eighteenth and early nineteenth centuries, led to the installation of a diving bell at the Royal Polytechnic Institute in London in 1838.[89] Although this diving bell submerged in a tank of water rather than in natural environments, Victorians gathered in throngs "to experience the thrill of breathing underwater, possibly dreaming of really being in the ocean, instead of central London."[90] The diving bell as an apparatus of transportation to another world thus resonates with the ballet *Ondine* and the naiad's surfacing from the depths of the sea in a shell. Coincidentally, a page from an 1843 edition of *The Examiner* grasped the idea of such a comparison. At the top of one page is an announcement of a benefit performance to be given by Cerrito and other famous ballerinas, including excerpts of *Ondine;* further down the same column is a listing of exhibits at the Royal Polytechnic Institute, advertising "the DIVER and DIVING BELL" (and microscope) among them.[91]

While reviews of the ballet made no such direct comparison between the shell and the diving bell, the theatricality of interactive exhibits affirmed what Aileen Robinson refers to as an "interplay among performance, technology, spectacle, and scientific demonstration."[92] This interplay is made explicit in Dion Boucicault's staging of a French *féerie* titled *Babil and Bijou* (1872) in London, in which the half-fairy Bijou traverses fairyland via diving bell to the bottom of the ocean and via balloon to the moon.[93] The interactions between scientific demonstrations and the theater are also apparent in Smith's *Natural History,* where he compares the collective movement of a *corps de ballet* to a lecture demonstration at the Polytechnic Institute that literally shocked its audience by having the spectators link hands as a machine passed a mild electric current through them.[94] Through this lens of interactivity, a reference to a nautilus shell as both a "fairy shell" and diving bell by British design theorist Charles Dresser in *Principles of Decorative Design* (1873) offers potential insight into a Victorian interpretation of Ondine's shell.[95] Though a nautilus and scallop are obviously different animals, retroactively extending a broader notion of a fairy shell/diving bell enables an association in which Ondine's ascent on the shell reflects an inverse of the descent of the diving bell, in which both signify a metaphorical crossing into another world. For Ondine, the shell brings her from a fantastical sea to the human world on land, while the diving bell simulates the human journey into the unfathomable depths, and back out.

Epilogue

Observing Ondine's *pas de l'ombre* evokes the practice of scientific study, as her efforts to capture her shadow could be seen as a metaphor for the Victorian naturalist's fascination with collecting specimens. An 1843 critique in *The Morning Chronicle* hints at the plausibility of such an interpretation: "The exquisite effect of Cerrito's dancing to her own shadow as though it were some strange and incomprehensible object, trying to follow its movements, and stooping down to catch it, is beyond description."[96] As Victorians watched her examine this unfamiliar creature, they also saw a self-reflexive image of their engagement with the natural world. However, the scientific method increasingly dictated the legitimacy of knowledge in the late nineteenth century, prompting the naturalist author John George Wood to write in 1868 that with modern technology, "the curious inquirer can see [aquatic creatures] without the assistance of either Nereid or Naiad."[97] Subsequently, Victorian interest in fairies waned by the turn of the century, primarily reducing them to themes for children, though the notorious Cottingley photographs by Elsie Wright and Frances Griffiths that fraudulently documented the existence of fairies in the 1920s again ignited controversy that captured the public's attention.[98]

In 1958, the water fairy resurfaced prominently in English ballet, in *Ondine,* choreographed by Frederick Ashton (1904–1988). Ashton's *Ondine* draws from the de la Motte Fouqué novel and pays homage to Perrot's ballet, both of which Ashton studied in preparation for his ballet.[99] While Ashton's *Ondine* is a twentieth-century work, his decision to set the ballet in Victorian England, an era he deeply admired, offers a novel and nostalgic view of the past that perhaps contains traces of Victorian values. I conclude this essay with a description of Ashton's *pas de l'ombre* as danced by Margot Fonteyn, to demonstrate a persistence of the Victorian sensibilities and their relationship to scientific discourse.[100] My interpretation of this choreography gestures toward the long reach of the nineteenth century and the enduring vestiges of the Victorian entanglement of fairies, science, and dance.

Fonteyn's Ondine emerges from a shimmering fountain in a courtyard, a raven-haired ingénue wearing a pale blue chiffon dress. She marvels at the new world around her, touching the foliage near the fountain before taking measured steps forward, pausing to feel the new sensation of air. A cascading harp melody approximates trickling water. Suddenly, she sees her shadow, first on a wall, and approaches it until it disappears in an empty doorway, and later reappears on the ground. She gives chase, darting about until she catches the shadow in her peripheral vision. She jubilantly leaps into the air before spinning to the floor, dropping onto her right hip with her legs folded to the side. In this position, one can transplant the image of the Victorian naturalist, seated by a tide-pool, recording her observations. She examines the shadow up close, making note of the way it imitates her lifting one arm, then both arms overhead, crossing her wrists and encircling her hands. She delights in the shadow's mimicry and studies it like a new species. Despite her temporal distance from the Victorian era, Ondine retains connections to science culture and nineteenth-century attitudes toward the mysteries of the aquatic realm.

Notes

1 de la Motte Fouqué, *Undine.*
2 Au, "The Shadow of Herself," 161–64; Guest, *Jules Perrot,* 99–104.
3 King, "Victorian Natural Science and the Seashore," 442–43.
4 For more on science and fairies, see Bown, *Fairies in Nineteenth-Century Art and Literature;* Forsberg, "Nature's Invisibilia"; Keene, *Science in Wonderland;* Silver, *Strange and Secret Peoples.*
5 Keene, *Science in Wonderland,* 19.
6 Keene, *Science in Wonderland,* 83.
7 Kelly, "Ponds in the Parlour," 55–56.
8 Auerbach, "Before the Curtain," 3.
9 Newey, "Fairies and Sylphs," 100.
10 Heath, *Beauties of the Opera and Ballet,* 70.
11 Gautier, *Gautier on Dance,* 15.
12 Lightman, "Victorian Science and Popular Visual Culture," 1.
13 McWilliam, *London's West End,* 47.
14 Forsberg, "Nature's Invisibilia," 644–45.
15 Smith, *The Natural History of the Ballet-Girl,* 17.
16 Nathanial Currier, *Fanny Ellsler [sic] in the Shadow Dance,* 1846, hand-colored lithograph.
17 Hazelton, "The Grieve Family," 31–32.
18 Hazelton, "The Grieve Family," 34.
19 Au, "The Shadow of Herself," 166–67; Guest, *Jules Perrot,* 98–99.
20 Au, "The Shadow of Herself," 167–68.
21 For more on the size and scale of fairies relative to humans, see Bown, *Fairies in Nineteenth-Century Art,* 66–82.
22 Bown, *Fairies in Nineteenth-Century Art,* 110.
23 Heath, *Beauties of the Opera and Ballet,* 65.
24 Seibold-Bultmann, "Monster Soup," 215.

25 Forsberg, "Nature's Invisibilia," 650.
26 Newey, "Fairies and Sylphs," 99.
27 Newey, "Fairies and Sylphs," 100.
28 "Her Majesty's Theatre." *Morning Post,* June 23, 1843. Italics in original. Perrot played Matteo, Ondine's beloved.
29 Guest, *Jules Perrot,* 103.
30 "The Story of Pantomime."
31 Morris, "The Other Giselles," 52–53.
32 Moncrieff, *Giselle,* 62.
33 Booth, *Theatre in the Victorian Age,* 90.
34 Guest, *Jules Perrot,* 105.
35 Burdekin, "Darkening the Auditorium in the Nineteenth Century British Theatre," 42.
36 "Theatrical Examiner," *Examiner,* June 24, 1843.
37 Guest, *Jules Perrot,* 103.
38 Shteir, *Cultivating Women, Cultivating Science,* 2.
39 *The Young Lady's Book* (1829).
40 Sheffield, *Revealing New* Worlds, 1–5.
41 Lightman, *Victorian Popularizers of Science,* 96.
42 Logan, *The Victorian Parlour,* 141.
43 Logan, *The Victorian Parlour,* 146.
44 Logan, *The Victorian Parlour,* 150–52; Keogh, *The Wardian Case,* 1; Rehbock, "The Victorian Aquarium in Ecological and Social Perspective," 526.
45 Rehbock, "The Victorian Aquarium," 522–31.
46 Rehbock, "The Victorian Aquarium," 531–33.
47 Kelly, "Ponds in the Parlour," 64–66.
48 Rehbock, "The Victorian Aquarium," 524–31.
49 Heath, *Beauties of the Opera,* 73. Italics in original.
50 Heath, *Beauties of the Opera,* 78.
51 Granata, *The Victorian Aquarium,* 5, 183–90.
52 Hibberd, *Rustic Adornments for Homes of Taste,* 52.
53 Armstrong, *Victorian Glassworlds,* 1–2.
54 Hall-Witt, *Fashionable Acts,* 29.
55 Kitchiner, *Practical Observations on Telescopes, Opera-Glasses, and Spectacles,* 26–61.
56 Turner, *Scientific Instruments, 1500–1900,* 98.
57 Graham, *"Gone to the Shops,"* 89.
58 Shears, *The Great Exhibition, 1851,* 60.
59 Seibold-Bultmann, "Monster Soup," 212.
60 Keene, *Science in Wonderland,* 83.
61 Armstrong, *Victorian Glassworlds,* 320.
62 Armstrong, *Victorian Glassworlds,* 320
63 Armstrong, *Victorian Glassworlds,* 317.
64 Forsberg, "Nature's Invisibilia," 655.
65 Seibold-Bultmann, "Monster Soup," 211.
66 Seibold-Bultmann, "Monster Soup," 214.
67 Forsberg, "Nature's Invisibilia," 642.

68 Forsberg, "Nature's Invisibilia," 638.
69 Cornwell, "Gothic and its Origins in East and West," 120. For a review of one of Cerrito's 1851 performances of *Ondine*, see "Her Majesty's Theatre," *Standard*, Aug. 25, 1851.
70 Seibold-Bultmann, "Monster Soup," 211.
71 Turner, *Fanny Cerrito as Ondine*, 1843.
72 Kosstrin, "Kinesthetic Seeing," 19.
73 Newey, "Fairies and Sylphs," 107–108.
74 Bown, *Fairies in Nineteenth-Century Art*, 55.
75 Bown, *Fairies in Nineteenth-Century Art*, 54.
76 "Her Majesty's Theatre," *Morning Post*, July 28, 1843, italics in original.
77 Freedgood, *Victorian Writing about Risk*, 75–76.
78 Granata, *The Victorian Aquarium*, 25.
79 King, "Victorian Natural Science," 454.
80 Marland and Adams, "Hydropathy at Home," 500–501.
81 "Her Majesty's Theatre." *Morning Post*, Aug. 14, 1843.
82 Guest, *Jules Perrot*, 107.
83 King, "Victorian Natural Science," 454.
84 Granata, *The Victorian Aquarium*, 27.
85 Granata, *The Victorian Aquarium*, 170.
86 Cohen, "Marine Bizarrerie," 169.
87 Granata, *The Victorian Aquarium*, 27–28.
88 Cohen, "Marine Bizarrerie," 169–70.
89 Robinson, "Knocking for Air," 43–45.
90 Granata, *The Victorian Aquarium*, 26.
91 *The Examiner, a Sunday Paper on Politics, Literature, and the Fine Arts. For the Year 1843*, 462.
92 Robinson, "Knocking for Air," 39.
93 Booth, *Victorian Spectacular Theatre 1850–1910*, 78.
94 Smith, *Natural History*, 47.
95 Dresser, *Principles of Decorative Design*, 19–20.
96 "The Italian Opera," *Morning Chronicle*, June 23, 1843.
97 Wood, "The Waves and their Inmates," 52. See also King, "Victorian Natural Science," 440.
98 Silver, *Strange and Secret Peoples*, 189–92.
99 Kavanagh, *Secret Muses*, 405.
100 For a film of a 1960 performance of Ashton's *Ondine*, see *The Royal Ballet*, DVD, directed by Paul Czinner.

Bibliography

Armstrong, Isobel. *Victorian Glassworlds: Glass Culture and the Imagination 1830–1880.* New York: Oxford University Press, 2008.

Au, Susan. "The Shadow of Herself: Some Sources of Jules Perrot's *Ondine.*" *Dance Chronicle* 2, n. 3 (1978): 159–71.

Auerbach, Nina. “Before the Curtain.” In *The Cambridge Companion to Victorian and Edwardian Theatre,* edited by Kerry Powell, 3–14. Cambridge: Cambridge University Press, 2004.

Booth, Michael R. *Theatre in the Victorian Age.* Cambridge: Cambridge University Press, 1991.

Booth, Michael R. *Victorian Spectacular Theatre 1850–1910.* London: Routledge & Kegan Paul, 1981.

Bown, Nicola. *Fairies in Nineteenth-Century Art and Literature.* Cambridge: Cambridge University Press, 2001.

Burdekin, Russell. “Darkening the Auditorium in the Nineteenth Century British Theatre.” *Theatre Notebook* 72, n. 1 (2018): 40–57.

Cohen, Margaret. “Marine Bizarrerie: The Imaginative Biology of the Underwater Frontier.” In *Coastal Cultures of the Long Nineteenth Century,* edited by Matthew Ingleby and Matthew P. M. Kerr, 169–85. Edinburgh: Edinburgh University Press, 2018.

Cornwell, Neil. “Gothic and its Origins in East and West: Vladimir Odoevsky and Fitz-James O’Brien.” In *Exhibited by Candlelight: Sources and Development in the Gothic Traditions,* edited by Valeria Tinkler-Villani and Peter Davidson, with Jane Stevenson, 117–28. Atlanta: Rodopi, 1995.

Currier, Nathanial. *Fanny Ellsler [sic] in the Shadow Dance,* 1846. Hand-colored lithograph, 26.9 by 35.7 centimeters. London, Victoria and Albert Museum. https://collections.vam.ac.uk/item/O106082/fanny-ellsler-sic-in-the-print-currier-n. Accessed Aug. 20, 2021.

de la Motte Fouqué, Friedrich. *Undine.* Translated by W. L. Courtney. London: William Heinemann, 1912.

Dresser, Christopher. *Principles of Decorative Design.* London: Cassell, Petter, Galpin & Co., 1873.

The Examiner, a Sunday Paper on Politics, Literature, and the Fine Arts. For the Year 1843. London, 1843.

Forsberg, Laura. “Nature’s Invisibilia: The Victorian Microscope and the Miniature Fairy,” *Victorian Studies* 57, n. 4 (Summer 2015): 638–66.

Freedgood, Elaine. *Victorian Writing about Risk: Imagining a Safe England in a Dangerous World.* New York: Cambridge University Press, 2000.

Gautier, Théophile. *Gautier on Dance,* edited by Ivor Guest. London: Dance Books Ltd, 1986.

Graham, Kelley. *“Gone to the Shops”: Shopping in Victorian England.* Westport, CT: Praeger, 2008.

Granata, Silvia. *The Victorian Aquarium: Literary Discussions on Nature, Culture, and Science.* Manchester, UK: Manchester University Press, 2021.

Guest, Ivor. *Jules Perrot: Master of the Romantic Ballet.* New York: Dance Horizons, 1984.

Hall-Witt, Jennifer. *Fashionable Acts: Opera and Elite Culture in London, 1780–1880.* Durham: University of New Hampshire Press, 2007.

Hazelton, Nancy J. Doran. “The Grieve Family: Patterning in Nineteenth-Century Scene Designs.” *Theatre Survey* 32, n. 1 (1991): 31–42.

Heath, Charles, ed. *Beauties of the Opera and Ballet.* London: David Bogue, 1844.

“Her Majesty’s Theatre.” *Morning Post.* Aug. 14, 1843. British Library Newspapers.

“Her Majesty’s Theatre.” *Morning Post.* June 23, 1843. British Library Newspapers.

"Her Majesty's Theatre." *Standard.* Aug. 25, 1851. Newspapers.com World Collection.

Hibberd, Shirley. *Rustic Adornments for Homes of Taste, and Recreations for Town Folk, in the Study and Imitation of Nature.* London: Groombridge and Sons, 1856.

"The Italian Opera." *Morning Chronicle.* June 23, 1843. British Library Newspapers.

Kavanagh, Julie. *Secret Muses: The Life of Frederick Ashton.* New York: Pantheon Books, 1996.

Keene, Melanie. *Science in Wonderland: The Scientific Fairy Tales of Victorian Britain.* Oxford: Oxford University Press. 2015.

Kelly, Ann. "Ponds in the Parlour: The Victorian Aquarium." *Things* 2 (Summer 1995): 55–68.

Keogh, Luke. *The Wardian Case.* Chicago: University of Chicago Press, 2020.

King, Amy M. "Victorian Natural Science and the Seashore." In *The Oxford Handbook of Victorian Literary Culture,* edited by Juliet John, 4339–457. Oxford: Oxford University Press, 2016.

Kitchiner, William. *Practical Observations on Telescopes, Opera-Glasses, and Spectacles,* 3rd ed. London: S. Bagster, 1818.

Kosstrin, Hannah. "Kinesthetic Seeing: A Model for Practice-in-Research." In *Futures of Dance Studies,* edited by Susan Manning, Janice Ross, and Rebecca Schneider, 19–35. Madison: University of Wisconsin Press, 2020.

Lightman, Bernard. *Victorian Popularizers of Science: Designing Nature for New Audiences.* Chicago: University of Chicago Press, 2007.

Lightman, Bernard. "Victorian Science and Popular Visual Culture." *Early Popular Visual Culture* 10, n. 1 (2012): 1–5.

Logan, Thad. *The Victorian Parlour.* Cambridge: Cambridge University Press, 2001.

Marland, Hilary and Jane Adams. "Hydropathy at Home: The Water Cure and Domestic Healing in Mid-Nineteenth-Century Britain." *Bulletin of the History of Medicine* 83 n. 3 (Fall 2009): 499–529.

McWilliam, Rohan. *London's West End: Creating the Pleasure District, 1800–1914.* New York: Oxford University Press, 2020.

Moncrieff, W. T. *Giselle; or, The Phantom Night Dancers: A Domestic, Melo-dramatic, Choreographic, Fantastique, Traditionary Tale of Superstition in Two Acts.* London: J. Limbard, 1842.

Morris, Mark. "The Other Giselles: Moncrieff's *Giselle; or, The Phantom Night Dancers,* Loder's *The Night Dancers,* and Puccini's *Le Villi.*" In *The Creation of iGiselle: Classical Ballet Meets Contemporary Video Games,* edited by Nora Foster Stovel, 51–88. Edmonton: University of Alberta Press, 2019.

Newey, Katherine. "Fairies and Sylphs: Femininity, Technology and Technique." In *Theatre, Performance and Analogue Technology: Historical Interfaces and Intermedialities,* edited by Kara Reilly, 97–116. New York: Palgrave Macmillan, 2013.

Rehbock Philip F. "The Victorian Aquarium in Ecological and Social Perspective." In *Oceanography, the Past,* edited by Mary Sears and Daniel Merriman, 522–39. New York: Springer-Verlag, 1980.

Robinson, Aileen. "Knocking for Air: The Diving Bell and Interactivity in Nineteenth-Century London." *Nineteenth Century Theatre and Film* 41 n. 1 (Summer 2014): 38–53.

The Royal Ballet. DVD. Directed by Paul Czinner. United Kingdom: Poetic Films Ltd., 2004.

Seibold-Bultmann, Ursula. "Monster Soup: The Microscope and Victorian Fantasy." *Interdisciplinary Science Reviews* 25, n. 3 (2000): 211–19.

Shears, Jonathan. *The Great Exhibition, 1851: A Sourcebook.* Manchester, UK: Manchester University Press, 2017.

Sheffield, Suzanne Le-May. *Revealing New Worlds: Three Victorian Women Naturalists.* New York: Routledge, 2001.

Shteir, Ann B. *Cultivating Women, Cultivating Science: Flora's Daughters and Botany in England, 1760–1860.* Baltimore: John Hopkins University Press, 1996.

Silver, Carole. *Strange and Secret Peoples: Fairies and Victorian Consciousness.* Oxford: Oxford University Press. 1999.

Smith, Albert. *The Natural History of the Ballet-Girl.* London: D. Bogue, 1847.

"The Story of Pantomime." Victoria and Albert Museum. https://www.vam.ac.uk/articles/the-story-of-pantomime. Accessed May 5, 2022.

"Theatrical Examiner." *Examiner.* June 24, 1843. British Library Newspapers.

Turner, G. A. *Fanny Cerrito as Ondine in 'Ondine, ou La Naïade' by Jules Perrot,* 1843. Oil on canvas, 59.7 by 49 centimeters. London, Victoria and Albert Museum. https://collections.vam.ac.uk/item/O60542/fanny-cerrito-as-ondine-in-painting-g-a-turner. Accessed Aug. 20, 2021.

Turner, Gerard L'Estrange. *Scientific Instruments, 1500–1900: An Introduction.* London: Philip Wilson, 1998.

Wood, John George. "The Waves and their Inmates; or, the Sea, the River, and the Lake." In *Routledge's Every Boy's Annual: An Entertaining Miscellany of Original Literature,* edited by Edmund Routledge, 51–58. London: George Routledge and Sons, 1868.

The Young Lady's Book: A Manual of Elegant Recreations, Exercises, and Pursuits. London: Vizetelly, Branston, and Co., 1829.

4

Phytology and Dance

The Impact of Plant Biology on Nineteenth-Century Flower Ballets

Alexander H. Schwan

Over the course of the early nineteenth century, botany—the leading science of the age of reason[1]—became immensely important for European culture through the publication of popular scientific writings. These helped to circulate new concepts of plants, new ways of categorizing them, and above all, a new understanding of how plants propagate and disseminate. Much of this popularized knowledge was further adopted in nonscientific literature, sometimes departing significantly from the original scientific approach. Through appropriations and transformations, principles of botany and their theoretical framework found their way into novels, poems, and new genres of plant-themed literature.[2] Phytology, or the science of plants, also influenced dance libretti, which were performed in a peculiar and entirely new dance genre: the flower ballet. Here, plants and flowers were performed as animated, moving, and dancing beings. No longer an additional ornament to costumes and stage design, plants entered the stage and danced.[3] They performed principles and ideologies of eighteenth-century biology, or at least those fragments of knowledge that had gone through the various adaptations and transformations of popular scientific literature.

In this essay, I unravel the influence of plant biology on the nineteenth-century flower ballet. My main focus is on the question of how historic phytology, its classificatory system, and its ideological background affected the imaginary of the flower ballet. Here, I will pay particular attention to the role of plant propagation and show how this was historically compared with human sexuality through an allegedly natural heteronormativity in both plant and human life.[4] This proposed linkage between biology and society made a significant impact on the flower ballet and led to dance performances where animated flowers danced the ideological norms of nineteenth-century Europe.

In analyzing this nexus of plant biology, popular science, and dance, I am aware of the methodological problem that much of the ideological burden of the flower ballet went unnoticed for the contemporary audience and dance critics. They saw and described dancers performing as various animated plants and posing in ornamental arrangement. But we might also consider how this choreography translated the scientific breakthroughs of early eighteenth-century biology into the very grounds for the exclusionary social codes of the flower ballet's time. What an audience today would interpret as the underlying racism, homophobia, and colonial mindset of European supremacy in these ballets were of course not mentioned in the historical writings on the performances. This reinforces the importance of analyzing the ideological burden retrospectively and with theoretical insights inaccessible in the historic discourse.

After a short introduction to the flower ballet as a genre, I explore how the phytological knowledge of the eighteenth century was interpreted liberally in visual and literary material that then directly influenced the aesthetics of the flower ballet. Through close readings of libretti, choreographic sketches, and contemporary dance critique, I show that remnants of biological thinking and their ideological interpretations were indeed openly visible on the ballet stage. My argument is scaffolded by four particular flowers that are either prominently represented in individual flower ballets or play a decisive role in our understanding of how scientific knowledge was adapted and transformed. The names of the flowers follow the binomial nomenclature, a naming system for organisms with two Latin or Latinized terms, of which the first describes the genus of the organism while the second functions as a specific epithet and distinguishes the species within the genus. The system was first introduced by Swedish botanist and physician Carl von Linné, or Linnaeus, in his work *Species Plantarum* (1753), and is still used today, with a postposed capital letter L. added to all names that follow the binomial nomenclature.[5]

The flower ballet is a subgenre of the *ballet fantastique* and as such belongs to the performance complex that, due to its thematic connection to Romanticism, was once called "the Romantic ballet,"[6] an expression in recent years increasingly replaced by the terminology of (early) nineteenth-century ballet. As part of the genre of the *ballet fantastique*,[7] flower ballets encompass entire dance pieces that follow a narrative structure related to dreams, states of intoxication, and para-reality. Their floral-themed narratives often center on the idea that a female protagonist has been transfigured into a flower and needs intervention by the male protagonist in order to transform back to her human nature.[8]

Gaining a remarkable popularity in Western and Eastern Europe in the mid- to late nineteenth century, the flower ballet genre also extended to smaller parts of larger ballets. These and choreographic highlights of entire *ballets fantastiques* were sometimes extracted from their original performance setting and incorporated into new revue-like structures. Within the Russian context alone, some of the more well-known floral choreographies are *Le Jardin animé* in *Le Corsaire*, the *Garland Waltz* and the *Rose Adagio* in *The Sleeping Beauty*, Marius Petipa's flower ballet *The Blue Dahlia* (1860), and his last, yet unperformed, ballet *The Romance of the Rosebud and the Butterfly*, and Lev Ivanov's *The Haarlem Tulip* (1887).[9]

As a fundamental underpinning of such flower performances, these choreographies shared the basic idea of showing plants as lively and moving entities performed by dancing human bodies. And, while representing the vivacity of plants though dance, flower ballets refrained from equating the ephemerality of flowers with that of movement. Their flowers never wilted but blossomed and bloomed in apparently endless repetitions and figurations.[10] Fitting to this eclipse of floral ephemerality, choreographies of the flower ballet relied mostly on choreographic strategies of arranging bodies in ornamental configurations that sometimes even resembled the flower performance by grouping dancers in petal-shape formations. Only in future theatrical dance movements would this flower-dance relation become less static and ornamental, including even the portrayal of wilting and decaying plants through dance.[11]

Calendula officinalis L.

In the flower ballet, then, the only way to represent an idea that exceeded the mere ornamentality of arrangement or movement was through an elaborate system of flower symbolism called floriography, or the language of flowers.[12] Here, species of flowers and plants, as well as trees, grass, and moss, were encoded with a message or linked to a multitude of metaphorical meaning. The floriographic system itself had only become possible through the rise of popular botany, as it relied on biological plant taxonomy with which plants could be differentiated from one another before being encoded with a symbolic meaning. These meanings were listed in floriographic dictionaries, or books on the language of flowers, with Charlotte de Latour's oft-reprinted *Le Langage des fleurs* (1819) the most prominent example.[13] Through floriographically encoded flowers, emotive messages could be communicated from one person to another. Provided that both partners had access to the same edition of a language-of-flower book, the nosegay or posey—a small bouquet of various flowers, each ascribed a different meaning—could even transmit complex messages, sentences, and entire poems.

Language-of-flower books were immensely popular throughout the nineteenth century and had a sustained influence on art, literature, and dance. It is likely that the creators of the flower ballets could count on the audience having this knowledge and thus chose the flowers for the choreography accordingly. Thus, whenever a specific plant appeared in a flower ballet—identified either by the libretto or through characteristic costuming of the dancers—this very plant carried at least one floriographic meaning that was, theoretically, decodable. However, since the meaning of each flower could differ from country to country and from one floriographic dictionary to another—and most importantly, since not all audience members might have been completely fluent in the language of flowers—the floriographic coding of the flower ballets was overdetermined: it was neither entirely arbitrary nor distinct, but rather a playfully polysemous reference to the system of floriography.[14]

Perhaps the most prominent example of this overdetermined usage of flower symbolism is the "Waltz of the Flowers" in *The Nutcracker*, first choreographed by Marius Petipa and Lev Ivanov in 1892 to the music of Pyotr Ilyich Tchaikovsky. In the original version of this *grand divertissement*, and quite typical for the flower ballet, the dancers did not embody just any flower, but performed a particular species. In its introduction to the ballet, the Marius Petipa Society states that "the flowers represented in the waltz . . . were marigolds."[15] But already with this information, the characteristic ambiguity of the flower ballet sets in, as it is not clear to which plant species the English name *marigold* refers, which undermines our ability to decode the embedded floriographic reference therein. Most commonly, *marigold* is the name for a ball-shaped flower of the genus *Tagetes* whose color spectrum oscillates between yellow and orange-red. But it could also refer to the completely different plant *Calendula officinalis*, a round-shaped, orange- to yellow-colored flower, also known as *pot marigold* due to its Christian-iconographic connection to Mary. The biographical explanation for Petipa's choice of the marigolds, as a tribute to his late daughter Evgenia (1877–1892), upon whose death marigold flowers allegedly blossomed in Petipa's garden, does not resolve the problem.[16] And further discombobulating the floriographic coding of this choreography, later versions of the *Waltz of the Flowers* skipped the reference to marigolds altogether and dressed the dancers in unspecific floral-themed costumes.[17]

Characteristic of the polysemous references in flower ballets, floriographic lexica translate both *Tagetes* and *Calendula officinalis* to grief, despair, and pain, but also to a variety of other emotions or topics.[18] Petipa likely chose the marigold because of its common floriographic encoding with emotions of grief in order to refer to the loss of his daughter. But the calendula could also symbolize completely different emotive conditions such as jealousy, cruelty, or an undefined uneasiness,[19] which reveals a basic idea symptomatic of the flower ballet in general. It shows the influence of floriography on nineteenth-century ballet, but simultaneously, that this influence did not generate an unambiguous meaning for the plants that were performed as animated beings. Between the artful arrangements of embodied flowers, the floral décor of stage and costumes, and the polysemy of any one species, the floriographic influence led to choreographies characterized by an excess of visual reference articulated through flowers. As a consequence, the ornamentality of the flower ballet did not simply amount to extraneous decoration; in fact, the flower ballets were nothing but their ornamentality. Overloaded with reference and with the singular flower bearing more than one meaning, these ballets were prime examples of an understanding of ornament as modus operandi in which ornamental elements were no longer reduced to superfluous additions, but could move, dance, and act.[20]

The principal idea of the flower ballet, that specific plants could be performed as animated beings, draws from a popular appropriation of earlier botanic insights. The ambiguity of the plants on stage was no contradiction to this, but a hint as to how loosely the original botanical knowledge had been adapted in popular culture. The same applied to the parallels between human and plant sexuality that stemmed from already popularized science and was then further adapted and transformed in literature and dance. To unravel this connection between botany and the flower ballet and to scrutinize its ideological burden, I am momentarily leaving the narrow field of nineteenth-century flower ballet. The following section goes back further in time to the end of the eighteenth century and to early publications of popular scientific writing to find the ideological basis upon which the flower ballets were created.

Andromeda polifolia L.

"Gentle Reader!—Lo, here a Camera Obscura is presented to thy view, in which are lights and shades dancing on a whited canvas, and magnified into apparent life!"[21] These are the first lines of Erasmus Darwin's prose poem *The Loves of the Plants,* the second part of his two-volume poem *The Botanic Garden* (1791). In this text, the English physician, natural philosopher, and poet Erasmus Darwin, the grandfather of Charles Darwin, explains the new botanic theory of his time for a primarily female readership. His work gained ample popularity in nineteenth-century Britain, particularly because of the elder Darwin's capacity to impart scientific knowledge through a vivid, almost terpsichorean combination of poetry and philosophical footnotes.[22] To circumvent the sessility of flowers, their missing ability to move about and touch one another, Darwin superimposes human behavior and feelings onto plants such that these fictionally animated entities are "magnified into apparent life."[23] And to provide a projection surface for the primarily female readership, Darwin portrays flowers as elegant women, "who may amuse . . . by the beauty of their persons, their graceful attitudes, or the brilliancy of their dress."[24] While this already makes flowers prominent and obvious subjects for dance interpretations, Darwin achieves his highest impact on the later flower ballets when he anthropomorphizes flowers as heterosexual men and women and portrays them with erotic desires: thus, to demonstrate that "Beaux and Beauties crowd the gaudy groves,/ And woo and win their vegetable Loves," he presents enamored flowers such as "The lovesick Violet" or "jealous Cowslips" and shows "How the young Rose in beauty's damask pride/ Drinks the warm blushes of his bashful bride."[25]

Behind all this stood another, even more seminal book: Linnaeus's *Systema Naturæ,* first published in Latin in 1735.[26] Here, the Swedish botanist and physician outlined his principles of a hierarchical classification of the natural word dividing it into the low-ranked mineral kingdom (*regnum lapideum*), then the plant kingdom (*regnum vegetabile*), and finally the highest level—that of the animal kingdom (*regnum animale*). Elaborating on previous systems of classification, Linnaeus further divided these kingdoms into a hierarchy of classes, orders, genera, and species. The taxonomy of botany, the plant kingdom, depended on the artificial and strict binomial distinction between the putatively "male" parts of the flower, the stamens, and the pistils as the "female" organs.[27] The number and proportions of the male parts accounted for the different classes, whereas the female parts were responsible for the hierarchically lower orders. Most relevant for the later influence on the flower ballet genre, Linnaeus described the different combinations of male and female organs in plants as various forms of marriages. In the words of Darwin's *The Loves of the Plants:* "Linneus, the celebrated Swedish naturalist, has demonstrated, that all flowers contain families of males and females, or both; and on their marriages has constructed his invaluable system of Botany."[28]

From the perspective of our current knowledge of plants and the diversity of sexualities involved in their propagation,[29] Linnaeus's turn to the metaphor of male-female marriage might seem like a prime example of what Emily Martin in her seminal essay "The Egg and the Sperm" has described as science's tendency to construct "romance based on stereotypical male-female roles."[30] And even more than this, Linnaeus's approach reveals an anthropomorphic heteronormativity based on a dual phenomenon of attributing human sexuality to plants and then attributing heterosexual marriage to them. His sexual system shows a general cultural willingness to look for and find heterosexual marriage in plant environments and prescribe it as the only legible form of sexuality. Consequently, his nuptial metaphor favored those combinations of plant organs as marriages where male and female organs met, while ignoring any possibility of same-sex pairings. And whereas Linnaeus disregarded combinations of several female organs with one or more male organs, his sexual system included polygamous constellations of one female organ subjected to up to twenty male organs.

Beyond the field of botany, the larger cultural dynamics of Linnaeus's heteronormative plant system can best be seen in how his Latin works were adapted in the English language. In the first attempt at such adaptations, James Lee's *Introduction to Botany* (1760), "'male' stamens are 'husbands,' 'female' pistils 'wives' and sexual union a 'marriage.' Flowers lacking stamens or anthers are termed 'eunuchs' and, not surprisingly, the removal of anthers is 'castration.'"[31] And even more explicit was the plant-pornography in Hugh Rose's *Elements of Botany* (1775), his translation of Linnaeus's *Philosophia Botanica* (1751), as it likened the fertilization process of plants with its union of pistils and stamens to "husbands and wives on their nuptial bed . . . the calyx then is the marriage bed, the *corolla* the curtains, the filaments the spermatic vessels, the *antheræ* the testicles, the dust the male sperm, the *stigma* the extremity of the female organ, the *style* the *vagina*, the *germen* the ovary."[32]

These texts formed the linkage between Linnaeus's *Systema Naturæ* and Erasmus Darwin's *The Loves of the Plants*, and as Linnaean thinking was further disseminated, the hetero-pornographic character of the human–plant analogy appears to have increased the more the system was adapted to British female readership. Darwin even extended Linnaeus's categorization of plant polygamy to scenarios of suggested group sexuality: "The freckled Iris owns a fiercer flame. / And *three* unjealous husbands wed the dame."[33] Thus, an already unresolved tension increased: on the one hand, most of the flowers—except for the stamen-less "eunuchs"—had both male and female organs. On the other hand, the analogy to wives and husbands portrayed them as having only one sex, which in the process of marriage (*vulgo* fertilization) unites with the opposite one. This heteronormative reading was even further bolstered when some flowers were portrayed as having a primarily female or male character.

Sometimes, Linnaeus's gendering of plants and his equation of flowers with women in particular also extended to binomial nomenclature. A prominent example was the small pink flower growing in the Lapponian bog and—

following a pre-Linnaean nomenclature—given the long Latin name *Erica palustris pendula, flore petiolo* (originally *petiolio*) *purpureo*. On a botanic field trip in 1732, Linnaeus became so fond of this flower with its delicate pink bell-shaped calyxes that he compared it to Andromeda, Cassiopeia's daughter in Greek mythology who was known for her outstanding beauty. Replacing the existing name for this plant from the older system of Johann Christian Buxbaum and Johann Bauhin, Linnaeus gave the flower a new name, calling it *Andromeda polifolia L.*[34]

Summarizing the nexus of conceptual and ideological notions that have emerged from botany and its popular adaptations, I point out three main ideas. First, in the attempt to understand and, even more, to explain the propagation of plants, herbal life was compared to human life, and plants were perceived through an anthropomorphizing perspective. Second, this often went along with a gendering of flowers as female and a portrayal of specific flowers as having a distinct humanlike character, such as modesty, selfishness, shyness, or pride. And third, the imposition of late eighteenth- and early nineteenth-century social norms on plants and the comparison of plant dissemination with human marriage and sexuality indexed a heteronormative perspective, which implicitly declared human homosexuality as unnatural. Established already in the eighteenth century, this conglomeration of botanical knowledge and ideological concepts established the background for the flower ballet.

Papaver somniferum L.

More than a hundred years later, and through further popular adaptations of botanical knowledge by James Lee, Hugh Rose, and Erasmus Darwin, Linnaeus's taxonomical system manifested in the flower ballet genre. Though their libretti may not have listed the incorporated flowers by their proper Linnaean names, the differentiations among these flowers had been consolidated by scientific botany. In their very structure, some flower ballets paid direct tribute to the taxonomical system that Linnaeus had introduced. An example is Henri Justament's naming his ballet *Le Royaume des fleurs* (The Kingdom of Flowers, Marseille 1845),[35] a reference to Linnaean taxonomy and its subdivision of nature into genera, orders, classes, and kingdoms.

More than a nod to Linnaeus was the first Russian flower-themed entertainment, *The Love of Roses, or The May Beetle and the Butterfly*, performed in the Mikhailovsky Theatre in St. Petersburg in November 1848. In line with taxonomical principles, it showed a sequence of three kingdoms: first, the kingdom of flowers, "in which 'floral performers' danced and sang,"[36] followed by the kingdom of butterflies, and, finally, the kingdom of vegetables. With this, the Linnaean differentiation between a plant kingdom (*regnum vegetabile*) and an animal kingdom (*regnum animale*) entered the stage, yet with a remarkable twist. Diminishing the Linnean taxonomy and his monarchical principals to kingdoms of flowers, butterflies, and vegetables, the choreography of *The Love of Roses, or The May Beetle and the Butterfly* was hardly a serious adaption of Linnaean's systems. Rather, it was an ironic pastiche, if not an open persiflage of his system.[37]

Most important here, however, was the impact of Linnaeus's and Erasmus Darwin's constructed parallels of heterosexual human and botanical marriages on the narrative and the choreographies of the flower ballets. The dancers in these performances did not embody pistils and stamens dancing on the stage, nor the marriage of "male" and "female" roses; rather, following a more general plant-human analogy, the performances were based on the construct of each dancer as a flower that was understood to represent one sex only. While Linnaeus had set up a complex heterosexual system of plant identification, the result onstage was a simplified yet no less heteronormative equation of plants with one sex—mostly, but not exclusively, the female one. Appropriate to the idea that a ballerina represented a putatively female and implicitly heterosexual flower, the same ballerina could wear a full body-costume of this flower, with the tutu as the calyx, and floral decorations of the same flower applied to the dress.[38] The ballerinas both danced as flowers and wore flowers, much like the botanic system conceptualized plants as women that simultaneously had small womanlike organs within them.[39]

Popular botany by authors such as James Lee, Hugh Rose, and Erasmus Darwin was particularly successful in the British context.[40] In speculating as to whether the historical audiences of flower ballets were aware of the botanical underpinning of a plant-themed choreography, it is worth having a detailed look at a flower ballet that arose from this British context. An excellent example is Paul Taglioni's *Théa, ou la Fée aux fleurs*, which was produced in 1847 for Her Majesty's Theatre in London, restaged in Berlin in the same year, and later reproduced in Milan in 1867.[41] Here, as in *The Nutcracker*'s "Waltz of the Flowers," dancers incorporated various flowers such as daffodils, daisies, delphiniums, fuchsias, tulips, nightshades, and violets, which formed floral-ornamental tableaus on the stage.[42] Together with the floral decoration of the costumes and the flower garlands or leaf-decorated sticks in the dancers' hands, this floral abundance was further complemented by the equally richly decorated libretto, as seen in figure 4.1, showing a page of the libretto with a summary of the ballet's narrative.

THEA.

Nous sommes dans un magnifique jardin de Bagdad. Au fond d'un kiosque oriental, repose le prince Hussein, qu'entourent plusieurs odalisques. Près du kiosque, Théa jette des regards passionnés sur le prince, que ne peuvent émouvoir, ni les airs de quelques esclaves, ni les danses voluptueuses auxquelles se livrent ses odalisques. Absorbé dans la contemplation des fleurs qu'il tient à la main, rien ne le tire de sa rêverie, et Théa gémit de tant d'indifférence. Hussein ordonne qu'on le laisse seul. Théa réclame la faveur de rester auprès de lui, mais le prince lui ordonne de s'éloigner. Elle feint d'obéir, puis se cache derrière une colonne. Hussein exprime sa joie d'être seul. Il va de fleurs en fleurs les contemple avec ravissement et Théa qui le suit, en éprouve une vive jalousie. Elle essaie encore de l'arracher à son extase, mais le prince semble n'avoir des yeux que pour ses fleurs. Enfin il s'éloigne.

La jeune fille se désespère; puis tout à coup une idée heureuse ramène le sourire sur ses lèvres. Elle s'approche de la statue de la Fée aux fleurs et la supplie de lui rendre le cœur de son amant. Puis elle tresse une couronne, la lance dans l'espace, et la couronne vient se poser sur la tête de la statue. Tout à coup Théa disparaît, ainsi que le kiosque, et à la même place s'élève un rosier entouré de mille touffes de fleurs. Le prince revient, et sa joie est extrême quand il apperçoit l'admirable changement qui s'est opéré. Il caresse toutes les fleurs, aspire leurs parfums et va cueillir une rose, quand derrière le rosier, apparaît la Fée. Celle-ci reproche à Hussein son audace, et le prince, tremblant, se jette à ses genoux et implore sa clémence. La Fée ne lui pardonnera que s'il épouse la fleur qui aura conçu de l'amour pour lui. Quelque surprise qu'il éprouve à cette proposition, Hussein y souscrit cependant. La Fée le fait se placer sur un banc de gazon, et pose sur sa tête une couronne de pavots. Le prince s'endort.

La nuit est venue. Une flamme bleue s'échappe de chaque touffe fleurie. Peu à peu toutes les fleurs s'animent, et deviennent des jeunes filles qui se sourient, marchent, essaient des pas et forment une guirlande vivante autour de la Fée. Pendant que les fleurs dansent, Théa, qui représente la rose, s'échappe, enlève au prince sa couronne de pavots et revient se mêler à ses compagnes. Le prince s'éveille; il est ébloui du tableau qui s'offre à ses yeux. Il s'élance après la rose objet de ses rêves et de ses vœux, tombe aux pieds de Théa et lui offre son amour et sa main. La reine des fleurs paraît et sanctionne leur union.

We are in a magnificent garden of Bagdad. Within an oriental kiosk, reposes the youthful Prince Hussein, surrounded by odalisks. Near the kiosk is Théa, regarding with passionate and tender looks the Prince, who, all unmoved by the music of his slaves and the voluptuous dances of his odalisks, is absorbed in the contemplation of some flowers which he holds in his hand. Nothing can withdraw him from his reverie, and Théa becomes dejected at his indifference. Hussein desires to be left alone. Théa entreats that she may remain; but the Prince commands her to leave him. She feigns acquiescence, but conceals herself behind one of the columns of the kiosk. Hussein appears delighted at his solitude, and goes with rapture from flower to flower, while Théa follows him with all the marks of jealousy. She again attempts to tear him from his ecstasy; but the Prince appears to have eyes for his flowers only. At last he quits the scene.

The neglected beauty weeps in despair; but a sudden thought brings smiles to her lips. She approaches the statue of the *Fée aux Fleurs*, and entreats the restoration of her lover's heart. She then forms a crown of flowers and throws it in the air; the crown lights on the head of the statue. Théa becomes motionless, and disappears, as does also the kiosk,—in the place of which arises a lovely rose-tree, surrounded by every variety of flower. The Prince returns, and sees with joy the change which has taken place. He kisses the flowers, aspires their perfumes, and is about to pluck a rose, when, behind the tree, appears the *Fée aux Fleurs*. She seems in anger, and reproaches Hussein with his audacity. The Prince throws himself on his knees and implores pardon; but the fairy will grant it only on condition of his marrying the flower which shall have conceived a passion for him. Surprised at this proposition, he nevertheless consents. The fairy conducts him to a bank of turf, places on his head a crown of poppies, and the Prince sleeps.

It is night; a blue flame issues from each floral group, the flowers become animated and are changed to nymphs, who smile, walk, dance, and form a living garland round the *Fée aux Fleurs*. While the flowers dance, Théa, who represents the rose, escapes, takes the poppy crown from the Prince's head, and returns to mix with her companions. Hussein awakes; he is dazzled by the *tableau* which presents itself, but seeks the rose, the object of his dream, falls at the feet of Théa, and offers her his heart and hand. The Queen of Flowers appears to sanction their union.

Figure 4.1. Program for Paul Taglioni's *Théa, ou la Fée aux fleurs* (1847). Her Majesty's Theatre, London. Published with permission of Harvard University Houghton Library.

Elaborating on Darwin's literary fantasy, the plot of *Théa* conceptualizes flowers not only as moving and dancing plants, but as potential partners for and rivals of romantic human relationships.[43] Similar to Henri Justament's equally orientalist flower ballet *Le Royaume des fleurs*, the plot of *Théa* is set in a harem. It centers around a male protagonist, Prince Hussein, who is libidinously interested only in flowers and spurns the advances of the odalisque *Théa*. Put into an intoxicated slumber by a crown of opium poppies (*Papaver somniferum L.*), Hussein falls in love with a rose he sees in his dreams, not knowing that this very flower is Théa, who has metamorphosed into a plant. After Théa transforms back into a woman and Hussein awakens from his sleep, they agree to marry.[44]

Resonating with the choreographic usages of the ballet position *arabesque*, the plot of *Théa* has an equally arabesque structure, where story lines bifurcate and narrative strands diverge like legs from the perpendicular line of the body.[45] Almost hidden by this arabesque-like narrative is the conceptual backdrop of ideas adopted from popular botany. The libretto not only anthropomorphizes flowers and genders them as female, but it also promotes heterosexuality as a norm, with the narrative even culminating in a noteworthy sexual "correction" in which the libidinous interests of the male protagonist are diverted from flowers to women.[46] Fittingly to this conversion therapy *avant la lettre*, the white opium poppies of Prince Hussein's crown not only bear the floriographic meaning of sleep, but also symbolize "my bane, my antidote."[47] The core idea of the plot twist—Prince Hussein's drug-induced change from an ignorance of the opposite sex to his willingness to marry Théa—demonstrates the ideological perspective of this flower ballet. It shows how heteronormative ideas that were already at work in Linnaeus's scientific innovation survived the various adaptations of this approach from scientific to popular botany, and from there to ballet.

And even more than this: not only were plants viewed through an anthropocentric perspective and an ideal of heterosexual marriage, but the performative visualization by the flower ballets gave visual credence to this ideology and afforded it an even greater distribution and public support. Through display on the ballet stage, the support for traditional and exclusive social norms became indisputable as these norms were presented and celebrated at the center of European nineteenth-century society. In a significant reverse argument, the unscientific imposing of heterosexuality on plants also implied a denaturalization of human homosexuality: when even in nature only female and male plants married, human same-sex desire would be seen as all the more unnatural.

At this point, the question arises as to what degree this nexus was perceptible by the historical audience. Were audience members of *Théa* in London, Berlin, or Milan, or those viewing any other flower ballet aware that they saw an illustration of popular botany based on the anthropocentric subjugation of botanic nature under exclusive social normativity? Taking the broad distribution and availability of popular botanic publications and further adoptions like the language-of-flower books into account[48]—especially in the context of the creation of *Théa* in London—a knowledge of botanical ideas, even if rudimentary, can be regarded as informing the perspective of the audience. Its members were versed in the language of anthropomorphism and the gendering of plants, which, in the course of the nineteenth century, had become part of the common imagination.[49]

Grandville's *Les Fleurs animées,*[50] published in 1847 shortly before the première of *Théa* in that same year, offers well-known substantiation for the popularity of this concept. In this book, the French caricaturist Jean Ignace Isidore Gérard, under the pseudonym Grandville, illustrates a text by Taxile Delord with lithographs that show flowers as animated beings and that portray them as fusions of plant and human bodies, dressed in contemporary Parisian fashion. The parallels with the costume design for *Théa* and its basic idea of animated plants are striking, and *Les Fleurs animées* is regarded as a direct inspiration for Taglioni's piece.[51] This substantiates the widely held idea, throughout Europe, of portraying flowers as womanlike beings, and how influential this portrayal was on visual culture.[52]

Consequently, the depiction of plants as humanized dancing entities met the horizon of audience expectations, and, far from any skepticism, dance reviews of the time celebrated the animation of flowers. On the occasion of the premiere of the Milan version of *Théa* in 1867, a contemporary review in *Il Corriere delle Dame* praised the *ballabile* of the finale and highlighted

> the seductive, charming effect of those forty or fifty girls dressed as roses, violets, sunflowers, geraniums, that turn round in a thousand whirling circles, forming delightful bouquets, and then unite to form a splendid bunch of living flowers: it's such an ingenious and singular thing that the applause explodes unwillingly from your lips and erupts in inarticulate sounds.[53]

Particularly noteworthy is how this review took for granted the gendering of flowers as female and connected femininity with the aesthetic discourse of ornamentality that was so characteristic of nineteenth-century ballet in general. With regard to the technical vocabulary, Susan Foster prominently noted that "the divergent vocabularies for male and female dancers symbolized a difference between the sexes far greater than the distinct styles of the eighteenth-century performers."[54]

By contrast, the historic evidence for the claim that the flower ballets performed an implicitly homophobic concept of human–plant relation is far thinner. We have no textual source that the creators and the contemporary reception were driven by any sort of ideological didacticism based on what was happening onstage, either as an explicit support of the performed exclusion or as an objection against it. But this does not mean connections cannot be drawn between the flower dances and contemporaneous practices of social erasure. While scholars have persuasively argued that nineteenth-century ballets like *La Sylphide* and *Giselle* can be read as making space for androgynous corporeality and queer spectatorship,[55] we might still find value in scrutinizing how these flower ballets utilized adaptations of botanical rhetoric to advance practices of exclusion. Based on this point, I return to the narrative of the different versions of *Théa* for a closer look at Taglioni's surviving sketches for this piece to show that ideology was indeed openly visible on the stage.

Helianthus annuus L.

In the 1867 Italian version, the plot of *Théa* was partly altered, with the protagonist now named Almanzor and the setting transferred from Baghdad to the Moorish court in Córdoba. Interestingly, a layer of persiflage was added to the piece that exaggerated and ridiculed the idea of dancing animated plants. This culminates in the finale, where a flower garden unites with a kitchen garden "represented by a rank of tomatos, beets, radish, mushrooms, that with the best of graces endeavour to court the roses and the violets,"[56] accompanied by "butterflies, hornets, bumble bees and other insects," all of which are embodied by dancers of the *corpo di ballo.* A historical review states how "here hilarity was added to admiration, and the audience bursts with laughter at that inimitable show."[57] Remarkable in this review is how it reveals the basic reception of *Théa* by a historical audience: hilarity, admiration, and laughter were caused by an extravaganza of dancing plants and insects.

However, the earlier Berlin version of *Théa* itself exaggerated its core idea of dancing flowers when it portrayed not only European flowers, but also trade plants, such as coffee, tobacco, cane, and tea.[58] This persiflage affirms the Orientalism of the piece and reveals how *Théa* as a typical example of the flower ballet genre was steeped in coloniality. The inclusion of colonial trade plants arguably signified to European viewers their own social and political world, which could have served to ground the orientalist narrative and its

Figure 4.2. *Soleil (Sunflower),* in Grandville, *Les Fleurs animées,* vol. 1, pl. 21. Public domain. Staatsbibliothek zu Berlin–PK, Manuscripts and Historic Prints Department, Signature: 4″LW7440–1 R.

harem setting in a sense of authenticity. Here again, the dancing plants are far from innocent fantasies of animated botanical life, but mirrors of sociopolitical reality and thus could be read as emblems of European supremacist ideology.[59]

Probably the most notorious flower representation in this context was the sunflower (*Helianthus annuus L.*), which was first domesticated in North and Central America and had arrived in Europe in the sixteenth century. Grandville in *Les Fleurs animées,* published in Paris 1847, draws on the local origin of the plant and, as seen in figure 4.2, uses racial stereotypes and exoticist fantasies to depict the sunflower as a kneeling indigenous man, holding his hands folded in a prayerlike pose and wearing a short leaf skirt around his hips and a crown of yellow petals around his face. The position of the kneeling and praying man bears a striking similarity to that of a Black slave on Josiah Wedgwood's famous anti-slavery medallion created in 1787 as a seal for the Society for the Abolition of the Slave Trade. Yet in exchanging the slave's metal chains with twines around the sunflower's arms and legs, it is more likely that Grandville's caricature refers to the abolitionist symbol in a mocking way. While Wedgwood's medallion shows a Christian Black slave, Grandville's lithograph depicts the flower as an indigenous sun worshipper, thus resonating with botanic knowledge about the heliotropism of the sunflower, whose head follows the position of the sun in the course of the day. As the most offensive image of *Les Fleurs animées,* the lithograph perpetuates primitivizing pictorial traditions to portray Indigenous people from North and Central America and ridicules Indigenous religion and spirituality. At the same time, this representation of the sunflower is in line with its ascribed non-European otherness, found earlier in popular writings like Darwin's *The Loves of the Plants* (1791), where the heliotropisms of the sunflower are orientalized as a kind of dervish dance: "Great Helianthus guides o'er twilight plains / In gay solemnity his Dervise-trains."[60]

The costume design for *Théa* seized on Grandville's sunflower images and combined them with an equally racist portrayal of African slaves. For yet another arabesque bifurcation from the main narrative, which was part of a revue-like sequence in the second image of the ballet, where dancers performed various identifiable flowers as animated beings, Taglioni choreographed a sunflower dance.[61] This was performed by one male dancer and a group of young male dance students, all of whom appeared in blackface and with a presumably yellow ruff. They were costumed in grass skirts with bangles around their arms and legs and with the illusion of dark skin, an effect likely created by either a full body tricot or black body paint. Figure 4.3 shows one of the surviving sketches for this costuming[62] as well as some of the key

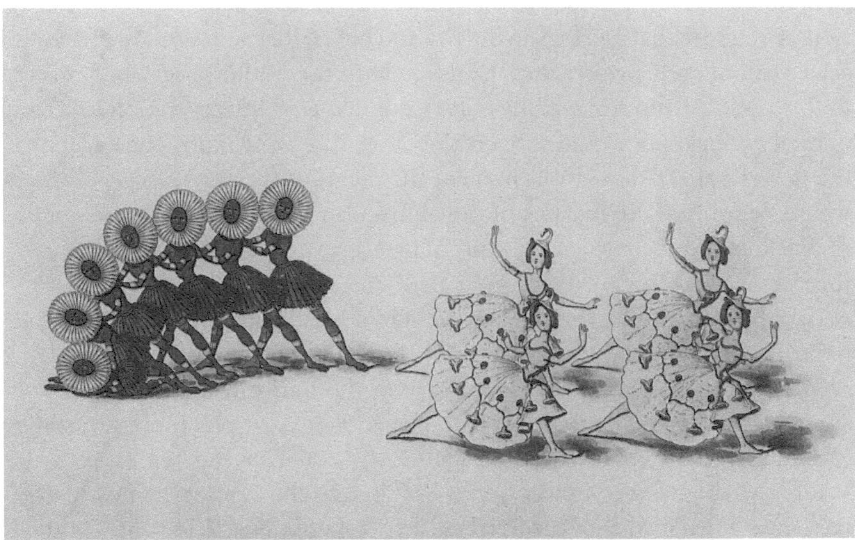

Figure 4.3. Sketch for Paul Taglioni's *Théa, ou la Fée aux fleurs* (1847–1867). Public domain. Reprinted in Lifar, *Le Ballet et la danse à l'époque romantique*, n.p.

dance steps of the scene: the dance of the sunflower men appears to have used a shoulder embrace reminiscent of folk dancing of unspecified origin. Their nonballetic and allegedly primitive motions are contrasted with the symmetric arrangement of a group of four female dancers embodying light-blue bell flowers.[63] Dressed in large, calyx-shaped tutus with appliqué bell flowers and a bell flower headdress, the female dancers perform European ballet motions. Thus, with elegantly elevated arms and outstretched feet, they represent the alleged supremacy of the European flora that contrast with the purportedly inferior sunflowers of non-European origin.

The depiction of the sunflower through the usage of racist clichés was only tangentially related to the plot of *Théa*. It was, however, a visualization for the homophobic subtext of the flower ballet. While most of the flowers were performed by female dancers, some plants, like the sunflower, had an ascribed masculine character and were danced by men.[64] Through the shoulder embrace, Taglioni's sketch not only alluded to clichés of folk or tribal dancing; it also shows a moment when the dance of the seven sunflower men appears to turn into stumbling, bringing their buttocks and pelvises close to one another. Judging from the sketch, the choreography at this point may be read as mimicking and mocking homosexual anal intercourse, thus visualizing the homophobic ideology that was already implicit in earlier botanic writings. The homophobia is not opposed to the erasure of homosexuality and the idealization of heterosexual pairings in botany and ballet, but rather follows from this selective and exclusive practice. Consequently, the sunflower group-sex scene in *Théa*—at least in my reading—does not invite an interpretation as a form of queer dance that would subvert the otherwise heteronormative agenda of the flower ballet. Rather, it is here, in the dance of the animated sunflowers, where primitivist stereotypes of non-European indigeneity, racist depiction of Blackness, and conspicuous hints of heterosexual fantasies of gay life come together.

To add on to this pictorial evidence of biologically informed heteronormativity in the flower ballet genre and its condensation in homophobic dance scenes, we must take into account that in the mid-nineteenth century, the sunflower itself—beyond its depiction as an animated plant in the tradition of Grandville—was floriographically already connected to same-sex desire. While the language-of-flower books ascribed to the sunflower mostly negative meanings such as "false riches" or "haughtiness,"[65] for the Aesthetic Movement in late nineteenth-century Britain, the sunflower stood for a queer lifestyle and was a self-chosen signifier for homosexual men. This, however, only spurred homophobic caricature, which depicted publicly known homosexual men, such as Oscar Wilde, together with sunflowers or even as sunflower-persons in the tradition of Grandville.[66] These caricatures indicate how widespread the floriographic identification of the sunflower with homosexuality in the nineteenth century was, and that Taglioni likely drew from this association for his creation of *Théa*. His choice for the sunflower built on the association of the flower with homosexuality, and—judging from his sketches—casting, costuming, and choreography added homophobia and racism as well.

Conclusion

The nineteenth-century flower ballet owed much of its conceptual ideas to earlier scientific developments in botany and their various adaptations. These included popular scientific writings as well as the further applications in floriographic literature and visual art. The very idea of creating entire ballets or smaller revue-like dance sequences of animated flowers—anthropomorphized and (mostly) gendered as female—stemmed from these sources. Most influential among them were Grandville's illustration of *Les Fleurs animées*, which paired the phantasma of plants as animated and humanlike beings with the idea of attributing each plant an individual character. Paul Taglioni's *Théa, ou la Fée aux fleurs*, a prominent example of the flower ballet genre, drew direct inspiration from these illustrations. From here, Taglioni adopted ideas for the costuming of the dancers as specific flowers and—as the example of the sunflower dance shows—approached the depiction of certain plants from a colonialist perspective.

The influence of botany on ballet also extended to the moral discourse and the social conceptions that the popular botanic publications imposed. Their simplifying and misleading equation of plant and human sexuality not only explained the propagation of plants to a wider and mostly female readership; it also made statements about human social norms, manifested and naturalized through the comparison with plants. In this perspective, *Théa*'s plot, which culminates in the apotheosis of a human male-female marriage with the support of a flower fairy, was a realization of what popular botany conveyed on a moral level. In particular, the ballet's disapproval of nonheterosexual romance and libidinous enthusiasm for flowers harkened back to Linnaean botany, where plant propagation was compared to heterosexual marriage.

The representation of anthropomorphized and gendered plants in parallel to—or, as in the case of *Théa*—as competitors of human relationships, romance, and partnership participated in the affirmation of existing social norms that accompanied popular botany. When heteronormative gender roles were projected onto plants, such norms received a naturalization that then reflected back upon human social life. Publicly shown on a ballet stage, the performed imaginary parallels between plant and human life contributed to this naturalization of existing norms and supported them further. The equation of plants and people developed as a performed discourse on social normativity with an implicit, or, in certain scenes, even explicit exclusion of otherness. From this perspective, the connection between eighteenth-century science and its popularization in dance was highly ambivalent. Science rendered the flower ballet its subject and led to performances of hitherto unseen animated floral ornamentality that carried the ideological burden of historic botany.

Notes

1 Foucault, *The Order of Things*, 162.
2 Kelley, *Clandestine Marriage*, 17–51.
3 Schwan, "Wilting Flowers in Dance," 264.
4 George, *Botany, Sexuality and Women's Writing*, 81–104.
5 Linnaeus, *Species plantarum*.
6 Garafola, *Rethinking the Sylph*.
7 On the *ballet fantastique* in general, see the study of Meglin, "Behind the Veil of Translucence."

8 Meisner, *Marius Petipa,* 222–23.
9 Meisner, *Marius Petipa,* 94–96; 300–18.
10 Schwan, "Wilting Flowers in Dance," 263–70.
11 Schwan, "Wilting Flowers in Dance," 271–73.
12 On floriography, see particularly Seaton, *The Language of Flowers;* Schwan, "Floriographie als Sprache der Emotionen."
13 Latour, *Le Langage des fleurs.*
14 Schwan, "Floriographie als Sprache der Emotionen," 219–21.
15 The Marius Petipa Society, "The Nutcracker." In Meisner, *Marius Petipa,* 229–33 and Wiley, *Tchaikovsky's Ballets,* 216, the flowers remain unidentified.
16 The Marius Petipa Society, "The Nutcracker." See also Wiley, *Tchaikovsky's Ballets,* 193–241, and *The Life and Ballets of Lev Ivanov,* 132–49; Mahiet, "The Aesthetics and Politics of Wonder in the First *Nutcracker.*" For the later cultural history of *The Nutcracker,* see particularly Fisher, *Nutcracker Nation,* 171–94.
17 The Marius Petipa Society, "The Nutcracker."
18 Seaton, *The Language of Flowers,* 184–85.
19 Seaton, *The Language of Flowers,* 184–85.
20 Bonne et al., "Y a-t-il une lecture symbolique de l'ornement?" 30; Schwan, "Wilting Flowers in Dance," 264, and "Arabesque Vision," 331.
21 Darwin, *The Loves of the Plants,* vii. See also Kaulbarsch, "Apparent Life."
22 Kaulbarsch, "Apparent Life," 167.
23 Darwin, *The Loves of the Plants,* vii. See also Schiebinger, *Nature's Body,* 11–39.
24 Darwin, *The Loves of the Plants,* ix.
25 Darwin, *The Loves of the Plants,* 2.
26 Linnaeus, *A System of Vegetables.*
27 George, "Botany in an English Dress," 2.
28 Darwin, *The Loves of the Plants,* 2.
29 The Tree of Sex Consortium, "Tree of Sex," 2–4.
30 Martin, "The Egg and the Sperm," 485.
31 George, "Botany in an English Dress," 2.
32 Rose, *The Elements of Botany,* 151.
33 Darwin, *The Loves of the Plants,* 8.
34 Kranz, "Zur Poetik der Pflanzennamen in der Botanik," 104.
35 Wittrock, *Arabesken,* 254.
36 Meisner, *Marius Petipa,* 95. See also The Marius Petipa Society, "The Blue Dahlia."
37 Hiner, "From Pudeur to Plaisir," 45.
38 For a contrasting reading of the ballerina, see Foster, "The Ballerina's Phallic Pointe."
39 Kaulbarsch, "Apparent Life," 171.
40 George, "Botany in an English Dress," 2.
41 On *Théa,* see particularly Toschi, "Seduction by Proxy." On P. Taglioni in general, see Oberzaucher-Schüller, "Bournonville, Paul Taglioni, and the Ballet of the Mid-Nineteenth Century," 473–79.
42 Taglioni, *Thea,* 3–4.
43 Toschi, "Seduction by Proxy," 316.
44 Toschi, "Seduction by Proxy," 316; Schwan, "Wilting Flowers in Dance," 265–67.

45 Schwan, "Arabesque Vision," 327. For a detailed, ten-page summary of the narrative, see the libretto of the Berlin version in Taglioni, *Thea,* 5–14.
46 Toschi, "Seduction by Proxy," 315–16.
47 Seaton, *The Language of Flowers,* 188–89.
48 Schwan, "Floriographie als Sprache der Emotionen," 220.
49 Hiner, "From Pudeur to Plaisir," 45.
50 Grandville, *Les Fleurs animées.*
51 Toschi, "Seduction by Proxy," 316; Wittrock, *Arabesken,* 203–208.
52 Kranz, Schwan, and Wittrock, "Einleitung," 18.
53 *Il Corriere delle Dame,* Mar. 1, 1867, 68–69, quoted in Toschi, "Seduction by Proxy," 330.
54 Foster, "The Ballerina's Phallic Pointe," 4.
55 Lee, "The Romantic Ballet."
56 *Il Corriere delle Dame,* Mar. 1, 1867, 68–69, quoted in Toschi, "Seduction by Proxy," 330 (orthography follows the original).
57 *Il Corriere delle Dame,* Mar. 1, 1867, 68–69, quoted in Toschi, "Seduction by Proxy," 330.
58 Taglioni, *Thea,* 4.
59 On the colonial aspects of Linnaean thinking, see Koerner, *Linnaeus,* 95–112; Schiebinger, *Plants and Empire,* 194–225.
60 Darwin, *The Loves of the Plants,* 23.
61 Taglioni, *Thea,* 4.
62 Lifar, *Le Ballet et la danse à l'époque romantique,* n.p. For a contextualization of Lifar's exhibition catalogue, published in German-occupied Paris in 1942, see Franko, *The Fascist Turn in the Dance of Serge Lifar,* 180–231.
63 Wittrock, *Arabesken,* 206–207.
64 Wittrock, *Arabesken,* 206–207.
65 Seaton, *The Language of Flowers,* 194–95.
66 Breward, "Aestheticism in the Marketplace," 200–205.

Bibliography

Bonne, Jean-Claude, Martine Denoyelle, Christian Michel, Odile Nouvel-Kammerer, Emmanuel Coquery, "Y a-t-il une lecture symbolique de l'ornement?" *Perspective* 1 (2010): 27–42.

Breward, Christopher. "Aestheticism in the Marketplace: Fashion, Lifestyle and Popular Taste." In *The Cult of Beauty: The Victorian Avant-Garde 1860–1900,* edited by Stephen Calloway and Lynn Federle Orr, 194–205. London: V&A Publishing, 2011.

Darwin, Erasmus. *The Loves of the Plants, with Philosophical Notes.* In *The Botanic Garden: A Poem, in Two Parts.* London: J. Johnson, 1791.

Fisher, Jennifer. *Nutcracker Nation: How an Old World Ballet Became a Christmas Tradition in the New World.* New Haven, CT: Yale University Press, 2003.

Foster, Susan Leigh. "The Ballerina's Phallic Pointe." In *Corporealities. Dancing Knowledge, Culture and Power,* edited by Susan Leigh Foster, 1–24. London: Routledge, 1996.

Foucault, Michel. *The Order of Things: An Archaeology of the Human Sciences.* New York: Vintage Books, 1994.

Franko, Mark. *The Fascist Turn in the Dance of Serge Lifar: Interwar French Ballet and the German Occupation.* New York: Oxford University Press, 2020.

Garafola, Lynn. *Rethinking the Sylph: New Perspectives on the Romantic Ballet.* Middletown, CT: Wesleyan University Press, 2011.

George, Sam. *Botany, Sexuality and Women's Writing, 1760–1830: From Modest Shoot to Forward Plant.* Manchester, UK: Manchester University Press, 2007.

George, Sam. "Linnaeus in Letters and the Cultivation of the Female Mind: 'Botany in an English Dress.'" *British Journal for Eighteenth-Century Studies* 28 (2005): 1–18.

Grandville, Jean-Jacques. *Les Fleurs animées,* introduction by Alphonse Karr, text by Taxile Delord. Paris: Gonet, 1847.

Hiner, Susan. "From Pudeur to Plaisir: Grandville's Flowers in the Kingdom of Fashion." *Dix-Neuf* 18, n. 1 (2014): 45–68.

Kaulbarsch, Vera. "'Apparent Life': Botanik, Visualität und Literatur bei Erasmus Darwin." *Literatur für Leser* 40, n. 2 (2017): 167–84.

Kelley, Theresa M. *Clandestine Marriage: Botany and Romantic Culture.* Baltimore: Johns Hopkins University Press, 2012.

Koerner, Lisbet. *Linnaeus: Nature and Nation.* Cambridge, MA: Harvard University Press, 1999.

Kranz, Isabel. "Zur Poetik der Pflanzennamen in der Botanik: Carl von Linné." *Poetica* 50 (2019): 96–118.

Kranz, Isabel, Alexander H. Schwan, and Eike Wittrock. "Einleitung." In: *Floriographie. Die Sprache der Blumen,* edited by Kranz, Isabel, Alexander H. Schwan, and Eike Wittrock, 11–30. Paderborn: Wilhelm Fink, 2016.

Latour, Charlotte de. *Le Langage des fleurs.* Paris: Andot, 1819.

Lee, Amanda. "The Romantic Ballet and the Nineteenth-Century Poetic Imagination." *Dance Chronicle* 39, n. 1 (2016): 32–55.

Lee, James. *An Introduction to Botany: Containing an Explanation of the Theory of that Science, and an Interpretation of its Technical Terms: Extracted from the Works of Dr. Linnæus.* London: Printed for J. and R. Tonson, 1760.

Lifar, Serge. *Le Ballet et la danse à l'époque romantique, 1800–1850.* Paris: Le Musée des arts décoratifs, 1942.

Linnaeus, C. [Carl von Linné]. *Species plantarum, exhibentes plantas rite cognitas, ad genera relatas, cum differentiis specificis, nominibus trivialibus, synonymis selectis, locis natalibus, secundum systema sexuale digestas.* Stockholm: Impensis Laurentii Salvii, 1753.

Linnaeus, C. [Carl von Linné]. *A System of Vegetables, According to Their Classes, Genera, Orders, Species with Their Characters and Differences . . . Translated from the Thirteenth Edition of the Systema Vegetabilium of the Late Professor Linnaeus and from the Supplementum Plantarum of the Present Professor Linnæus . . . By A Botanical Society at Lichfield.* 2 vols. Lichfield: printed by John Jackson for Leigh and Sotheby, London, 1783.

Mahiet, Damien. "The Aesthetics and Politics of Wonder in the First *Nutcracker.*" *19th-Century Music* 40, n. 2 (2016): 131–58.

The Marius Petipa Society. *The Nutcracker.* https://petipasociety.com/the-nutcracker/. Accessed Apr. 18, 2022.

The Marius Petipa Society. *The Blue Dahlia*. https://petipasociety.com/the-blue-dahlia. Accessed Apr. 18, 2022.

Martin, Emily. "The Egg and the Sperm: How Science Has Constructed a Romance Based on Stereotypical Male-Female Roles." *Signs* 16, n. 3 (1991): 485–501.

Meglin, Joellen A. "Behind the Veil of Translucence: An Intertextual Reading of the Ballet Fantastique in France, 1831–1841. Part One. Ancestors of the Sylphide in the Conte Fantastique." *Dance Chronicle* 28, n. 1 (2005): 67–142.

Meisner, Nadine. *Marius Petipa: The Emperor's Ballet Master.* New York: Oxford University Press, 2019.

Oberzaucher-Schüller, Gunhild. "One Out of Many? Bournonville, Paul Taglioni, and the Ballet of the Mid-Nineteenth Century." *Dance Chronicle* 29, n. 3 (2006): 471–83.

Rose, Hugh. *The Elements of Botany . . . Being A Translation of the Philosophia Botanica, and Other Treatises of Linnæus.* London: T. Cadell [etc.], 1775.

Schiebinger, Londa L. *Nature's Body: Sexual Politics and the Making of Modern Science.* London: Pandora, 1993.

Schiebinger, Londa L. *Plants and Empire: Colonial Bioprospecting in the Atlantic World.* Cambridge, MA: Harvard University Press, 2004.

Schwan, Alexander H. "Arabesque Vision: On Perceiving Dancing as *Écriture Corporelle* in William Forsythe's *The Vertiginous Thrill of Exactitude.*" In *Vision in Motion. Streams of Sensation and Configurations of Time,* edited by Michael F. Zimmermann, 317–33. Zurich: diaphanes, 2016.

Schwan, Alexander H. "'Blumen müssen oft bezeigen, was die Lippen gern verschweigen:' Floriographie als Sprache der Emotionen." In *Gefühle Sprechen: Emotionen an den Anfängen und Grenzen der Sprache,* edited by Viktoria Räuchle and Maria Römer, Maria, 199–221. Würzburg: Königshausen & Neumann, 2014.

Schwan, Alexander H. "Wilting Flowers in Dance: Choreographic Approaches to Floral Ephemerality." In *Floriographie. Die Sprache der Blumen,* edited by Isabel Kranz, Alexander H. Schwan, and Eike Wittrock, 259–85. Paderborn: Wilhelm Fink, 2016.

Seaton, Beverly. *The Language of Flowers. A History.* Charlottesville, VA: University of Virginia Press, 1995.

Taglioni, Paul. *Thea, oder: Die Blumenfee. Ballet in 3 Bildern.* Berlin, 1847.

The Tree of Sex Consortium. "Tree of Sex: A Database of Sexual Systems." *Scientific Data* 1, n. 140015 (2014). doi:10.1038/sdata.2014.15.

Toschi, Andrea. "Seduction by Proxy: *Thea* by Paolo Taglioni and Constantino Dall'Argine." In *Souvenirs de Taglioni,* Vol. II: *Bühnentanz in der ersten Hälfte des 19. Jahrhunderts.* Edited by Gunhild Oberzaucher-Schüller, 315–32. Munich: Kieser, 2007.

Wiley, Roland John. *The Life and Ballets of Lev Ivanov. Choreographer of The Nutcracker and Swan Lake.* Oxford: Clarendon Press, 1997.

Wiley, Roland John. *Tchaikovsky's Ballets: Swan Lake, Sleeping Beauty, Nutcracker.* Oxford: Oxford University Press, 1991.

Wittrock, Eike. *Arabesken: Das Ornamentale des Balletts im frühen 19. Jahrhundert.* Bielefeld: transcript, 2017.

5

New Sensations

Response to "Science under the Surface: Victorian Science in the Ballet *Ondine*" by Steven Ha and "Phytology and Dance: The Impact of Plant Biology on Nineteenth-Century Flower Ballets" by Alexander H. Schwan

Whitney Laemmli

In the autumn of 1844, the center of a growing hubbub in London society was neither a royal scandal nor a brewing war, but a book. Titled *Vestiges of the Natural History of Creation,* the text's anonymous author argued that all entities currently in existence—from the white cliffs of Dover to the Dover sole to humanity itself—had evolved from earlier forms. In purple prose, it traced the history of the universe from the emergence of the solar system from a gaseous "Fire-mist" to the present, upsetting both religious and scientific orthodoxies in the process. The first edition of the text sold out within days, and by early 1845, Prince Albert was known to be reading it aloud to Queen Victoria each afternoon. But while *Vestiges* may have started out as the latest fascination of the wealthy and elite, its explosive claims quickly reached beyond the upper crust. Fueled by the new availability of cheap print, affordable editions of the book also found a receptive audience among the poor and the burgeoning middle classes in cities and towns across the country.

The story of *Vestiges* is told by historian James Secord in his book, *Victorian Sensation.*[1] Among Secord's claims is the argument that it was the cheap and scandalous *Vestiges,* rather than Charles Darwin's dignified 1859 tome *On the Origin of the Species,* that conditioned nineteenth-century Britons to think in evolutionary terms. Previous historians, Secord contended, had been overawed by Darwin's intellectual contributions and thus neglected to pay attention to the other—and perhaps more significant—ways in which people in the past came to think about science. When asking, "How did Victorians understand evolutionary theory?" Secord responded not with an explication of any author's set of ideas but with the rejoinder "through books."

The contributions by Steven Ha and Alexander Schwan to the current volume make similar interventions, although they would answer the question not just with "through books," but also "through dance." In "Phytology and Dance," Schwan explores how plant biology influenced the construction of nineteenth-century flower ballets—and how, in turn, these ballets may have changed the ways their audiences conceived of the natural and the social worlds. He focuses first on popular botanical collecting and floriography, arguing that the then-ubiquitous Victorian floral dictionaries shaped onstage imagery as well as audiences' reading of the ballets' messages. Schwan then moves to a simultaneous consideration of Erasmus Darwin's 1791 prose poem *The Loves of the Plants* and of Linnean taxonomy writ large, contending that these authors' tendencies to write about plant reproduction in anthropomorphic terms naturalized the idea of a ballet's female protagonist *as* flower. He concludes by suggesting how this framing might have reinforced conservative ideas about gender and sexuality.

Ha's "Science under the Surface" takes a comparable approach, using the 1843 ballet *Ondine* to investigate Victorians' new "fascination and anxiety about the marine realm." The story of a naiad who pursues a human man, the production had particular resonance at a moment in which English audiences were strolling the seashore searching for specimens, descending in diving bells, and peering through microscopes at previously unseen—and sometimes unnerving—worlds. As Ha points out, the aquatic Ondine was both alluring and threatening. As such, she functioned simultaneously as an embodiment of Victorians' fears about the microscopic organisms now apparently lurking in every drop of water and as a tool for integrating these new discoveries into existing imaginative structures. The result, Ha ventures, may have been a growing comfort with and even excitement about the study of the natural world. (Audiences were also perhaps reassured by the fact that Ondine, who withers when too-long removed from her natural environment, is ultimately unable to displace her human rival, Giannina.)

Schwan's and Ha's interventions should be important reminders to historians of science that natural knowledge has long been communicated in forms beyond books, lectures, and images. While the field has become adept at attending to the cross-pollination between science and literature, the visual arts, and music, the study of science and dance has lagged behind.[2] There are undoubtedly a variety of reasons for this lacuna, including the historical tendency to associate dance with science's others: the embodied, irrational, feminine, and racialized rather than the intellectual, objective, masculine, and white. There are, however, also other unique challenges that studying dance in the past presents, including dance's relative ephemerality. While historians like Secord have been able to scour surviving copies of books like *Vestiges* for marginalia suggesting how different groups of readers responded to the text, analogous artifacts rarely exist for dance. Reviews, letters, and diary entries can, of course, provide valuable insights, but are often difficult to track down, leaving exactly how a piece was interpreted—and how those interpretations may have varied across social groups—frustratingly opaque.

Both Ha and Schwan tackle this obstacle head-on, acknowledging the fact that certain facets of the historical audience's experience remain fuzzy. Schwan, for example, notes that a ballet's creator could never be sure that the intent motivating the choice of a particular flower would be understood by the audience, as the meaning of each flower "could differ from country to country and from one floriographic dictionary to another." Moreover, though floriography was a common pastime, it was not one in which all audience members were equally expert. Schwan's analysis of the gender and sexual politics of the flower ballet raises similar questions. Though some audience members may have indeed come away from a ballet like *Théa* convinced of the naturalness or superiority of heterosexual marriage, the history of popular engagement with Linnean botany suggests other possibilities as well. In fact, as historians like Londa Schiebinger and Stefan Müller-Wille have demonstrated, both Erasmus Darwin and Linnaeus were regularly attacked for the ways in which their depictions of plant reproduction highlighted nonnormative forms of sexuality, "from monogamy and polygyny, to homosexuality, miscegenation and incest."[3] The coupling of a single pistil with multiple stamens—"twenty or more husbands in the same bed with a woman"—was not exactly the kind of "straightlaced middle-class monogamy" to which social conservatives aspired.[4] A number of well-known figures raged loudly against this sexual imagery, calling it "loathsome harlotry," "unsuitable to the laws and manners of our people," and a shock to female modesty that went "far beyond all decent limits."[5] To some, Darwin was even more suspect, given his radical politics and advocacy of free love.

This is not to say that a heterosexist reading of the flower ballet was impossible, but rather that at least some audience members may have interpreted a piece like *Théa* quite differently. They might, for example, have focused not on the final redirection of Prince Hussein's affections to an appropriate female object, but on the initial depiction of his infatuation with flowers—as well as the fact that he falls in love with Théa only after she appears to him in the guise of a rose. The ballet's setting in a harem is also significant, given widespread Orientalist tropes about the sexual proclivities of non-Westerners. In this light, the floral motifs may have served not as a public call to order and normalcy, but rather as an acknowledgment of the existence of "the underworld of exuberant sexual drives and desires" then often associated with plant life.[6]

Ha too faces the methodological hurdle, in the absence of a large body of textual material, of imagining how Victorian audiences saw *Ondine*. To do so, he turns in part to material culture, considering how viewers' engagement with objects like hot air balloons, seashells, Wardian cases, home aquaria, and gas lighting conditioned their understanding of the onstage spectacle. One of Ha's most interesting propositions is built on the fact that opera glasses and microscopes relied on analogous technology, could be purchased in the same settings, and produced comparable ways of seeing. Both tools, he writes, engendered a "disconnect from one's immediate surroundings," beckoning the viewer into strange and wonderous realms "encircled by darkness." Here, Ha suggests that dance brought nineteenth-century Britons into the world of science not just in terms of content, but via practice, giving viewers a literal glimpse of what it might feel like to approach the natural world through a new lens. The resulting experience, he reveals, mixed the objective and the magical, the scientific and the imaginative. Ha's attention to hobbies and the home also speaks to a growing body of scholarship that has shown how the history of science can be enriched by paying attention to the influence of previously ignored domains like private and domestic life.[7]

In recent years, historians have done much to undermine long-standing ideas about the binary opposition between science and Romanticism, demonstrating that embracing the scientific or the technological did not mean welcoming disenchantment. Instead, they have shown, the two often built on one another in enormously productive ways. The nineteenth century was a time in which audiences lined up to feel electricity with their own bodies, when photography was used to document both spirits and scientific specimens, and when lively machines could enact utopian visions.[8] Adding dance to this picture not only enriches but may also complicate our understanding of the period, suggesting new questions for exploration. Future research, for example, might explicitly consider how the medium of dance—the human body—mattered to its reception. Did dance engage the public with science in a way that was fundamentally different from a poem, play, or painting?

In 2008, *Science* and the American Association for the Advancement of Science began sponsoring an annual "Dance Your Ph.D." competition, inviting doctoral candidates in the sciences to present their research through "interpretive dance."[9] The contest drew widespread attention relatively quickly, in part because the concept seemed, well, funny. What could dance possibly have to do with climate science, genetics, or nanotechnology? (And would all those serious scientists even be willing to attempt something so silly?) Though many who have participated in the competition have done so with great sincerity and joy—and have reportedly come to appreciate the ways in which movement can serve as a powerful form of thinking and communicating—the enterprise is still infused with an air of the transgressive.[10] Along with the other texts in this volume, the essays by Ha and Schwan should help put any giggles to rest. Not only do their essays demonstrate that dance has long served as a medium for communicating science, but they also reveal how training historical attention on this oft-neglected art form can open our own eyes to formerly invisible worlds.

Notes

1 Secord, *Victorian Sensation.*

2 See, for example, Brain, *The Pulse of Modernism;* Clarke and Henderson, *From Energy to Information;* Henderson, *The Fourth Dimension and Non-Euclidian Geometry in Modern Art;* Hui, Kursell, and Jackson, "Music, Sound, and the Laboratory from 1750–1980"; Jones and Galison, *Picturing Science, Producing Art;* McCray, *Making Art Work.*

3 Müller-Wille, "The Love of Plants," 268.

4 Schiebinger, *Nature's Body,* 24.

5 Quoted in Schiebinger, *Nature's Body,* 30.

6 Müller-Wille, "Linnaeus and the Love Lives of Plants," 318.

7 Coen, "The Experimental Multispecies Household"; Tonn, "Laboratory of Domesticity." See also Opitz et al., *Domesticity in the Making of Modern Science.*

8 Lightman, *Victorian Popularizers of Science;* Tresch, *The Romantic Machine;* and Tucker, *Nature Exposed.*

9 "Announcing the Annual Dance Your Ph.D. Contest."

10 Myers, "Performing the Protein Fold," in *Simulation and Its Discontents,* 171–202.

Bibliography

"Announcing the Annual Dance Your Ph.D. Contest." *Science.* https://www.science.org/content/page/announcing-annual-dance-your-ph-d-contest. Accessed June 9, 2023.

Brain, Robert. *The Pulse of Modernism: Physiological Aesthetics in Fin-de-Siecle Europe.* Seattle: University of Washington Press, 2015.

Clarke, Bruce, and Linda Dalrymple Henderson, eds. *From Energy to Information: Representation in Science and Technology, Art, and Literature.* Stanford, CA: Stanford University Press, 2002.

Coen, Deborah R. "The Experimental Multispecies Household," *Historical Studies in the Natural Sciences* 51, n. 3 (2021): 330–78.

Henderson, Linda. *The Fourth Dimension and Non-Euclidian Geometry in Modern Art.* Princeton, NJ: Princeton University Press, 1983.

Hui, Alexandra, Julia Kursell, and Myles W. Jackson, eds. "Music, Sound, and the Laboratory from 1750–1980," *Osiris* 28, n. 1 (2013).

Jones, Caroline A., and Peter Galison, eds. *Picturing Science, Producing Art.* New York: Routledge, 1998.

Lightman, Bernard. *Victorian Popularizers of Science: Designing Nature for New Audiences.* Chicago: University of Chicago Press, 2007.

McCray, W. Patrick. *Making Art Work: How Cold War Engineers and Artists Forged a New Creative Culture.* Cambridge, MA: MIT Press, 2020.

Müller-Wille, Staffan. "Linnaeus and the Love Lives of Plants." In *Reproduction: Antiquity to the Present Day,* edited by Nick Hopwood, Rebecca Flemming, and Lauren Kassell, 305–18. Cambridge: Cambridge University Press, 2018.

Müller-Wille, Staffan. "The Love of Plants," *Nature* 446 (Mar. 15, 2007): 268.

Myers, Natasha. "Performing the Protein Fold." In *Simulation and Its Discontents,* edited by Sherry Turkle, 171–202. Cambridge, MA: MIT Press, 2009.

Opitz, Donald, Staffan Bergwik, and Brigitte van Tiggelen, eds. *Domesticity in the Making of Modern Science.* Basingstoke: Palgrave Macmillan, 2016.

Schiebinger, Londa L. *Nature's Body: Gender in the Making of Modern Science.* New Brunswick, NJ: Rutgers University Press, 2004.

Secord, James. *Victorian Sensation: The Extraordinary Publication, Reception, and Secret Authorship of "Vestiges of the Natural History of Creation."* Chicago: University of Chicago Press, 2000.

Tonn, Jenna. "Laboratory of Domesticity: Gender, Race, and Science at the Bermuda Biological Station for Research, 1903–30." *History of Science* 57, n. 2 (2019): 231–59.

Tresch, John. *The Romantic Machine: Utopian Science and Technology after Napoleon.* Chicago: University of Chicago Press, 2012.

Tucker, Jennifer. *Nature Exposed: Photography as Eyewitness in Victorian Science.* Baltimore: Johns Hopkins University Press, 2006.

II

Dancing Ideologies

Nation, Sexuality, Sciencing

6

Dr. Louis Véron, Medical Philosophy, and Medical Practice at the Paris Opera

ELIZABETH CLAIRE

Louis Véron and the Gendering of Ballet History

Dr. Louis-Désiré Véron (1798–1867), managing director of the Académie Royale de Musique de Paris from 1831 to 1835, looms large in ballet history.[1] He has at once been considered responsible for the widescale commercialization of French ballet and influential in enabling the elite prostitution of "ballet-girls"[2] via the development of a particular form of sociability in the Green Room.[3] Recent investigation, however, has engaged archival research to revisit women's contributions to nineteenth-century French ballet, questioning aspects of these long-held assumptions.[4]

Vannina Olivesi's analysis of the Paris Opera's accounting records (*registres comptables*) and contract negotiations for dance soloists demonstrates that Véron's management during the July Monarchy was not uniquely responsible for the "feminization" of ballet and the rising influence of a class of *vedette* (celebrity) ballerinas the likes of Thérèse Elssler, Fanny Cerrito, and Marie Taglioni.[5] Olivesi establishes that a gradual feminization of the ballet workforce began to take root already in the mid-eighteenth century, unfolding in a gendered battle for recognition by dancing masters and *vedette* ballerinas alike, who sought remuneration as legitimate artists and authors in the theater arts.[6]

However, only men could benefit from the right to be remunerated for diverse professional functions at the Paris Opera—*maître, professeur, compositeur*—in addition to the role of *interprète* (dancer) and thus to accumulate earnings for performing distinct tasks. This had been the case since 1669, when the founding documents of the Opera disqualified women from a particular category of employment: that of ballet master.[7] Adding to the artistically frustrating fact that women were generally excluded from composing full-scale ballets, these institutionalized restrictions meant that women were less well-paid than their male counterparts, struggled to make late-stage career transitions into teaching and choreography, suffered from lower retirement pay, and had fewer options for dance-related work while pregnant, injured, or otherwise unable to perform. A few *vedettes* achieved public recognition as choreographers by employing alternative strategies with support from local, familial, and international professional networks, with mitigated success.[8]

Emmanuelle Delattre-Destemberg explains how the professional tensions and ambitions Olivesi examines intersected with an institutional culture of neglect at the Opera's dance school. This negligence was allayed by direct involvement of young dancers' mothers who did not hesitate to write to the opera administration to negotiate their daughters' advancements.[9] Delattre-Destemberg's discussion of the "omnipresent mothers" of the "opera rat" confirms Véron's lax oversight of visitors in the *coulisses de l'Opéra,* and depicts a complexly gendered system of unequal social relations where youth and poverty collided in an intense competition between ballet students. Fathers also negotiated with the opera administration: a touching letter drafted in 1805 by one Monsieur Petit, employee of the War Ministry whose three children danced in the corps de ballet (the eldest of whom died from an injury sustained at the Opera), demonstrates parents' detailed knowledge of their children's working conditions.[10] Such family dynamics, situated in the interstice between ambition, suffering, and care, alert us, as Delattre-Destemberg's dissertation does, that ballet bodies—whether belonging to elite *vedettes* or the more burdened dancers of the *corps de ballet* with their contiguous "mothers"—were, above all, bodies that labored.[11]

Madison Mainwaring examines another kind of women's work, discovering a heretofore unknown genre of French dance journalism in which women writers responded to the Romantic ballet. Female spectatorship dialogues in important ways with medical philosophy and emergent French feminisms, including the thorny question of dance's role in the contagion of revolutionary or feminist thought itself.[12] Physicians were disposed to attribute the cause of certain sexual *maladies des femmes* to female imaginations perceived as dominated by fiction and "the monopoly acquired by female legs [i.e., ballerinas] over female existence."[13] Mainwaring affirms that not only did ballerinas "look back" at their audience, they also looked at (and up to) one another in ways that moved women in the audience. Ballerinas danced not only for an ogling male literati in the parterre but also for the critical and "phantasmagorical gazes" of their female spectators and patrons, who were inspired to imagine "proto-feminist alternatives to their existing lives" in the ballets women performed.[14] While female-centered networks of labor, care, and patronage were crucial to the ballerina's career advancement, they also benefited ballet's new class of bourgeois benefactresses, and some took pen to paper to express it.

Much theorization of Romantic ballet has focused on the idea of aesthetic and corporeal submission by ballerinas to a dominant and pervasive "male gaze." This argument, however, much like the male journalism the research was based on, conflates dancers' corporeality on stage with the promise of sexual availability of their bodies, ultimately ascribing to Véron and prostitution a critical role in the feminization of ballet that neither he nor the Green Room entirely deserve. Compounding this problem, French historiography has largely focused on analyzing artistic oeuvres in dance, reproducing hagiographic biographies of famous dancers, and minimizing the question of theater management.[15]

Little has thus been written that examines how Véron's intellectual training, professional networks, and mastery of medicine, journalism, and marketing impacted his vision of himself as an employer of *dancing bodies that mattered.*[16] This must be part of the project to reframe nineteenth-century ballet as more than an artistic vocation with a gendered aesthetic history, but rather as a category of labor with an economic and corporeal history whose particularity was its predominantly female population. Accounting for the fact that Véron held a legitimate medical diploma, was trained in the emerging science of public hygiene, and had clinical experience in the hospitals and anatomy theaters provokes the following questions: How did the feminization of the opera ballet's workforce determine the concerns of medical practitioners at the Paris Opera? And, can we ascertain if Véron's medical training influenced his managerial approach to this question?[17]

Medical Professionals and Public Hygiene at the Paris Opera

As early as 1780, records show that the Paris Opera had a surgeon appointed to its administration.[18] In 1803–1804, the year the Napoleonic Code was formalized, the Opera's "in-house" surgeon Dr. Delatour requested the appointment of an assistant, generating the first administrative dossier filed under the category of "health" at the Opera.[19] While the institution began to consider how medical risks interfered with their employees' availability to work, this didn't mean the Opera was prepared to anticipate or prevent injuries and illness by adaptions in the workplace. Delatour's request did result in the appointment of an assistant, one Dr. Dudanjon, a contributing author of the Napoleonic Code.[20] However, as Delattre-Destemberg points out in reference to "health officers" (*officiers de santé*) at the Opera's ballet school, naming an assistant to the in-house surgeon did not imply that physicians were regularly available to personnel, but rather that these health officers had exclusive authority to intercede administratively on behalf of Opera employees.[21]

In 1806, a decree from the Préfet du palais announced that subsequent to three months of sick leave for a chronic illness, artists' salaries would be reduced by half.[22] The Opera employed physicians to check up on absent dancers, certify the veracity of their illnesses, and assess their readiness to return to work. Delattre-Destemberg cites the example of an undated report addressed by Dr. Delatour to the Opera administration in which he visited the domicile of Mademoiselle Saulnier, "dance artist," whom the doctor found "sick in bed with a sore throat that has affected her for some days" as well as a "rheumatic pain in her knee" caused by "atmospheric variations." He concluded that her illness would produce no "vexing consequences" that might prohibit her from returning to the stage.[23]

This preliminary administrative shift toward regulating dancers' fitness to work aligns with a culture of institutional reform that began under Napoleon in response to nascent conceptions about the role of state government in what was to be called "public health." Following the liberalization and restructuration of the medical profession under the Revolution, an 1802 decree formalized a proposal that had been debated by Mirabeau, Talleyrand, and Condorcet to reconstitute the medical profession inside the new Institut national de France under a department that would oversee a new category of "moral and political sciences." In the same year Napoleon also created the Conseil de salubrité de Paris, composed of doctors, pharmacists, chemists, veterinarians, and industrialists under the direction of the Paris Police Prefect. Henceforward, physicians no longer limited themselves to treating individual illnesses but sought to keep an "intimate relationship with social organization" by participating in the administration of the State and maintaining "public health and hygiene."[24]

While Napoleon's government reinforced the emerging science of "public hygiene" by providing a bureaucratic structure for its implementation, decades would pass before these ideas were generalized as a public service. The major thrust of initial medical regulations at the Opera, for example, served merely to guarantee the financial security of the institution. Physicians were solicited to assist in the evaluation of how to "reform employees," that is, dock an artist's pay, fire them, or refuse to renew their contract after absences due to illness. In 1804, Opera regulations stipulated that an absence due to pregnancy constituted "an interruption of service that implied a full suspension of remuneration, as was the case with any illness."[25]

Medical thinking about public health and hygiene developed slowly within French government institutions like the Opera and the Paris Medical School. Despite interruptions and structural changes that reorganized the teaching institution after the 1789 Revolution, a survey of thesis defense juries during the 1790s and early 1800s illustrates broad continuity in teaching faculty at the Académie de médecine de Paris.[26] At the Paris Opera, a similar stability can be observed before and after the Revolution regarding the presence and promotion of professional dancers in important positions of authority: only half of the dance soloists under contract at the Opera in 1796–1797 (*An V Républicain*) were recruited after 1789, while roughly one-quarter of them had been employed there for over twenty years.[27] Pierre Gardel (1758–1840), whose father and brother were also professional ballet masters, maintained a long and successful career from 1771 to 1829 as both soloist and choreographer. He was promoted to *maître de ballet* in 1788, directed the Opera ballet throughout the 1790s and even daringly experimented with the Opera's first comic ballet in 1800, where he staged a fashionable "public health" concern about the phenomenon of post-Revolutionary dance mania with his "folie-pantomime" *La Dansomanie.* This extremely popular domestic-themed ballet with a shrewd political subtext about revolution connected the phenomena of dance hysteria to a comic suggestion that young bourgeois might use the charms of dancing (instead of the guillotine) to overturn the misguided constraints of paternal (and paternalistic governmental) authority.[28]

Within this continuity, institutional development both at the Opera and at the Académie de médecine did slowly integrate medical practice. One important improvement after 1794, resulting in part from the medical school's greater financial stability, was the implementation of "anatomo-clinical" teaching methods including hands-on experience for students in Parisian hospitals and opportunities to participate in autopsies. Véron benefited from these pedagogical innovations when he studied there. Practical work in the hospitals and anatomy theaters allowed students to study the effects of disease in patients and in dissected cadavers. Such experiences were not part of eighteenth-century instruction of nosology (the classification of diseases), which focused solely on theory and on symptoms of living patients. The Paris medical school's dissection practicums were unique in France in the early nineteenth century and required employing a new category of personnel. These new employees comprised most notably a chief of "anatomical work" and various assistants, including a position for a "surveying physician," who followed students' clinical work in the hospitals.[29] Knowledge of anatomy was thus adopted by a new generation of students in the study of pathology and would begin to inform new thinking about public hygiene.

The presence of such personnel at the Paris Opera meant its artists were increasingly subjected to the surveillance of physicians called upon to assist other types of officers (*pompiers, police, militaires,* et cetera.) during theatrical events. In 1825, for example, the Département des Beaux-Arts of the Maison du roi, which oversaw the Opera administration, engaged a small team of physicians to attend performances and dress rehearsals at a selection of the most important state theaters in Paris, including the Académie royale de musique, the Théâtre-Français, Théâtre-Italien, and Odéon Theater. These doctors, were not remunerated full-time for their work in the theaters but received free access to performances and organized soirées, and shared (among nine practitioners) an annual stipend of three thousand francs, enough to pay one month's rent.[30] In short, no physician could make his living performing these duties, but the practice did allow for a privileged backstage presence of (male) physicians in the theater, a genre of access to performances that Véron the medical student may have observed intimately long before he was in a position to capitalize upon it.

Dr. Louis Véron and Medical Philosophy

Louis Véron serves as our point of entry here, not only because of the outsized and contested role he has occupied in the historiography of French ballet, but also because his medical thesis demonstrates that his thoughts on corporeality were aligned with a larger Enlightenment project to bring the science of sensation and the philosophy of perception to bear on understandings of the art of expression.[31] His interest in the theater may have influenced his choice to study how the theory and anatomy of sensation informed medical thinking about the body's capacity to convey ideas. We will examine his thesis in order to assess how contemporary medical theories of corporeality may have impacted the perception and treatment of dancers' bodies at the Opera.

Véron's medical dissertation, *General Considerations About Sensation, followed by some Medical Proposals* (1823), recapitulated Enlightenment medical philosophy about the "science of sensibility": Véron's thesis examined human sensation in relation to imagination, expression, and idea formation.[32] He wrote about a spectrum of sensations ranging from the pleasurable to the painful, citing the idea that heightened experiences of sensation could ultimately produce pathology. In one cryptic passage, Véron suggests that "pleasurable sensations" are merely the "beginning of pain," implying that pain resulted from an "excess" of sensation. The question of how excess sensation impacted physiological functions of the body was a recurrent topic in medicine, particularly with respect to clinical observations about mental pathologies and experimental attempts to cure them; the question was also explored by writers and painters of the Romantic movement and in philosophies of the sublime.

Véron's conclusion echoed emergent medical notions about public hygiene. Habits formed by frequent recourse to behaviors that produced physical sensation—including especially activities that risked heating the imagination (i.e., reading fiction, waltzing at balls or, as Mainwaring's research implies, watching ballets)—led to experiences of excess sensation. Over time, such experiences rendered the human body "insensible" to the familiar "impressions" that had before been "agreeable" or even "disturbing."[33] The spectrum of sensation Véron described followed a standard pattern discussed elsewhere in medical discourse: the patient begins by performing a behavior that produces a sensation of pleasure; repeated recourse to the behavior generates a need for deeper "impressions" or greater stimuli for the body to experience a pleasant sensation; the frequent repetition of pleasureful sensations hence becomes excessive and causes disturbing sensations (ex., vertigo in the case of waltzing); this overstimulation of the imagination causes a physiological breakdown (*lesion*) of the sensory system which ultimately leads to the onset of various pathologies.[34]

Conventional examples of these pathologies taken from influential eighteenth-century medical texts include Tissot's theories on *Onanism* and Bienville's treaty on *Nymphomania*.[35] The elucidation that sexual manias belonged to a category of diseases of the imagination influenced awareness among medical professionals that sexually transmitted diseases must be controlled as part of the emerging question of "public health." The immediate solution of keeping women's libido under wraps resonated with the moral imperatives of the 1804 Napoleonic Code and physicians' new implication in surveying "public hygiene." The majority of medical texts on hygiene from the early part of the century confirm a moralizing emphasis on "habits," and those treating the question of the imagination suggested both prevention (avoid young women's partaking in desire-stimulating activities), and marriage as a cure for these nervous diseases.[36]

When Véron evoked the function of the nervous system in his medical thesis, he predictably cited the president of his doctoral defense jury, M. Richerand, chief surgeon *adjoint* at the Saint-Louis Hospital in Paris and professor of pathology at the *Faculté de Médecine de Paris*. Richerand supervised medical students in coordination with Philippe Pinel, chief physician at the Salpêtrière hospital.[37] Pinel's early experiments with clinical psychology, what the French then called *médecine morale* (including notably melancholia, hysteria, and their treatment), preceded and influenced Charcot's arguably better-known and more spectacular work with hysterics, which is evoked frequently in feminist dance history scholarship on scopophilia.[38] In his discussion of how "habits" influence sensation, Véron also cites Jean Noël Hallé, who taught hygiene at the medical school. Hallé was an important physician for Napoléon: knighted *chevalier* of the Légion d'honneur, named to the new Institut de France and the imposing Collège de France, Hallé was ideally situated to promote the new science of public hygiene and its pedagogy. Véron was thus steeped in a teaching culture in which hygiene, pathology, and mental medicine were at the top of the hierarchy of concerns.

Véron also references the anatomist François-Xavier Bichat when seeking to explain the importance of symmetry in the human body. Bichat was not only a qualified anatomist; he also practiced surgery, pathology, physiology, histology, and therapy, and maintained a keen interest in *médecine morale*. Bichat was hailed as a medical genius, one who quite romantically died at the age of thirty-one from a feverish dedication to his writing and experimental work. Bichat was author of the well-received *Recherches physiologiques sur la vie et la mort* (1797), a work composed of an initial theoretical section followed by an experimental one reporting on Bichat's extensive anatomical dissections. This two-part structure may have inspired the atypical organization of Véron's medical thesis. Bichat's most renowned publication was his best-selling four-volume *Anatomie Générale appliquée à la physiologie et à la médecine* (1801), based on the more than 600 autopsies he supposedly performed in the year 1800 alone.[39] As Secretary of the prestigious Société de médecine de Paris, founded in 1800 under Napoleon's auspices, Bichat's influence would have been significant at the Paris School of Medicine when Véron studied there. The connection of Bichat's idea of corporeal symmetry to an abstract idealization of the ballet dancer's healthfulness would ultimately, if not immediately, influence the recruitment process at the ballet school of the Paris Opera. It would be interesting to know more about the impact of Bichat's anatomical research on Véron's thinking about the dancing body. Unfortunately, Véron does not discuss dancing specifically in his thesis.

Between Medical Philosophy and Aesthetics

If Véron was renowned in his youth for an insatiable appetite for the theater, it may come as no surprise that he chose to study the science of sensation and expression when obliged by his father to pursue a medical degree.[40] Physical sensation and expression, idea formation and its connection to the mental faculty of the imagination—all notions referenced in Véron's medical thesis—were also key topics for many of the Enlightenment philosophers and ballet masters debating the artistic value of dance in the eighteenth century. Louis de Cahusac and Jean-Georges Noverre had written extensively on the art of ballet, arguing for its reform and for dance's recognition as a poetic art.[41] Both Cahusac and Noverre evoked the powers of the imagination in the physiological function of sensation as a means to justify the poetic nature of ballet's eloquence, which they defined against the improvised ballet *entrées* in lyric operas and dance practices. The authors qualified such *entrées* improvised by solo *interprètes* (who were increasingly female) as neither aesthetic nor poetical, but rather as detrimental to the artistry, authority and auctoriality of ballet masters.[42] Hector Berlioz would mobilize similar ideas about the imagination and sensation in his philosophical writings on music and acoustics to discuss the relationship between sound and an audience's emotional response to a musical composition.[43]

Véron's medical thesis discussed the relationship between human sensation and the discovery of perspective in the visual arts, notably painting. He analyzed how the impressions bodies received through various sensations (especially sight and hearing) could lead to erroneous perceptions if taken as isolated information. He cited the way a ventriloquist's performance tricks its audience as to the source of a sound. Humans, he argued, receive and react to numerous and "varied impressions" through their sensations, which constitute the "materiality of thought, of intelligence."[44] It is for this reason that when we are preoccupied, we might fail to perceive certain things: our sensations must be "directed" by "an act of intelligence, by our will."[45] Véron defended the notion that human intelligence and the "faculties of the soul [*facultés de l'âme*]"—here evoking the imagination—exist to "survey and judge" the operations of perception, to combine the impressions from the various senses and determine an "exact and true" understanding of what we witness and feel.[46]

Véron argued that sensations do not make or give humans intelligence; rather, human thought is generated by the faculties of the soul that survey and exploit sensations in order to determine the relationship established between us and all of nature.[47] Véron's examination of the science of sensation concludes with an example from comparative anatomy—perhaps the most obvious moment in which the incongruity of the two sections of his dissertation finds some resolution—in which he recapitulates the work of Pierre-Jean-Georges Cabanis on sensations, distinguishing animal instinct from human intelligence.[48] Véron wrote: "Sight in animals allows them to recognize objects; in humans it provides the symbols for our thoughts. Hearing offers to animals the sensation of sounds; it gives men the very expression of ideas. In a word, animals see, humans read. Animals hear sounds, humans listen to the expression of thoughts."[49] It would perhaps be an act of over-interpretation to suggest that Véron subtly references the form of lyric opera when he writes that humans listen not to "sounds" but to "the expression of thoughts." However, the connection he makes between human intellection and imagination's role in the art of painting recalls Immanuel Kant's notion of "aesthetic judgment" as well as Noverre's references to Denis Diderot's *tableaux* and Cahusac's defense of the art of French ballet that likewise evoked painting and the imagination. Véron's discussion of human sensation resonates with German philosophical discussions of aesthetics (Kant et al.), and although Véron does not directly cite Georg Wilhelm Friedrich Hegel (whose trip to Paris did not occur until 1827), Véron would have been studying medicine at the same time that Hegel was delivering his now-famous Berlin lectures on aesthetics.[50] More certain is that by ending his dissertation with a gesture toward comparative anatomy, Véron concluded with a scientific endorsement of the human art of eloquence, one that echoed the Enlightenment science of aesthetics in which the expression of ideas through the poetic arts was a distinguishing factor of human exceptionalism.

Given Véron's extraordinarily brief career in medical practice and the fact that he did not continue to publish medical research after receiving his diploma, turning instead to the lucrative marketing of pharmaceuticals and the rewarding vocation of literary publishing, it is difficult to evaluate how Véron's "medical thinking" evolved while he managed the Paris Opera.[51] There seem to be two opposing forces at work in his dissertation, as evidenced in the unrelated sections: one theoretical, on the science of sensation, and one anatomical, offering practical "medical proposals." In the theoretical passage, one distinguishes with difficulty Véron's medical philosophy from a discussion of aesthetics. In his *Mémoires*, Véron seems to apply his theory of sensation directly to his study of how to market ballets. He writes:

> I gathered that ballets founded on dramatic action would never be assured a great success. I studied the receipts earned by all the old works [predating his tenure at the Opera], and I ascertained that those ballets that had succeeded most were *Les Filets de Vulcain* and *Flore et Zéphyre*. . . . Above all, the public demands that a ballet have a striking and varied score, new and unusual costumes, contrasting and novel sets, and a simple, easily understood plot, but one in which the dance develops naturally out of the dramatic situations. One must add to all this the seductiveness of a beautiful young artist, who dances differ-

ently from, and more successfully than, those who preceded her. *When one speaks neither to the mind nor to the heart, one must speak to the senses, and above all to the eyes.*[52]

This excerpt could easily lead contemporary readers to think that Véron was interested solely in marketing scopophilia, but this would be anachronistic. Véron's emphasis on the critical power of visual sensation to engage imaginations (in this case of potential spectators) was a central tenet of the science of sensation Véron explored in the theoretical section of his dissertation; the concept connected back to medieval philosophy concerning the powers of the imagination itself and was unrelated to sexual predation.[53] Véron was just as likely to have been assessing how best to cater to an important new cadre of bourgeois *female* Opera patrons.

While it does not explain his implementation of legislation as managing director of the Opera, reading Véron's dissertation allows us to concur that he may have been a more complex figure for ballet than a simple arbiter of a commercialized male gaze that set the stage for André Levinson's 1929 declaration that Marie Taglioni had "evicted" the male dancer with her point shoes, to the ultimate detriment of the French art of ballet.[54]

Dr. Louis Véron and Medical Practice

Appended to his dissertation on sensation, Véron published an additional and unrelated set of observations concerning childhood diseases, reflecting the results of his own extensive anatomical dissection of child cadavers during his practical training under Jacques-François Baron, vice president of the Société anatomique de Paris, and chief physician at the Hospice des Enfants-Trouvés. Proposing separate and incongruous theoretical and anatomical sections in a dissertation was an unusual choice for a medical thesis of this period. Possibly, Véron was unsure how he would make use of his diploma and thus attempted to cover many bases, from medical theory on sensibility to the practical work of anatomy instruction, hoping to increase his chances on a job market still dependent on elite social networking. The division was also a reflection of the dual nature of evolving pedagogy at the Paris medical school: theoretical work leading to a dissertation and a diploma, and new practical training including autopsies and hospital experience that was not fully integrated into the theory.

After graduating, Véron put most of his professional energy not into medical work but into journalism, as a theater critic and founder of the successful literary magazine *La Revue de Paris,* hailed for publishing writers like Victor

Hugo and Honoré de Balzac. While his dissertation on sensation had arguably little impact on medical research, Véron's literary journal flourished, remaining active until 1970. Ultimately, he made his medical fortune marketing a pharmaceutical product, *la pâte Regnauld,* invented by a Parisian friend and neighbor.[55]

In 1831, thanks in large part to friendships with composers and *hommes de lettres* like Rossini and Eugène Scribe, Véron was named "Director-entrepreneur" of the Académie de musique, a position he occupied until 1835. During this brief time, he oversaw a radical restructuring and deregulation of employment procedures at the Opera, as well as the temporary closure of the theater due to the seven-month cholera epidemic that ravaged Paris in 1832. Véron also created a position for a *Médecin des théâtres* whose objective was to "ensure the security of the increasingly larger audiences."[56] As a *directeur-entrepreneur,* he was called upon personally to compensate financial losses incurred during the cholera epidemic, and was responsible for authorizing a number of soloists to leave France for temporary contracts at the English opera house where, as Véron wrote, "they even performed *Robert le Diable,*" the very lucrative opera created under Véron's supervision that he had been forced to cancel on April 7 when the epidemic broke out in Paris.[57]

Les Cancans de l'Opéra: Journal d'une habilleuse—a manuscript written not, in fact, by a female costume assistant as the title suggests but by one Louis Gentil, hired as materials manager at the Paris Opera in 1831—details the comings and goings of various protagonists who interacted with the Opera dancers during Véron's tenure.[58] Delattre-Destemberg's research corroborates the constant circulation of people inside what she describes as an "anthill of diverse activities" where dancers were expected to practice their trade, highlighting the erratic and permissive surveillance practiced under Véron's management. The detrimental effects of this circulation, she argues, were compounded by the fact that the dance school was frequently moved to temporary and alternative locations to make way for construction projects and other institutional priorities.[59]

Véron's decision to open a number of transitional backstage spaces to an elite group of male subscribers placed a new burden of surveillance on mothers who chaperoned the young corps de ballet dancers. Prior to Véron's arrival, under the reign of Charles X, "the four coulisses to the right and the left [of the stage] were guarded by eight *factionnaires* from the Royal Guard, as well as six *laquais* who surveyed the corridors."[60] As a result, performing artists left their dressing areas only to take their place on stage, and for this they were also accompanied by *huissiers* in black suits. After 1831, according to the *Dictionnaire des coulisses,* published by an anonymous balletophile, the backstage corridors became a place where "everyone could penetrate with a bit of audacity; it's the rendezvous for the actors, the theater physicians, the journalists and the mothers."[61] This citation, which lumps together journalists and theater physicians as the potential source of young dancers' seduction and negotiation of a *protectorat,* lends a certain skepticism as to the effectiveness of the medical presence Véron introduced.

Again, we are reminded that young dancers were especially vulnerable and dependent on their chaperones. Gentil, the faux *habilleuse* and an ally of Véron concerning tensions with the oversight committee—the *commission de surveillance* appointed by the Ministère de l'intérieur[62]—suggested that the institution relied on regulations in order to limit how mothers might seek the favor of wealthy men for their daughters. However, Delattre-Destemberg's depiction of the maternal figure suggests someone much more preoccupied with the simple survival of her charge. It should be noted that the initial 1827 regulations (reinforced again in 1838 and 1846), which excluded mothers from specific spaces within the Opera in an attempt to limit their "omnipresence," were adopted well before Véron was hired. Most striking, however, is how the maternal figure seems to perform a quasi-medical role, attending to her daughter's every physical need while dressing and undressing her before escorting "like a wooden crutch" the dancer's fatigued body as they exited the theater at the end of a long day and night perhaps stopping at a soirée chez Véron before finally returning home.[63]

Much remains to be studied about how families negotiated the safety, health, and prosperity of their young charges. Distinguishing high-end prostitution from practices "connected with gallantry" that were integral to a "system of *protectorat*" and also involved privileged access to the Green Room by *abonnés,* remains challenging, since both activities (high-end prostitution and gallantry) coexisted in the same spaces.[64] Even where the comparatively privileged *vedette* ballerina's strategies for professional recognition and survival were concerned, research is still necessary to determine how their careers were impacted by pregnancies, illnesses, physical injuries, and work-related deaths. How were such events managed by the Opera administration before, during, and after Véron's time there? Anecdotal sources suggest, for example, that Marie Taglioni's 1835 "knee injury" (a dissimulated pregnancy[65]) required her to endure onerous treatment prescribed by the Opera physicians—the application of forty leeches—so that she could keep dancing and maintain her monthly *appointements* despite the Opera management's skepticism about her condition. Taglioni, however, returned directly to the stage as *La Sylphide* after the birth of her child, and the episode did not ultimately interfere with her career as a performer.[66] In contrast, the death of Taglioni's protégée Emma Livry, who perished in a stage fire while rehearsing *La muette de Portici*, had an arguably devastating effect on Taglioni's career as a choreographer (Livry performed the title role in Taglioni's initially successful *Le Papillon*).[67]

And After Véron . . . ?

Despite Véron's brief tenure (1831–1835), which ended in his resignation eighteen months before the end of his contract, his influence on the institution, as we have noted, has been assigned much import in dance historiography, thanks in large part to the enduring legacy of *La Sylphide* and Marie Taglioni, and the outsized influence of literary sources including Véron's own *Mémoires*. Yet aside from his creation of the post of *médecin des théâtres*, his granting of leave to important soloists, and his ultimate closure of the theater during the height of the cholera epidemic, there is little evidence that Véron prioritized either public health or employee health when he restructured employment at the Opera. As we have noted, the social question of public health was already influential in the French capital as early as 1802. However, *hygiénisme* as a widespread institutional reform movement did not permeate French workplace culture until the 1840s when political mobilization and philanthropy increased public awareness of the effects of industrial development and led to the widespread institutionalization of workers' rights.[68]

Evidence that the Paris Opera was affected by the labor movement in France is attested by an 1842 report issued by one Deputy Guillaume in which the inspector deplored the "lack of hygiene and the decline [*dépérissement*] of the state of the Opera's dance conservatory."[69] But it was not until 1852 that a first *Arrêté préfectoral* (May 2, 1852) mandated establishment of a medical dispensary in all Parisian theaters, to be set up in an "orchestra stall" or balcony and managed by the *surveillant des coulisses*, a surgeon who, in addition to his scrutiny of the theater during performances, was thereafter made available to treat accidents or illnesses on site. Such a dispensary existed at the Paris Opera, which probably served as a model for this legislation. The dispensaries were to be "comfortably outfitted, heated and well-lit," and contain a small pharmacy paid for by the Ministère de l'Intérieur. The physicians' operation of these dispensaries was "under the surveillance" of the Conseil de salubrité invented in 1802 under Napoleon.[70] Stunningly, a clause in the legislation indicated that the Paris Opera was exempt from following the new decree, implying that the privileged institution need not submit to the new modalities of surveillance by the Paris Police Prefect, and could maintain the status quo of their operations.[71]

Seeking preventive solutions to the medical problems dancers experienced as a result of their work required thinking about human anatomy as well as public hygiene and how the two medical domains should be applied to the daily practice and performance of ballet. However, Delattre-Destemberg's research states clearly that in the 1850s the Paris Opera, unlike the case of La Scala in Milan, did *not* call upon members of the medical profession to assist their own ballet masters in assessing the physiological aptitude of young candidates seeking entrance at the ballet school. The medical innovation at La Scala, inspired by Carlo Blasis's theorization of the art of dance, was executed according to a division of professional expertise in the audition process: ballet masters first examined "the physiognomy and corporeal disposition of the young aspiring dancers, looking to establish a regularity in this regard"; and the La Scala surgeon(s), in collaboration with the theater's "house physician," then inspected the "physical constitution and corporeal aptitude" of each candidate, searching for physical defects that would make admission to the school impracticable.[72] Delattre-Destemberg suggests that the notions of "physical aptitude" and "anatomical conformity" were probably adopted at the Paris Opera after the working model developed in Milan: in a moment of economic and institutional crisis at the ballet school and with the intention to implement reforms in the 1850s, the Paris Opera procured a copy of the statutes of the Milanese ballet school that remains in their archives today.[73] A later 1859 French dance treatise by ballet master G. Léopold Adice (*professeur de la classe de perfectionnement*) at the Opera, followed up by publishing an anatomical treatise on injuries specific to the professional dancer.[74]

It has yet to be confirmed whether Véron read Carlo Blasis's *Traité élémentaire théorique et pratique de l'art de la danse* (1820) or attempted to apply his own anatomical knowledge to the question of recruitment. However, as Delattre-Destemberg demonstrated, anatomical examinations by physicians within the audition process were not institutionalized at the Paris Opera until well after Véron's tenure; not until 1860 did the Paris Opera *Comité* overseeing auditions solicit medical professionals to participate in juries, asking physicians to judge the strength, health and physical conformity of aspiring young dancers.[75] To this end, the Opera administration created a statute that "defined stricter modalities of recruitment, notably concerning the physical capacities and anatomical potential of the candidates."[76]

Conclusions and Suggestions for Further Research

While we may hypothesize that Véron's attachment to the science of aesthetics and his lifelong love of the theater influenced and was influenced by his training in the humanist philosophy of sensation and the anatomy of childhood diseases, it remains difficult to assess whether this prompted him to introduce a vision of the Paris Opera as an institution destined to improve the health and working conditions of its ballet dancers. The evolution of medical legislation at the Paris Opera was driven by a slow but growing awareness that dancing bodies were bodies that suffered. However, medical notions of hygiene and public health had only a minor impact on the institution before midcentury, despite the presence of Dr. Véron and his experience directing the theater during the cholera epidemic. At best, Véron's *Mémoires* suggest a paternalistic approach to questions of medical leave and a focus on profitability that would explain the Opera physicians' management of Marie Taglioni's "knee injury." The Opera preserved its special status and immunity to scrutiny under new legislation even as the labor movement against industrial abuses gained ground in France.

A systematic study of medical reports addressed to the Opera administration and medical certificates issued to dancers who missed work due to illness, pregnancy, or accident, as well as an analysis of the impact of sick leave on the career trajectories and remuneration of dance artists, remains necessary before we can fully determine how medical philosophy and practice impacted dance artists. While Véron sought to foster a culture of dance art in the French capital, his contentious relationship with the structures of governance to which he was contractually bound suggest it is unlikely that he pushed for government hygiene inspections or mandates for preventive measures in the workplace.

Whether *vedette* artists or "*petits rats*" from the École de Ballet, Opera dancers would have to wait until 1892 before medical professionals were required to enforce government legislation aimed at improving their health and safety. A legal limit was then set on working hours and routine inspections by physicians ensured that hygiene problems were addressed. These could include basic issues like those identified in Dr. Gaches-Sarraute's 1893 report to Opera management: a lack of potable drinking water, problems of malnutrition due to a scarcity of available food and the artists' habit of "drinking at the buvette on an empty stomach," or even the moral dangers resulting from unstructured leisure time when dancers, too far from their homes to return for the mid-day meal, "roamed the corridors" of the theater between morning training sessions and evening performances.[77] To this day, the Opera runs its own cafeteria where artists can eat meals at government-subsidized prices.

Acknowledgments

I am grateful to Emmanuelle Delattre-Destemberg and Vannina Olivesi whose archival research provides the foundation for much of what I can report about medical and employment practices at the Paris Opera. Without permission to consult their unpublished dissertations (and our many years of academic collaboration), I could not have deciphered how medical philosophy was (and was not) applied to the art and labor of ballet at the Paris Opera.

Notes

1 The institution changed names regularly while at rue Le Peletier: *Académie royale de musique* (1821–1848), *Académie nationale de musique* (1848–1851), *Académie impériale de musique* (1852–1870), *Académie nationale de musique* (1871–1873). In 1873, a fire destroyed "Le Peletier" and the Opera moved to the Palais Garnier. I will use simply "Paris Opera."

2 See Albert Smith's illustrated physiognomy *The Natural History of the Ballet-Girl.*

3 McCarren, *Dance Pathologies;* Garafola, *Rethinking the Sylph;* Foster, *Choreography and Narrative;* Foster, "The Ballerina's Phallic Pointe," in *Corporealities;* Marquié, *Histoire et esthétique de la danse de ballet au XIXe siècle.*

4 Delattre-Destemberg, *Les enfants de Terpischore;* Olivesi, *'Vedettes' et 'artistes';* Mainwaring, *Reclaiming the Silences of Dance.*

5 Olivesi, *'Vedettes' et 'artistes';* regarding contract negotiations, see Olivesi, "Entre discrétion et quête de reconnaissance," in *Giovanni Coralli, l'autore di Giselle.* Marian Smith demonstrated how a few resonant contributions to discourse on French ballet had an outsized influenced on historiography; she cites André Levinson's 1929 declaration that "Marie Taglioni had 'evicted' men from the stage" and connects this back to male dance criticism during the July Monarchy, cf. Smith, "The Disappearing Danseur." For further discussions decentering the Paris Opera's influence on the gendered history of ballet, Sarah Gutsche-Miller's *Parisian Music-Hall Ballet, 1871–1913;* Oberzaucher-Schüller, "Multi-Layered Shifting Processes."

6 Olivesi, *Vedettes' et 'artistes';* see also Glon, Olivesi, and Vallejos, "Penser l'auteur en danse au XVIIIe siècle"; Glon, Olivesi, and Vallejos, "Writing (for) the Ballet in the Eighteenth Century"; Lilti, *Figures publiques.*

7 Olivesi, "Between Pleasure and Censure," 45; see also McCleave, "Marie Sallé."

8 On the gendered division of labor and accumulation of "*emplois qualifies* [qualified work]" at the Paris Opera, see Olivesi, '*Vedettes' et 'artistes,*' 65–66; see also Olivesi, "Between Pleasure and Censure"; Olivesi, "Archéologie du vedettariat chorégraphique à l'Opéra"; see also Mainwaring, "Dancing in Pants."

9 Delattre-Destemberg, *Les enfants*, especially "*Des mères omniprésentes à l'Opéra,*" and "*Des mères entremetteuses,*" 435–51, 475–97. The maternal figures referenced in the literature and the archives could be close kin—aunts, sisters, or even neighbors—acting as chaperones who accompanied the young dancers during morning classes, evening performances, and post-performance social gatherings. One source cites a soirée at Dr. Véron's house attended by nineteen women including eleven dancers and five "mothers," cf. Delattre-Destemberg, 443.

10 Archives nationales (AN), Series AJ^{13} 64, *lettre de Monsieur Petit,* 1805, cited in Delattre-Destemberg, *Les enfants,* 452–55.

11 Delattre-Destemberg, *Les enfants;* Sabee, *The* Rat de l'Opéra.

12 Claire, "Le corps en revolution," and "*Waltzmania* in the Paris Pleasure Gardens."

13 Daubié, *La femme pauvre au XIXe siècle,* 304; for a comparison between Adice's legitimizing anatomical discourse about a gender-neutral dancers' body in need of medical protection and journalistic scopophilia promoting a moralizing physiognomy of eroticized ballerina legs, see Delattre-Destemberg, *Les enfants,* 263–68.

14 Mainwaring, *Reclaiming the Silences; Le regard féminin;* and "Dancing in Pants"; for the initial discussion of the "ballet-girl" who "looks back," see McCarren, *Dance Pathologies,* 99–102.

15 Goetschel and Yon, *Directeurs de théâtres XIXe–XXe siècles;* Lacombe's *Histoire de l'opéra français* discusses Opera administration and briefly treats dance, but not the medical dimension.

16 Claire, "Dance Studies, Gender and the Question of History." For French historiography of Paris Opera administration under Véron: Drysdale, *Louis Véron and the Finances of the Académie Royale de Musique;* Monnier, *L'Opéra de Paris de Louis XIV au début du XXe siècle;* Juin, *1831–1835: l'Opéra de Paris sous la direction Véron;* Ehrhard, "L'Opéra sous la direction Véron (1831–1835)." These works do not address the question of public health.

17 The impetus for this essay emerged after a scholarly exchange with Laura Smith, whose doctoral research (University of California-Los Angeles, World Arts and Cultures) takes a medical humanities approach to French ballet.

18 Thuillier, *Les pensions de retraite des artistes de l'Opéra (1713–1914),* cited in Delattre-Destemberg, *Les enfants,* 259.

19 AN, AJ^{13} 63, *officiers de santé,* an XII; cited in Delattre-Destemberg, *Les enfants,* 260.

20 AN, AJ^{13} 63, *officiers de santé,* an XIII; cited in Delattre-Destemberg, *Les enfants,* 260.

21 Delattre-Destemberg, *Les enfants,* 259; for the Opera's management of medical personnel see, "*La santé au travail: La médecine à l'Opéra,*" 259–63; see also Monnier, *L'Opéra de Paris,* 538–41.

22 AN, AJ^{13} 74, *arrêté du préfet du palais,* Nov. 20, 1806; cited in Delattre-Destemberg, *Les enfants,* 260.

23 AN, AJ^{13} 64, *rapport du docteur Delatour, chirurgien titulaire de l'Académie impériale de musique;* cited in Delattre-Destemberg, *Les enfants,* 260.

24 Following the 1793 abolition of the *Académies royales,* the Thermidorian Convention decided in 1795 [*Article 300 de la Constitution de l'An III*] to re-authorize scientific societies; on the creation of a department of "*Sciences morales et politiques,*" at the new

Institut national see *Annales d'hygiène publique et de médicine légale, tome I, 1re partie* (1829), v.

25 AN, AJ[13] 72, 1804; this remained the case as late as 1846 at the height of the French labor movement, AN, AJ[13] 193, *règlement du service de la danse, clauses communes,* 1846, cited in Delattre-Destemberg, *Les enfants,* 268–69; 449.

26 The *École de Santé* (Paris, 1794) established on the site of the Royal Surgery Academy (rue des Cordeliers), became the *École de Médecine de Paris* in 1798, and the *Faculté de Médecine de Paris* in 1808. After 1813 (completion date for Gondoin's architectural work) the school remained essentially unchanged until 1876; see Le Roux des Tillets, Huard and Imbault-Huart, "Structure et fonctionnement de la Faculté de Médecine de Paris en 1813," 139–40.

27 AN, F[17] 1298; see Bourdin, "Les fantômes de l'Opéra ou les abandons du Directoire," *Annales historiques de la Révolution française.*

28 Claire, "*La Dansomanie.*"

29 Le Roux, Huard and Imbault-Huart, "Structure et fonctionnement," 156–57, esp. fn. 15.

30 Archives départementales, AJ[13] 109, *règlement sur le service de santé,* Oct. 24, 1825, cited in Delattre-Destemberg, *Les enfants,* 260.

31 Véron, "Considérations générales sur les sensations, suivies de quelques propositions médicales." All translations are my own.

32 On the science of sensibility and the emergence of aesthetics during the Enlightenment see Vermeir and Funk, *The Science of Sensibility.*

33 Véron, *Considérations générales,* 12–13.

34 Claire, "A Moral Defence of the Regency Ballroom"; Claire, "Monstrous Choreographies"; Claire, "Inscrire le corps révolutionnaire dans la pathologie morale."

35 Tissot, *Ouvres complètes de Tissot;* de Bienville, *La nymphomanie.*

36 Vitet, *Le médecin du peuple,* vol. XI, 471.

37 Le Roux, "Structure et fonctionnement," 157.

38 For reasons related to discourses on ballet aesthetics in French literature, these essays focused on the late nineteenth century, without considering the chronology for scientific developments in mental medicine; this has led dance scholars erroneously to consider hysteria as a nineteenth-century disease.

39 Paul Ganière, "Bichat, François-Xavier (1771–1802), Médecin," *Dictionnaire Napoléon* I, 225–26.

40 Véron, *Considérations générales.*

41 Vallejos, *Les philosophes de la danse;* Pfister, *La danse à la conquete du statut d'art.*

42 My current book project, *Imagination Embodied,* contains a chapter on the role of imagination and poetic eloquence in eighteenth-century dance.

43 Véron was in good company as a medical student: Hector Berlioz, also obliged by his father, attended the *Faculté de Médecine de Paris* from 1820 to 1822. See Raz, "Hector Berlioz's Neurophysiological Imagination."

44 Véron, *Considérations générales,* 14.

45 Véron, *Considérations générales,* 15.

46 Véron, *Considérations générales,* 10.

47 Véron, *Considérations générales,* 15.

48 Cabanis, *Rapports du physique et du moral de l'homme* I, 133–35.

49 Véron, *Considérations générales,* 17.

50 The circulation of literati between Berlin and Paris and work of translators like Élise Voïart suggest that Véron could have been aware of Hegel's work before 1827.

51 Véron published one medical work in 1825, *Observations sur les maladies des enfants* and served briefly as a physician for the *Musées royaux* but had already renounced a medical career by 1828; see Grégoire, *Supplément* to the *Dictionnaire encyclopédique d'histoire,* 56.

52 Huckenpahler, "Confessions of an Opera Director," 73, emphasis mine.

53 Delaurenti, "La fascination et l'action à distance."

54 Smith, "The Disappearing Danseur," 33–57, esp. fn. 4.

55 Sergent and Bouvet, "Quelques documents iconographiques sur le docteur Véron," 136–39.

56 Planchon, "Le Fabuleux Destin du Docteur Véron," 22. Planchon does not elaborate on the nature of this medical presence.

57 Huckenpahler, "Confessions," 82.

58 Tamvaco's critical edition of Gentil's manuscript provides artists' biographies that allow researchers to identify dancers' social networks; see *Les Cancans de l'Opéra.* Robin-Challan is responsible for our knowledge of the manuscript, cf. *Danse et danseuses à l'Opéra de Paris, 1830–1850.*

59 Delattre-Destemberg, *Les enfants,* 31.

60 Tamvaco, *Les Cancans,* 143; see also, Delattre-Destemberg, *Les enfants,* 483.

61 Anon., *Dictionnaire des coulisses,* 34; also cited in Delattre-Destemberg, *Les enfants,* 482.

62 Concerning tensions between Véron and the *commission de surveillance,* see Véron, *Mémoires* III, 197; see also Juin, *1831–1835,* 76–84.

63 Delattre-Destemberg, *Les enfants,* 446.

64 Quijuano-Gonzalez, *Capital de l'amour,* 156–57; see also Delattre-Destemberg, *Les enfants,* 491–92.

65 De Boigne, *Petits mémoires de l'Opéra,* 52, 274–75.

66 Dr. Le Guise, the opera physician in charge of her treatment, resigned after a second consultation with Taglioni who apparently refused either to admit her pregnancy or to repeat the leeching treatment (De Boigne, *Petits mémoires de l'Opéra,* 275).

67 Olivesi, *'Vedettes' et 'artistes,'* 299–301.

68 Villermé, *Tableau de l'état physique et moral des ouvriers.* For a discussion of early hygienists, see Kudlick, "The Culture of Statistics and the Crisis of Cholera in Paris, 1830–1850"; La Berge, *Mission and Method.* On the politics of philanthropy during the cholera epidemic, see Kudlick, "Giving Is Deceiving."

69 Bibliothèque-Musée de l'Opéra, Arch. 19/282, *rapport du député Guillaume,* 1842, cited in Delattre-Destemberg, *Les enfants,* 262.

70 *Arrêté du préfet de police, Service medical,* 2 mai 1852, reprinted in, Bouchard, *La langue théâtrale,* 351–52. The legislation specifies that physicians would not be remunerated for running the dispensaries.

71 *Arrêté du préfet de police, Service medical,* 2 mai 1852, reprinted in, Bouchard, *La langue théâtrale,* 352.
72 AN, AJ^{13} 479, *règlement intérieur de l'École de danse de Milan, article 1,* cited in Delattre-Destemberg, *Les enfants,* 165–66.
73 Blasis, *Traité élémentaire théorique et pratique;* AN, AJ^{13} 479, *règlement intérieur de l'École de danse de Milan,* cited in Delattre-Destemberg, *Les enfants,* 163–64.
74 Adice, *Théorie de la gymnastique de la danse théâtrale;* on this treatise, see Delattre-Destemberg, *Les enfants,* 30–31, fn. 105.
75 AN, AJ^{13} 1186, *règlement du Conservatoire de danse du 15 mars 1860, article 37;* see also Bouchard, *La langue théâtrale,* 351; and Delattre-Destemberg, *Les enfants,* 80, 261.
76 Delattre-Destemberg, *Les enfants,* 501.
77 AN, AJ^{13} 1269, *rapport du médecin de l'Opéra, le docteur Gaches-Sarraute, adressé au directeur de l'Opéra, le 25 avril 1893,* and the *Loi du 2 novembre 1892,* cited in Delattre-Destemberg, *Les enfants,* 87–88.

Bibliography

Archival

Archives nationales

AJ^{13} 109, *règlement sur le service de santé,* Oct. 24, 1825.
AJ^{13} 63, *officiers de santé,* an XII.
AJ^{13} 63, *officiers de santé,* an XIII.
AJ^{13} 64, *lettre de Monsieur Petit,* 1805.
AJ^{13} 64, *rapport du docteur Delatour, chirurgien titulaire de l'Académie impériale de musique.*
AJ^{13} 72, *Règlement pour les théâtres du 25 avril 1807 [?], titre premier, article 1* [1804?].
AJ^{13} 74, *arrêté du préfet du palais,* Nov. 20, 1806.
AJ^{13} 193, *règlement du service de la danse, clauses communes,* 1846.
AJ^{13} 479, *règlement intérieur de l'École de danse de Milan et dispositions fondamentales du règlement de l'École de danse de Milan, 1er août 1850.*
AJ^{13} 1186, *règlement du Conservatoire de danse,* 15 mars 1860, art. 37.
AJ^{13} 1269, *rapport du médecin de l'Opéra, le docteur Gaches-Sarraute, adressé au directeur de l'Opéra, le 25 avril 1893,* and the *Loi du 2 novembre 1892.*
F^{17} 1298, *État sommaire des versements.*

Bibliothèque nationale de France (BNF Opéra).

Arch. 19/282, *rapport du député Guillaume,* 1842.

Books and Articles

Adice, G. Léopold. *Théorie de la gymnastique de la danse théâtrale, avec une monographie des divers malaises qui sont la conséquence de l'exercice de la danse théâtrale.* Paris: N. Chaix, 1859.

Annales d'hygiène publique et de médicine légale, tome I, 1re partie. Paris: Gabon, 1829.

Anon., *Dictionnaire des coulisses, ou Vade-Mecum à l'usage des habitués des théâtres . . . suivi d'une Notice historique sur chaque Théâtre.* Paris: Dezauche, 1832.

Bienville, D. T. de. *La nymphomanie ou Traité de la fureur uterine: dans lequel on explique avec autant de claret que de méthode, les commencements de cette cruelle maladie, dont on développe des différentes causes . . .* [ed. 1778]. Paris: Office de librairie, 1886.

Blasis, Carlo. *Traité élémentaire théorique et pratique de l'art de la danse.* Milan: Joseph Beati et Antoine Tenenti, 1820.

Bouchard, Alfred. *La langue théâtrale: vocabulaire historique, descriptif et anecdotique des termes et des choses du théâtre, suivi d'un appendice contenant la législation théâtrale en vigueur.* Paris: Arnaud et Labat, 1878.

Bourdin, Philippe. "Les fantômes de l'Opéra ou les abandons du Directoire." *Annales historiques de la Révolution française* 379, n. 1, (2015): 109–29.

Cabanis, Pierre-Jean-Georges. *Rapports du physique et du moral de l'homme,* vol. 1. Paris: Crapelet, 1805 [2nd ed.].

Claire, Elizabeth. "Dance Studies, Gender and the Question of History." *Clio. Women, Gender, History: Dancing* 46, n. 2 (2017): 161–88.

Claire, Elizabeth. "*La Dansomanie,* une expression symptomatique, entre ballet et bal au Théâtre de la République et des Arts." In *Les arts de la scène à l'épreuve de l'histoire,* edited by Roxane Martin and Marina Nordera, 357–71. Paris: Honoré Champion, 2011.

Claire, Elizabeth. "Le corps en revolution." In *Histoire de la danse en Occident. De la préhistoire à nos jours,* edited by Laura Cappelle, 90–101. Paris: Éd. du Seuil, 2020.

Claire, Elizabeth. "A Moral Defence of the Regency Ballroom—*vide Wilsons Rooms.*" In *European Drama and Performance Studies No. 8: Danse et morale. Une approche généalogique,* edited by Marie Glon and Juan Ignacio Vallejos, 199–235. Paris: Éd. Classiques Garnier, 2017.

Claire, Elizabeth. "Monstrous Choreographies: Waltzing, Madness & Miscarriage." In *Studies in Eighteenth-Century Culture* 38, edited by Linda Zionkowski, 199–235. Baltimore: Johns Hopkins University Press, 2009.

Claire, Elizabeth. "*Waltzmania* in the Paris Pleasure Gardens." In *Cultures of Contagion,* edited by Béatrice Delaurenti and Thomas LeRoux, 95–101. Cambridge, MA: MIT Press, 2021.

Daubié, Julie-Victoire. *La femme pauvre au XIXe siècle.* Paris: Guillaumin, 1866.

de Boigne, Charles. *Petits mémoire de l'Opéra.* Paris: Librairie Nouvelle, 1857.

Delattre-Destemberg, Emmanuelle. "Les enfants de Terpsichore: histoire de l'École et des élèves de la danse de l'Académie de musique (1783–1913)." PhD diss., Université Versailles-Saint-Quentin-en-Yvelines/Université Paris-Saclay, 2016.

Delaurenti, Béatrice. "La fascination et l'action à distance: questions médiévales (1230–1370)." *Médiévales* 50 (spring 2006): 137–54.

Drysdale, John D. *Louis Véron and the Finances of the Académie Royale de Musique.* New York: Peter Lang, 2003.

Drysdale, John D. "Louis Véron and the Finances of the Académie Royale de Musique, 1827–1835." PhD diss., University of Southampton, 2000.

Ehrhard, August. "L'Opéra sous la direction Véron, 1831–1835." *Revue Musicale de Lyon* (1907–1908): 49 pp.

Foster, Susan Leigh. "The Ballerina's Phallic Pointe." In *Corporealities: Dancing Knowledge, Culture and Power,* edited by Susan Leigh Foster, 1–24. London: Routledge, 1996.

Foster, Susan Leigh. *Choreography and Narrative. Ballet's Staging of Story and Desire,* 2nd ed. Bloomington: Indiana University Press, 1998.

Ganière, Paul. "Bichat, François-Xavier (1771–1802), Médecin." *Dictionnaire Napoléon,* vol. 1, 2nd ed. Paris: Éditions Fayard, 1999.

Garafola, Lynn (ed.). *Rethinking the Sylph. New Perspectives on the Romantic Ballet.* Hanover, NH: Wesleyan University Press, 1997.

Glon, Marie, Vannina Olivesi, and Juan Ignacio Vallejos. "Penser l'auteur en danse au XVIIIe siècle." In *Chemins de la création. Auteur, autorité et pouvoir dans la musique et les arts du spectacle,* edited by Vincent Cotro and Catherine Douzou, 21–42. Paris: Editions Kimé, 2017.

Glon, Marie, Vannina Olivesi, and Juan Ignacio Vallejos. "Writing (for) the Ballet in the Eighteenth Century." *European Drama and Performance Studies* 2, n. 9: *Écrire pour la scène (XVe-XVIIIe siècle)* (2017): 263–79.

Goetschel, Pascale, and Jean-Claude Yon, eds. *Directeurs de théâtres XIXè-XXè siècles, histoire d'une profession.* Paris: Publications de la Sorbonne, 2008.

Grégoire, Louis. *Supplément, Dictionnaire encyclopédique d'histoire, de biographie, de mythologie et de géographie.* Paris: Garnier Frères, 1874.

Gutsche-Miller, Sarah. *Parisian Music-Hall Ballet, 1871–1913.* Rochester, NY: University of Rochester Press, 2015.

Huckenpahler, Victoria. "Confessions of an Opera Director: Chapters from the Mémoires of Dr. Louis Véron, Part One." *Dance Chronicle* 7, n. 1 (1984): 50–106.

Juin, Guillaume. "1831–1835: l'Opéra de Paris sous la direction Véron." Master's thesis, Université de Paris Sorbonne IV, 2006.

Kudlick, Catherine J. "The Culture of Statistics and the Crisis of Cholera in Paris, 1830–1850." In *Recreating Authority in Revolutionary France,* edited by Bryant T. Ragan Jr. and Elizabeth A. Williams, 98–124. New Brunswick, NJ: Rutgers University Press, 1992.

Kudlick, Catherine J. "Giving Is Deceiving: Cholera, Charity, and the Question for Authority in 1832." *French Historical Studies* 18, n. 2 (Autumn 1993): 457–81.

La Berge, Ann Elizabeth Fowler. *Mission and Method: The Early Nineteenth-Century French Public Health Movement.* Cambridge: Cambridge University Press, 1992.

Lacombe, Hervé. *Histoire de l'opéra français. Du Consulat aux débuts de la IIIe République.* Paris: Fayard, 2020.

Le Roux des Tillets, J.-J. P. Huard, and M.-J. Imbault-Huart. "Structure et fonctionnement de la Faculté de Médecine de Paris en 1813." *Revue d'histoire des sciences* 28, n. 2 (Apr. 1975): 139–68.

Lilti, Antoine. *Figures publiques. L'invention de la célébrité, 1750–1850.* Paris: Fayard, 2014.

Mainwaring, Madison. "Reclaiming the Silences of Dance: Women and Ballet in Nineteenth-Century France." PhD diss., Yale University, 2023.

Mainwaring, Madison. "Dancing in Pants: Drag and Agency at the Paris Opera, 1858–1889." In *Méthodes en movement. Vers une histoire décentrée de la danse,* vol. 1, edited by Marie-Hélène Delavaud-Roux, Florence Poudru, and Aude Thuries, 93–111. Paris: L'Harmattan, 2024.

Marquié, Hélène. *Histoire et esthétique de la danse de ballet au XIXe* siècle. Nice: Université de Nice-Sophia Antipolis, 2014.

McCarren, Felicia. *Dance Pathologies. Performance, Poetics, Medicine.* Stanford, CA: Stanford University Press, 1998.

McCleave, Sarah. "Marie Sallé, a Wise Professional Woman of Influence." In *Women's Work: making dance in Europe before 1800,* edited by Lynn Matluck Brooks, 160–82. Madison: University of Wisconsin Press, 2007.

Monnier, Franck. "L'Opéra de Paris de Louis XIV au début du XXe siècle: régime juridique et financier." PhD diss., Université Panthéon-Assas, 2012.

Oberzaucher-Schüller, Gunhild. "Multi-Layered Shifting Processes: Changes of Conceptions in Nineteenth-Century Stage Dance." In *Times of Change: Artistic Perspectives and Cultural Perspectives in Nineteenth-Century Dance,* edited by Irene Brandenburg, Francesca Falcone, Claudia Jeschke and Bruno Ligore, 163–76. Bologna: Massimiliano Piretti Editore, 2022.

Olivesi, Vannina. "Archéologie du vedettariat chorégraphique à l'Opéra. Talents, célébrité genres de danse et hiérarchisation sexuée des rémunérations dans les *Tablettes de renommée* (1785) de Mathurin Roze de Chantoiseau." In *European Drama and Performance Studies* 18, *Molière and After. Aspects of the Theatrical Enterprise in 17th- and 18th-Century France,* edited by Sabine Chaouche, Jan Clarke, Paris, Classiques Garnier (2022–2021): 249–76.

Olivesi, Vannina. "Between Pleasure and Censure: Marie Taglioni, Choreographer During the Second Empire," ed. and trans. Elizabeth Claire. *CLIO. Women, Gender, History: Dancing* 46, n. 2 (2017): 43–64.

Olivesi, Vannina. "Entre discrétion et quête de reconnaissance: Jean Coralli à l'Opéra, une ascension professionnelle au défi de la standardisation du ballet (1830–1854)." In *Giovanni Coralli, l'autore di Giselle,* edited by José Sasportes and Patrizia Veroli, 123–231. Rome: Aracne Editrice, 2018.

Olivesi, Vannina. "'Vedettes' et 'artistes,' une histoire de la féminisation du ballet de l'Opéra, 1830–1860." PhD diss., École des Hautes Études en Sciences Sociales (EHESS), Paris, 2021.

Pfister, Béatrice. "La danse à la conquete du statut d'art. Apologie et théorie du ballet dans les textes français et italiens de la fin du XVIe à la fin du XVIIIe siècle." PhD diss., Paris Sorbonne-Nouvelle, 2020.

Planchon, Claude A. "Le Fabuleux Destin du Docteur Véron." *Vesalius: Acta Internationalia Historiae Medicinae,* 10, n. 1 (June 2004): 20–24.

Sergent, Louis, and Maurice Bouvet. "Quelques documents iconographiques sur le docteur Véron: directeur de l'Opéra et spécialiste." *Revue d'histoire de la pharmacie* 38, n. 128 (1950): 136–39.

Quijuano-Gonzalez, Lola. *Capital de l'amour. Filles et lieux de Plaisir à Paris au XIXe siècle.* Paris: Vendémiaire, 2015.

Raz, Carmel. "Hector Berlioz's Neurophysiological Imagination," *Journal of the American Musicological Society* 75, n. 1 (2022): 1–37.

Robin-Challan, Louise. "Danse et danseuses à l'Opéra de Paris, 1830–1850." PhD diss., Université Paris Diderot, 1983.

Sabee, Olivia. "The *Rat de l'Opéra* and the Social Imaginary of Labour: Dance in July Monarchy Popular Culture." *French Studies: A Quarterly Review* 76, n. 4 (October 2022): 554–75.

Smith, Albert. *The Natural History of the Ballet-Girl.* London: D. Bogue, 1847; reprint Dance Books, 1996.

Smith, Marian. "The Disappearing Danseur." *Cambridge Opera Journal* 19, n. 1 (2007): 33–57.

Tamvaco, Jean-Louis. *Les Cancans de l'Opéra: Chroniques de l'Académie royale de musique et du théâtre à Paris sous les deux Restaurations,* 2 vols. Paris: CNRS, 2000.

Thuillier, Guy. *Les pensions de retraite des artistes de l'Opéra (1713–1914).* Paris: Comité d'histoire de la Sécurité sociale, 1999.

Tissot, M. P. *Ouvres complètes de Tissot, Docteur et Professeur en Médecine, . . . Précédée d'un Précis historique sur la vie de l'Auteur, et accompagnée de Notes. Par M. J. N. Hallé, Docteur et Professeur en Médecine, de l'École de Paris, etc. etc.,* vol. I, 1761, 1st ed. Paris: Chez Allut, 1809.

Vallejos, Juan Ignacio. "Les philosophes de la danse. Le projet du ballet pantomime dans l'Europe des Lumières (1760–1776)." PhD diss., EHESS, Paris, 2012.

Vermeir, Koen, and Michael Funk Deckard, eds. *The Science of Sensibility: Reading Burke's Philosophical Enquiry.* Dordrecht: Springer, 2012.

Véron, Louis. "Considérations générales sur les sensations, suivies de quelques propositions médicales; Thèse présentée et soutenue à la Faculté de Médecine de Paris le 23 août 1823, pour obtenir le grade de Docteur en médecine." Paris: Didot le Jeune, 1823.

Véron, Louis. *Mémoires d'un bourgeois de Paris* vol. 3. Paris: Gabriel de Gonet Éditeur, 1853.

Villermé, Louis-René. *Tableau de l'état physique et moral des ouvriers employés dans les manufactures de coton, de laine et de soie.* Paris: J. Renouard, 1840.

Vitet, L., *Le médecin du peuple, ou Traité complet des maladies dont le peuple est communément affecté: ouvrage composé avant la révolution française, par L. Vitet, Médecin professeur,* tome XI, "Maladies des femmes, Accouchemens." Lyon: Frères Perisse, 1804.

7

Imagination, Sensation, and Habits

Medical Rhetoric and Popular Literature in Perceptions of the Paris Opera

Response to "Dr. Louis Véron, Medical Philosophy, and Medical Practice at the Paris Opera" by Elizabeth Claire

Olivia Sabee

Much has been made of Dr. Louis Véron's tenure as managing director of the Paris Opera, largely in reference to the privatization of the Opera, Véron's emphasis on profits, and the prominence of the Jockey Club.[1] More recent studies have underscored the variety of experiences available to women through their connections to the Opera and their agency within the greater system. Elizabeth Claire's essay in this volume adds another dimension to consideration of the Opera in its examination of how Véron's medical thesis and his medical philosophy affected the practice and perception of dance during this moment of rapid change in the intertwined areas of medicine, workers' rights, gender roles, and spectatorship.[2]

Véron's views on sensation intersected with trends affecting dance practice and spectatorship during and just after his tenure at the Opera. As cited in Claire's essay for this volume, Véron's evocation of the "faculties of the soul," or the imagination, aligns with links among imagination, contagion, and dance made by writers for the popular press in describing the allure of dance careers for working-class women. Beginning in the 1830s, the career of the *rat de l'opéra* (young dancer training at the Paris Opera) took on new prominence in the public imagination by means of *physiologies* (moral caricatures highlighting individuals within social groups) and *livrets* for popular theater and dance performances, as well as more generally within the popular press.[3] "Since 1770," Philibert Audebrand writes in the upscale *physiologie, Les Français peints par eux-mêmes,* "few young girls from the working class have known how to resist the desire, ignited in them like a fever, to appear in public, among the . . . splendors of a ballet."[4] Audebrand's language mimics the rhetoric of feverish contagion associated with the waltz, yet he attaches this language to the desire of working-class women for careers in dance. These girls, Audebrand writes, allow their imaginations to run away with them, dreaming "a pink and golden dream" of the riches and fame that a career at the Opera might bring them (were they to escape its lower echelons).[5]

The parallel between early nineteenth-century medical discourse and depictions of *rats de l'Opéra* in popular literature does not end with the power of the imagination as associated with fantasies of stardom on the stage. Rather, the association in Véron's medical thesis of "pleasurable sensations" with the "beginning of pain" finds echoes in then-popular depictions of dance. The dancer Lélia in Albéric Second's *Les petits mystères de l'Opéra* describes the tribulations of dance training: "If you knew, sir, everything that was necessary in a young girl: courage, patience, resignation, and constant work; if you knew that she had to endure horrible torture and swallow silent tears even to become a mediocre dancer, you would be moved and afraid all at once."[6] Lélia describes as "torture" the stretching practices and contraptions that forced the dancer's body into external rotation.[7] In these fictionalized accounts, dancers follow their imaginations, seeking transformation, fame, and fortune. It is unclear whether dance brings these aspiring professional dancers pleasure (other popular literature explicitly links women dancing recreationally with leisure, if not pleasure), but such grueling work apparently feeds their imagination of their futures.

Claire proposes the value of further research into the work of anatomist François-Xavier Bichat as potentially influencing Véron's medical thesis and implications for practices at the Opera. In addition to Bichat's emphasis on symmetry, Bichat also emphasizes the relationship between sensation and habits[8] in a way Véron would later echo—and that finds echoes in popular literature of subsequent decades. "Habits blunt all relative pleasures," Bichat writes, and over time, "the more the years accumulate sensations, the less we are provided with new [sensations] to feel, and they exhaust sources of happiness."[9] Véron's thesis expresses a similar sentiment: "Habits finish by making almost imperceptible all impressions that were first pleasant or painful."[10] Later in that passage, Véron employs the same verb used earlier by Bichat, "émousser," "to blunt."

Louis Couailhac's *Physiologie du théâtre* emphasizes the hard, thirteen-hour daily labor of the *figurante*.[11] Similarly, Philippe Dumanoir and Hippolyte Cogniard's tableau-*vaudeville, Les Danseuses à la classe* (1835), emphasizes the dancer's daily training in a song ("To flutter/With a light step,/ First thing in the morning/ We get started").[12] Although neither Couailhac, nor Dumanoir and Cogniard bring up the idea of sensation, they and other writers of popular literature emphasize the concept of habit in dance training. Further research might illuminate relationships between sensation and bodily training in Opera practices in the years following Véron's tenure. Could an understanding of sensation and its relationship to the daily repetition of bodily labor have delayed moves toward medical practices at the Opera driven by worker's rights movements and a desire for workers' improved physical health? Did readers and viewers of these works assume, based on then-current conceptions of sensation and habit, that dancers would cease to perceive the physical pain of their labor over time? Later, as Claire underscores, physical examinations were instituted to ensure employment of more perfect bodies from the earliest stages of training—a reflection of Bichat's ideas concerning symmetry and health. Yet prior to implementation of these examination practices or later health-oriented measures for dancing, emergent medical notions about sensation influenced popular perceptions of dancers and their labor. During this moment of change in the worlds of medicine and performing arts, the intersection of Véron's medical training and his work at the Paris Opera ultimately served to shape ballet training, performance, and spectatorship.

Notes

1 For example, Guest, *The Romantic Ballet in Paris,* 191–94; and Homans, *Apollo's Angels,* 144–45. Olivesi amends this narrative, arguing that tactics employed by Véron had been in place during the tenure of Sosthène de la Rochefoucauld (Olivesi, "Vedettes' et 'artistes'" 95–152).

2 In addition to citations Claire provides on perceptions of masculinity, see Murray, "Dancing While Male," esp. 267–91.

3 Jarrasse, *Les Deux corps de la danse,* 93–101; Coons, "Artiste or coquette?" 140–64; Sabee, "The *Rat de l'Opéra,*" 568–73; Mainwaring, "Ballet and Celebrity," 36–52.

4 Audebrand, "La Figurante," v. 1, 113. All translations are my own.

5 Audebrand, "La Figurante," v. 1, 113.

6 Second, *Les Petits Mystères,* 149.

7 Second, *Les Petits Mystères,* 150.

8 Bichat, *Anatomie Générale, v.* 2, t.4, 755.

9 Bichat, *Anatomie Générale,* v. 2, t.4, 755.

10 Véron, *Considérations générales sur les sensations,* 13.

11 Couailhac, *Physiologie du théâtre,* 45.

12 Dumanoir and Cogniard, *Les Danseuses à la classe,* 30.

Bibliography

Audebrand, Philibert. "La Figurante." In *Les Français peints par eux-mêmes: encyclopédie morale du dix-neuvième siècle,* ed. Léon Curmer, v. 1, 113–20. Paris: L. Curmer, 1841–1843.

Bichat, François-Xavier. *Anatomie Générale appliquée à la physiologie et à la médecine.* Paris: Brosson, 1801, 4 vols.

Coons, Lorraine. "Artiste or coquette? Les petits rats of the Paris Opera Ballet." *French Cultural Studies* 25 (2014): 140–64.

Couailhac, [Louis]. *Physiologie du théâtre, par un journaliste.* Paris: J. Laisné, 1841.

Dumanoir, [Philippe] and [Hippolyte] Cogniard. *Les Danseuses à la classe, tableau-vaudeville en un acte* [*L'École de danse, ou la matinée d'une danseuse*]. Paris: Barba, 1835.

Guest, Ivor. *The Romantic Ballet in Paris.* Alton: Dance Books, [1966] 2008.

Homans, Jennifer. *Apollo's Angels: A History of Ballet.* New York: Random House, 2010.

Jarrasse, Bénédicte. *Les Deux corps de la danse: Imaginaires et représentations à l'âge romantique.* Paris: Centre national de la danse, 2017.

Mainwaring, Madison. "Ballet and Celebrity at the Paris Opera, or the Dreams of the *Rat.*" *Nineteenth-Century French Studies* 52, n. 1–2 (2023–2024): 36–52.

Murray, Colin. "Dancing While Male: Theatrical Sovereignty in the French Romantic Ballet." PhD diss., Temple University, 2023.

Olivesi, Vannina. "'Vedettes' et 'artistes.'" PhD diss., École des hautes études en sciences sociales, 2021.

Sabee, Olivia. "The *Rat de l'Opéra* and the Social Imaginary of Labour: Dance in July Monarchy Popular Culture." *French Studies* 76, n. 4 (2022): 554–75.

Second, Albéric. *Les Petits Mystères de l'Opéra.* Paris: G. Kugelmann, 1844.

Véron, Louis. *Considérations générales sur les sensations, suivies de quelques propositions médicales.* Paris: Didot le Jeune, 1823.

8

The Paradox of the "Subtle Body"

Dance, Tantra, and Science

Pallabi Chakravorty

The construction of a universal Indian dancing body emerged within a complex interaction between East and West in the nineteenth and twentieth centuries. The colonial birth of the Indic body emerged from a transcultural and transdisciplinary dialogue across science, art, and religion, which materialized through several key figures. In this essay I will explore this deeply sedimented relationship between the colonial project of scientific inquiry and classification and Indian native belief systems in reformulating the very ontology of the "ideal body." I will show how the process brought yoga, dance, spirituality, and medical science onto the same "stage" through a multitude of braided trajectories often not identifiable through simple chronological recounting. The literal staging of science in colonial India was ideologically imbricated within the revamping of Indian dance, where notions of the "subtle body" and the "medical body" cohered. The emergence of a universal body brought forth through this interaction of art and science included both a body of knowledge and a physio-cultural body.[1]

In this essay I will explore the following: first, the inception of the subtle body in nineteenth-century discourses around yoga, body, and theosophy; second, the intersections of Western science and Indian spirituality through international figures such as Helena Blavatsky, Vivekananda, and others; third, the refashioning of Indian temple and courtly traditions as national/classical/contemporary dance by Rukmini Devi and later by Chandralekha as the ideal Indian/universal body; and fourth, the simultaneous melding of the yogic subtle body with American modern and postmodern dance, health, somatics, and neurobiology. I will show how the intersections among yoga, science, and classical Indian dance were forged through theosophy, Hindu philosophy, Sanskritic aesthetics, and Western dance, which together created a powerful cross-cultural dialogue with the science and art of the body.

The Inception of the "Subtle" Body in Colonial India

During the eighteenth and nineteenth centuries, the various sects of yogis and fakirs (holy persons belonging to a multitude of castes and Muslim communities) came to be associated with the anthropological "primitive other."[2] These primitive yogis/yoginis were snake charmers, healers, and miracle workers who indulged in various penances and bodily mutilations like lying on nails, walking over fire, or standing on one leg for days. They engaged in various *tapas* (forms of meditation) that appeared to endow them with magical powers of healing, levitation, and immortality. Kirin Narayan astutely observes how during this time, when Sanskrit texts were being translated into European languages, these practitioners were differentiated from the yogis whom people read about in sacred texts such as Bhagavat Gita.[3] Associated as they were with spiritual transcendence, these disembodied yogis of the texts became separated from the living yogis in colonial India. The transcendental yogis inhabited the "subtle" body, which came to be enunciated through a combination of colonial attitudes toward precolonial yoga (Hatha Yoga) and the doctrines of bodily purification associated with Vedantic traditions of *sukhya deha* (perfected body/divine body) or *siddha deha* (liberated body).[4]

Today, the literary heritage of the Hatha Yoga tradition is enormous. The texts are written in Sanskrit and various other north Indian languages.[5] The Nath yogis or the Nath sect are the primary practitioners of Hatha Yoga; they connect their lineage to the teachings of Goraksthana (in Hindi, *Goraknath*) at the present time. The term *Nath* can mean a protector or master, or it can simply mean Shiva, the *Adi-Nath,* the primeval lord. The Nath *Panths* (sects) are a diverse group with complex and competing ideas, practices, and theologies. They have been associated with different religious movements, from Tantric schools of diverse traditions (including Kashmiri Shaivism, Shakta, and Buddhist practices) to Islam and Jainism. Nath poetry often conceptualizes identities through bodies, as we see in several verses by Kabir, a fifteenth-century Indian mystic poet and saint.[6] The Nath tradition diffused, from roughly the eleventh century to the eighteenth century, into different trajectories depending on regional differences, combining aspects of Vaisnavism and Sufism, often expressed through singing *kirtans* (devotional songs), dancing, or beating drums.[7]

These diverse texts reveal that there are different kinds of yogic bodies that strive for and attain immortality, or the state of *Shiva*. The ideas of immortality are condensed in the concept of *siddhi*, or liberation, associated with various stages of yogic ascendance encapsulated in *siddha deha*, or immortality. The ideal of siddhi or siddha connects intense yogic *sadhana* (practice) to spiritual liberation.[8] The initial state of the body (*divya deha*) corresponds to the initial stage of the practice, where the body does not have any spiritual character and can be manipulated through *asanas, mudras, mantras* (chants), and meditation. These ideas of bodily manipulation and stages are connected to the Kaula tantra tradition that flourished in the first millennium AD, which is associated with the visualization of Kundalini (a notion of a coiled snake identified with feminine power/consciousness) rising through the *chakras*, or bodily stations. The various yoginis or tantric goddesses associated with magic, possession, and divination rites form a vast corpus that also includes Sufi practices.[9] These different practices crisscrossed and adhered to the eleventh-century Hatha Yoga to construct the power of *siddha deha*.

All these ideas and embodied practices are derived from an even earlier Bindu-oriented system, according to the Sanskrit scholar and yoga practitioner James Mallinson.[10] Here, the conceptualization of the *deha* or body is not the physiological body of Western science, but a psychophysical body consisting of *chakras* (stations), *nadis* (veins), and other physiological or anatomical elements. *Prana* (breath or bodily wind) is held to be responsible for the biological functions of the body, with *prana* flowing through the *nadis*. These subtle structures can be manipulated by *mantras* (chants) and deities—primarily female ones—as well as by *asanas* (postures) and *mudras* (symbolic gestures or poses), the practice of which is called *sadhana*. The Kundalini (consciousness/feminine power/*Shakti*) rises through the central channel, or *susuma*, of the subtle body, catalyzing *moksha/siddhi*, or liberation/immortality (associated with the Hindu belief of reincarnation and a release from the cycle of birth). The *dehas* (bodies) of Hatha Yoga and Kashmiri tantric traditions correspond with the three-bodies framework of the Vedantic traditions, consisting of *sthula sarira* (gross body), *suksma sarira* (subtle body), and *karana sarira* (casual body).

The term "subtle body," derived from *suksma sarira*, appeared first in the late 1800s in the writings of prominent theosophical figures such as Helena Blavatsky, Charles W. Leadbeater, and Annie Besant.[11] The Theosophical Society was an organization founded in the United States in 1875, dedicated to the pursuit of esoteric and occult knowledge. The members were interested in Eastern traditions such as Hinduism and Buddhism for the creation of new occult practices that sprang from creolized Asian cultural worlds and American discourses.[12] Helena Blavatsky's ambitious text, *The Secret Doctrine*, propagated this theosophical idea of a hybrid and complex cosmology and belief system, which fused the various traditions represented in yoga with mesmerism and universalized them.[13] She produced a mythical narrative of human evolution combining yogic principles with terminology like *energy* and *vibrations* drawn from modern physics.[14] Rather than the extreme physical asceticism associated with Hatha Yoga traditions, Blavatsky recommended "mystic meditation" that connected the cultivation of the subtle body to yoga and mesmerism.[15] By invoking medical science, however questionable, through mesmerism, and combining it with the "ancient science" of the Vedas, she discarded colonial India's "primitive yogis" and their Hatha yoga practice from the emergent scientific yoga discourses of the nineteenth century.

The Theosophical Society developed its practice of meditation by combining yoga and mesmerism in the cultivation of the "subtle body." Madame Blavatsky and her colleague Colonel Henry Olcott believed that mesmerism, which forms an important part of the practice's magic, is related to yogic meditation techniques. In *Isis Unveiled* (1877), Blavatsky promoted mesmerism as a modern counterpart to yoga. Following this trajectory, in the introduction to the theosophical edition of The Yoga Sutras, titled The Yoga Philosophy (published in 1882), Olcott introduced Kundalini Yoga in a mesmeric way, proclaiming yoga as "self-mesmerism."[16]

The Intersection of Western Science and Indian Spirituality

By framing Indian spirituality in a Western episteme, a new organization of knowledge emerged in the late nineteenth and twentieth centuries. This deep transformation of native "objects" for native "eyes" through the terms of science was a dual process of presentation and representation, whether it be of yoga or *Jati* (the Indian caste system). Gyan Prakash writes that this enunciation of "scientificity" happened through the process of representing science as Western, but at the same time, it also meant that science was "going native." "The enunciation of colonial science, therefore, was a profoundly ambivalent process that, while formulating scientific knowledge, articulated western science in native objects," writes Prakash.[17] Throughout the nineteenth century, various displays and exhibitions helped establish the credibility of science to colonial subjects; this was predicated upon science's staging and its visual confirmation through the notion that "seeing is believing." Among these displays was the presentation of the ethnological congress of all races (1869–1870) in India, following colonial racial typologies and classifications. The ethnological congress of all races included gathering, classifying, and exhibiting actual people in stalls. At the same time, such public displays among both the Europeans and Indians also helped to establish the scientific claims of mesmerism. Dr. James Esdaile, an exponent of this treatment, established a "Mesmeric Hospital" in Calcutta in 1846, which became popularly known as "*jadoo* hospital," a house of magic, among the lower-class native population. Interestingly, Dr. Esdaile himself used the term *belatee muntur/mantra* (European chant) while claiming it as a science. To quote Prakash again: "Indeed it was in the public display of its magical effect that mesmerism emerged as science, perched precariously in between cold scientific scrutiny and superstition of its wildest and most absurd forms."[18] Through this rearticulation, science and magic overcame their opposition and united in a unique relationship between "wonderous science and knowledge seeking wonder."[19] This union seamlessly fused yoga with mesmerism.

The integration of science and magic in the pursuit of knowledge about the subtle body is perhaps best enunciated through the merging of Darwinian evolution and Hindu avatarism by the theosophists, and later by Indian social reformers and nationalists like Keshab Chandra and Vivekananda. The evolutionary conceptualization of rebirth and *moksha* (in the Sankhya philosophy of *The Yoga Sutras*) is later found in the *Puranas* (ancient Sanskrit texts) in the depiction of the avatars or *dashavatars* (ten incarnations of Vishnu, the god who protects the world). The *dashavatars* document the evolution of consciousness from lower forms to higher forms: fish, tortoise, boar, man-lion, dwarf, Parasurama, Rama, Krishna, Buddha, and Kalki. In *Isis Unveiled,* mentioned earlier, Blavatsky first introduced the evolutionary view of the *dashavatar* as foreshadowing the modern theory of evolution from simple to complex life through different bodily stages. She believed that the *avataras* of the Hindu texts were an allegorical interpretation of Darwinian evolution (millennia before Darwin conceptualized it). Her interpretation influenced the British orientalists and important nationalists like Vivekananda, who were trying to reform Hinduism and show the inherent rationalism in the religion. Even preeminent British scholars like Monier Monier-Williams wrote that Hindus were Darwinians long before Darwin was born. Keshab Chandra Sen, a pioneering Hindu reformist and cofounder of the Bramho Samaj, wrote in 1882, "The Hindu Avatar rises from the lowest scale of life through the fish, the tortoise, and the hog up to the perfection of humanity. Indian Avatarism is, indeed, a prime representation of the ascending scale of divine creation. Such precisely is the modern theory of evolution."[20]

During the nationalist period, Vivekananda revived and reclaimed the foundational narrative of Samkhya/Sankhya Yoga (not unlike classical Indian dance, as I show in the next section) through yogic texts such as Patanjali's

The Yoga Sutras (written in approximately 200 BCE). He understood the ultimate liberation of the subtle body in the context of rebirth as propounded in Vedic philosophy. Vivekananda had been influenced by Brahmo Samaj[21] leaders like Keshab Chandra Sen and the theosophists, but he became a seeker of direct experiential knowledge through living Brahmin mystics, especially Ramkrishna Paramahamsa. He repositioned the idea of avatarism as stages of consciousness and believed that the evolution of consciousness can be achieved in a single lifetime through the practice of yoga, ultimately arriving at divine consciousness or *siddhi* (liberation). This idea became encapsulated in the movement of the gross to the subtle body. Vivekananda argued that Patanjali, the writer of *The Yoga Sutras,* was the true father of evolution, both physical and spiritual.[22] Thus, the gradual removal of yoga from the lived practices of *sadhus* and *fakirs* (which included low-caste Hindus and Muslim Sufis) and its promotion through Vivekananda's *Raja Yoga,* published in 1896, not only made yoga a transcendental spiritual pursuit of human consciousness, but elevated it to a high intellectual and scientific status. This elevated, transcendental status of yoga could therefore no longer include tribal and marginalized castes in India, nor the Muslim practitioners, who were lumped together as the colonial "primitive other."[23]

Swami Vivekananda became a living icon of how Hinduism existed not only in ancient scriptures and holy texts, as the British would have one believe, but also as a living force, transmitted by current teachers such as Ramkrishna Paramahamsa. In a famous speech in Chicago in 1893 to the World Parliament of Religions (held in conjunction with the Columbian Exposition that celebrated the four-hundredth anniversary of Columbus's arrival in America), Vivekananda boldly argued against the weaknesses of Western morals, as evidenced in colonialism, the West's lack of religious spirituality, and what he perceived as Western fanaticism. These remarks came in response to accusations that he was a "half-educated theological student who had come to instruct wise and erudite Orientals."[24] In some exchanges, he was also identified as a Brahmin monk, even though he belonged to the Kshatriya caste by birth. Vivekananda's self-representation as a monk in saffron garb (*swami*) became symbolically significant later for the revival of Brahminical Hinduism and bodily practices, such as yoga, as integral to Indian identity, nationhood, and moral pursuit. This particular confluence of yoga, morality, and embodiment had far-reaching consequences in modern India for fashioning national bodies and identities. The subtle body, now universalized as science, came to be associated with Brahminical purity and Sanskritic ideology. In the next section, I show that the entanglement of yoga with classical Indian dance revivalism through figures such as Rukmini Devi helped to create a hegemonic dominance of Brahminical bodily attitudes in Indian dance, erasing lower-caste or Muslim aesthetics.

The Intersection of Indian Concert Dance and Yoga in Constructing the "Subtle Body" as Universal

The revival of classical Indian dance and yoga are overlapping subjects of appropriation and erasure. The magnificent iconography of the dancing Shiva, invoked by Rukmini Devi Arundale and other cultural revivalists, brought the two traditions together, where dance became yoga and yoga became the dance of life. Rukmini Devi Arundale established her school, Kalakshetra, in Madras (Chennai) in 1936, and it continues to be one of the most recognized institutions for Indian dance, not only in India but throughout the world. In 1994, the government of India recognized Kalakshetra as an "Institute of National Importance," a title given to premier institutes of higher education in India that makes them eligible for government funding. Rukmini Devi's approach to the reconstruction of Bharatanatyam from the local Sadir (the dance practiced by the "devadasis") became a model for how regional dances such as Odissi and Kathak were redefined as national classical dances in modern India. Rukmini Devi claimed that Bharatanatyam was the mother of all regional dances in India and that other styles were merely aspects of its interpretation.

In search of new aesthetics, Rukmini Devi connected Bharatanatyam to the Natyashastra, a Sanskrit treatise on the performing arts considered the fifth Veda and ascribed to the sage Bharata sometime between 200 BCE and 200 CE. She invoked the archetypal images of women from India's mythical past that reside in the images of "warrior" and "mother" as Hindu goddesses.[25] She fused the mother goddess and Shiva (drawing on Kundalini as the female power/Shakti) but sanitized and Brahminized the dance through Sanskrit texts to form the bedrock of the classical Indian dance of Bharatanatyam. Rukmini Devi's influence was not limited to the revival of Bharatanatyam and the invention of the Vedic philosophical moorings of Indian dance; she was also an influential figure in the international theosophical movement. She imbued theosophy with yogic philosophical ideals and propagated her work through the Theosophical Society of India (established in 1882).

During the early twentieth century, Rukmini Devi emerged globally as an influential Indian figure of both culture and politics. The construction of classicism in India, such as shaping Bharatanatyam along the lines of theosophy, became crystallized through the interaction of Rukmini and Annie Besant, the renowned political and spiritual leader who became engaged with the theosophical movement in 1889, and as part of her theosophy-related work, traveled to India. There, Besant became the president of the Indian National Congress and fought for home rule against the British; she also became the president of the Theosophical Society in 1907, and spearheaded theosophical thought in India.[26] She recruited Rukmini Devi, the pioneer of India's dance revival, to head the society after her and Besant made Rukmini the leader of the worldwide theosophical movement, hailing her as the "World Mother."

The intersections of theosophy, yoga, and classical Indian dance philosophy that Rukmini Devi created and continued are of utmost significance. They have a common orientation to transcendence through practices of the yogic body. Further, the overlapping ideas of these practices included vegetarianism and *ahimsa* (nonviolence), the latter of which translated into practices of noninjury to animals, in Rukmini Devi's interpretation. In fact, Devi's efforts contributed to the Prevention of Cruelty to Animals Act established in 1960. The animal rights and human rights ideals (*ahimsa* and *satyagraha*) also cohered with Gandhian philosophy. Gandhi himself never became a theosophist but was influenced by its ideals. His approach to the body as the moral and spiritual repository of a culture and nation manifested in his lifelong preoccupation with his own body. He entered the cultivation of his body by tapping into the spiritual power of *Shakti* and embodied it through his personal and collective *Swaraj* (self-rule).[27] Theosophical doctrines brought Gandhi and Rukmini together, and although Gandhi never danced (that we know of), he disciplined his body like a dancer and a yogi.

A. K. Coomaraswamy, the renowned Sri Lankan/Tamil art historian and revivalist, brought dance within the ambit of the revival of the arts in India during anti-colonial nationalism. His famous book *The Dance of Shiva* (1924) invoked Shiva as the foundational deity of Indian dance (as the Natyashastra was said to have been spoken by Shiva to the sage Bharata), thus making dance an aspect of Hindu religion (read: Brahminical practice), and therefore a sacred pursuit.[28] Coomaraswamy wrote extensively about the importance of the Shiva symbol in art and Hindu philosophy in *The Dance of Shiva*. Following him, the all-encompassing symbol of Shiva as the creator and destroyer of the universe—wherein the whole cosmos is his theater, and he is both the audience and the actor/performer—became the mantra of many writers/scholars and practitioners alike. The mantra was repeatedly chanted in dance books, lectures, demonstrations, and program notes, and the practice continues today. The refinement of the body through classical Indian dance that was seen as an iteration of yoga was theorized by foundational dance scholars such as Kapila Vatsyayan, in such phrases as, "the body can and does move from the physical to the metaphysical, and from the grossest to the subtlest, from the time actual to the time transcendental."[29] Although Coomaraswamy discussed three different kinds of Shiva dances represented in poetry and iconography—delineating regional differences through texts and iconography (such as Shiva's image as the Bhairava/Yogi in Shiva Shakti)—he emphasized the Nataraja imagery in the golden hall of Chidambaram, or *Tillai,* as his supreme image. The dance of Shiva in Chidambaram (which is imagined to be the center of the universe) was established as the quintessential symbol of the lord of the dance. This iconography is widely represented in the South Indian bronze and copper images of Nataraja. The image of Shiva as both a yogi and a dancer was thus forever etched into Indian dance discourses through Coomaraswamy's and later Vatsyayan's influential writings.

Coomaraswamy also argued for the superiority of the Hindu way of thinking and for the contributions of Brahmins to Indian civilization as the highest form of spiritual, aesthetic, and scientific knowledge. The purpose of his writings was to glorify colonized India's cultural and civilizational heritage to the West. His mission was to rewrite the colonial history of decay and inferiority regarding India and Hinduism propagated by historians such as James Mill (especially in the hugely impactful book *The History of British India,* first published in 1817). Much has been written about the narrative of the golden Hindu past as fuel for nationalists to ignite a cultural renaissance (*nabajagoron*) in Bengal and India.[30] In the context of performance, the narrative that the body was "corrupted" during colonialism but could be recovered, refined, and cleansed by the "right" kind of practice became the classical doctrine for both dance and yoga. Thus, the body had to be reclaimed through the process of yogic rehabilitation and refinement to gain spiritual high ground (the subtle body). Reclaiming the dancing body through yogic formulations logically became the hallmark of dance revivalism, so much so that all of the eight classical dances (Bharatanatyam, Kathak, Odissi, Manipuri, Kathakali, Kuchipudi, Sattriya, Chhau) came to be homogenized under a single philosophy. Rukmini Devi pioneered this revitalization of the body through her reform of the local Sadir into Bharatanatyam. In the process, the body of the "ideal" dancer changed from the low-caste traditional practitioners ("devadasis") of colonial India to the Brahmin elite women in modern India. Both in dance and yoga, the importance of establishing a golden age and classical (read: Brahminical) heritage became the defining arc for the dancing body, which aimed for yogic transcendence through *moksha,* as mentioned above.

There has been an outpouring of scholarship since the 1980s on the reconstruction of Bharatanatyam from Sadir and on Rukmini Devi's role in displacing the living tradition of the "devadasi" art practices. Occupying a large part of global dance scholarship emanating primarily from England and

America, this scholarship on Bharatanatyam has (almost) come to stand for all Indian classical dance, thus making the entire history of Indian concert dance a neo-Brahminical category.[31] This, in turn, has impacted various dance styles in India in multiple ways. For instance, Kathak dance has been Hinduized under powerful gurus like the late Birju Maharaj at national institutions such as Kathak Kendra. This hegemonic Kathak narrative has given rise to a new dance called Sufi Kathak, which attempts to revive the *tawaif* (courtesan) and Islamic aspects of the dance, thus compartmentalizing elements of Hindu and Muslim syncretism and its secular history. But such male-dominated nationalist narratives of classical dance in India have been nuanced by contemporary dance trailblazers, such as Chandralekha. The resurrection of *Shakti,* or the feminine energy, through *prana,* has yoked yoga to the classical dance of Bharatanatyam once again within her powerful choreographies, albeit in a different mold than that of Rukmini. Chandralekha Prabhudas Patel, popularly known as Chandralekha, received her early training in Bharatanatyam from guru Ellappa Pillai. She was also influenced by women pioneers such as Rukmini Devi and Balasaraswati, but as I discuss below, she changed her focus to make yoga central to her choreographies.

Return of the Psychophysical Body through Physical Culture, Globalized Dance, and Health

In this section, I explore how the physicality of a strong and healthy yogic body that Chandralekha invoked through her vision of Indian dance and Bharatanatyam echoed an earlier movement of body culture. That is, I show that the discourse on the subtle body also intersected with the physical culture movement during Indian nationalism. By drawing on that earlier movement, I argue that Chandralekha resurrected the subtle body in contemporary India to refashion a new corporeal feminist practice that was both indigenous and universalist.

The intricate relationship between yoga, dance, health, and body that informs the ideological and affective bedrock for Indian nationalism is a complicated and fascinating subject. The revival of Postural Yoga or Hatha Yoga is connected to the nationalist physical culture movement that developed in the late nineteenth century to display the prowess of strong muscular bodies. Mark Singleton argues that to a large extent, what is practiced today as *asanas,* or Postural Yoga, emerged in the first half of the twentieth century.[32] It was a hybrid product of colonial India's exchange with and assimilation of the worldwide physical culture movement, which included the launching of a popular self-instruction genre. It is noteworthy that the staging of the first modern Olympics was also around the same time as the publication of Vivekananda's *Raja Yoga* in 1896.[33] Vivekananda's call for national awakening through the teachings of Vedanta and yoga in colonial India also included the reconstruction of masculine vigor among the native population. Indian nationalists like him posited that yoga was going to develop the bodies, characters, and souls of the Indian people, which had been emasculated by colonial rule. The development of the physical body with national character echoed across many chambers, and even women such as Sarla Devi, who was Rabindranath Tagore's niece, campaigned for the building of health and vigor among young women and men through martial arts, yoga, and physical culture.

The scientific claims of yoga regarding the building of stronger and healthier bodies were gaining ground and became part of both the international physical culture movement and the international spiritual movement, especially Transcendentalism, around the turn of the twentieth century. The key figures in making yoga part of the physical culture movement in India were Manik Rao and Tiruka.[34] The latter studied with the renowned figure of the yoga renaissance in India, Swami Kuvalayananda, who founded the Kaivalyadhama Yoga Institute and Research Center in Lonavala Maharashtra in 1924. The former was Kuvalayanda's first yoga guru, with whom he studied the Indian system of physical education called Yog-Vyayam in Baroda at the Vyayamshala (school of physical culture). At Kaivalyadhama, Kuvalayananda carried forward his scientific research on yoga with human subjects to show the efficacy of asanas in curing diseases, bringing yoga into direct contact and alignment with Western medical science.[35]

The internationalization of yoga, physical culture, and health also included figures such as Indra Devi (Eugenie V. Peterson) (1899–2002), another trailblazing figure who spread yoga globally, including among many celebrities in Hollywood. One could argue that she was one of the pioneering figures in creating the global, commodified yoga body, decontextualized from its Indian roots and practiced as a New Age panacea. She was the first woman to be trained by the famous yoga guru Sri Tirumalai Krishnamacharya, who was in the court of the Maharaja of Mysore.[36] Other students of Krishnamacharya, such as B. K. S. Iyengar and K. Pattabhi Jois, also became iconic gurus who established renowned institutions and styles of modern Postural Yoga in India.

A parallel narrative to the internationalization of yoga by Western women pioneers like Indra Devi can be found in the work of Ruth St. Denis, considered a mother of American modern dance, who also brought Indian dance to the US. St. Denis promoted the revival of classical Indian dance in the early part of the twentieth century, in conjunction with figures like Rukmini

Devi Arundale and Madame Menaka, who was associated with the revival of Kathak from the nautch dancers/tawaifs/baijis during the 1930s. In fact, the turn-of-the-century "oriental" dance practiced by St. Denis and others was a "synthesis of Asian inspired techniques from Transcendentalism, Theosophy, Vedanta, and Yoga," writes Mark Singleton.[37] Singleton quotes Matthew Allen to explain this symbiosis of yoga and Indian dance with Western modern dance, in which dance pioneers like St. Denis participated in "a drama of appropriation and legitimation with a pan South Asian framework of nationalist aspiration and cultural regeneration."[38] St. Denis founded the Society for the Spiritual Arts in New York and the New York School of Natya with American dancer La Meri (Russell Meriwether Hughes) in 1938. Her approach to spirituality and physicality was grounded in her study of the Delsarte technique. François Delsarte, investigating how emotions are physically expressed in the body, categorized them in charts and diagrams. His technique was later named "harmonic gymnastics" and popularized by Genevieve Stebbins. Stebbins performed her own interpretations of ancient dances on stage, which St. Denis saw as a child and found inspiring. Stebbins's book, *The Delsarte System of Expression* (1885), became hugely successful in the international market. Singleton writes that some of Stebbins's deep breathing techniques are connected to *pranayama*, or yoga breathing, as practiced by Brahmins.[39] Through Stebbins, yoga became intricately intertwined with Western modern dance, which St. Denis later took up and combined with Indian dance. Ruth St. Denis's choreographies, such as *East Indian Nautch Dance* (1910, 1932), *Radha* (1906), and *The Yogi* (1908), are expressions of such orientalist imaginings.

Dance scholars such as Kapila Vatsyayan, mentioned in the previous section, have theorized about the antiquity of classical Indian dance movements, postures, and gestures by analyzing temple sculptures. The dancing figures in these temple sculptures show that many of the postures resemble yoginis, especially figures such as *yakshis* (nature spirits) and *salabhanjikas* (decorative female figures, also symbols of fertility).[40] On the other hand, contemporary scholars of yoga have argued that the proliferation of postural asanas we see today in yoga practice is a development of recent times. As a dancer myself, I have often seen the similarities of postures like Virabhadrasana (the warrior pose) with the movements and grounded stances represented in the iconographies of the goddesses Kali or Durga in their images of Shakti. One might argue this to be the classical yogini culture of yoga and tantra as we find in modern Odissi dances (especially in the Odissi of guru Surendranath).[41] Despite the debates about the antiquity of asanas that are practiced today, what is relevant here is their continuous reinvention and dialogue with dance and approaches to the body, spirituality, health, and science.

Chandralekha reinvented the yogic tradition of the subtle body (although she did not use the term) for contemporary India through her discourse on healing and revitalizing the body, which she claimed had been broken by mechanization and deracination. She critiqued and departed from Rukmini Devi's model of Bharatanatyam, yoga, and tantra by creating a new vision of dance in contemporary India through feminine power and sexuality. She sought to reject the excessive ornamentation in Bharatanatyam and its superficial preoccupation with spirituality and transcendence. The yogic ideal of the body that she reimagined through her stunning choreographies such as *Angika* (1985) and *Sharira* (2000) (both Sanskrit terms meaning "bodies"), once again draw from antiquity. She wanted to go back to the philosophical roots of Bharatanatyam and yoga/martial arts and reaffirm the body's sexuality and corporeality. She proposed that the mechanization of modern life had weakened the spine, which had to be reinvigorated through introspection of the body. While rejecting the divine origins of Indian dance, she envisioned prana/breath/energy to be embodied in the symbol of Shakti, the female principle of tantra and Kundalini in a new incarnation.[42]

However, by positing the female principle in Sanskritic yoga-tantra to be universal, Chandralekha abstracted the historicity of these practices yet again. Although she foregrounded the urban body decayed by capitalism and industrialization, she went back to Patanjali's *The Yoga Sutras* (*Purusha* and *Prakriti*), much as Vivekananda did, without any indication of its heterogeneous past and its development by Sufis or Pirs (Muslim spiritual guides, holy men), as I have discussed above. By situating the human body in this universalist stance and introducing *asanas* and dance as the path to its rehabilitation, Chandralekha once again contextualized Indian dances within a Vedic/Brahmin cosmology. This becomes apparent if we look at her choreographies such as *Navagraha*, where she used renowned Carnatic musician Muthuswami Dikshitar's composition based on the praise of the nine planets of ancient *jotisha sastra* (Vedic astrology) and *mantras* (Vedic chants). Similarly, in *Lilavati*, she used the ancient treatise on mathematics from the Vedic tradition by Bhaskara Acharya. In both cases, she invoked ancient scriptures and knowledge systems to establish the ancestry of Indian dance movements in order to create contemporary work. She claimed the "body as mandala" to be the core of Indian aesthetic tradition, thereby eschewing other possibilities of imagining the body, especially as conceived in non-Sanskritic, vernacular, or Islamic sources.[43] In the process, she universalized the bodies of contemporary Indian dancers, who became unmarked as a category in Indian dance discourses, despite their elite or upper-caste positions.[44]

Asana practice has gained visibility in globalized India, but it is differently situated than in the past. It has aligned with a universal and scientific understanding of bodies and health decontextualized from its local roots, while at the same time, it has been appropriated by Hindutva groups (rightwing Hindu nationalist groups) to perpetuate a virulent form of Hinduism. However, in this new and globalized landscape, yoga is in a position to have a dialogue with postmodern or contemporary Western dance and their universal claims. Modern American dance, which broke from ballet in the early part of the twentieth century, was itself rejected in the 1960s by a new movement called postmodern dance. The relationship between dance and the body and their union in somatics emerged as a productive ground where yoga intersected with phenomenology and neurobiology.[45] The union of yoga and dance as "movement" folded into a new universal understanding of the body, where these practices were placed as alternative therapeutic practices to Western medical science. In this new understanding of corporeality, where science and the Indian subtle body merged, a hybrid, universal body arose. Here, the medical anatomical body for therapeutic rehabilitation completely replaces the ancient knowledge of the subtle body in terms of *chakras, kundalini,* and *sushuma.* Joseph Alter compares this new reconfiguration of yoga and the medical body to a cyborg, drawing on Donna Haraway's conceptualization of a hybrid figure who crosses the boundary between human and machine.[46]

Conclusion

This essay has charted the heterogenous construction of the "subtle body" in colonial and postcolonial India. In this interdisciplinary construction of the body as both art and science, yoga and dance were yoked together within the ancient Hindu philosophy of Sankhya and tantra. The inherent mutability of the body in Hindu philosophy, I show, allowed for its adaptation in various discourses and practices, from Sufism and Transcendentalism to neurobiology. In all these discursive trajectories that connected the orient and the occident, there was a continuous reinvention of the ideals of the body that merged rationalism and science with aesthetics and spirituality. The merging of yoga and dance with health and science helped to universalize the heterogenous Indian body through colonialism and nationalism. Contemporary discourses of the "subtle body," including therapeutic concepts of mind-body and somatics, have broken new ground, as the yoga-dance-science merging discussed in this essay remains active through globalization, commodification, and Hindutvization.[47]

Notes

1 Chakravorty, "Artistry and the Moral Body," was an earlier version of this research.
2 Hindu was a label given to a multitude of sects who had a multitude of caste identities by the British to distinguish them from the Muslims for administrative purposes.
3 Narayan, "Refractions of the Field at Home," 490.
4 Ondracka, "Perfected Body, Divine Body and Other Bodies in the Natha-Siddha Sanskrit Texts," 210–32.
5 Mallik, *Siddha-Siddhanta-Paddhati and Other Works of the Natha Yogis,* 27.
6 See Hess and Singh, *The Bijak of Kabir.*
7 Lorenzen and Muñoz, "Introduction," xv–xviii.
8 Kapstein, "An Inexhaustible Treasury of Verse."
9 Ernst, "Being Careful with Goddess," and D'Silva, "Bodies in Translation."
10 Mallinson, "Yoga: Haṭha Yoga," 770–81. Bindu means a point or a dot, indicating a stage in the process of yogic transcendence.
11 Samuel and Johnston, "General Introduction," 2.
12 Albanese, "Metaphysical Asia," 336, and White, "From Liberation to Self-Realization."
13 Albanese, "Metaphysical Asia," 336; and White, "From Liberation to Self-Realization."
14 Samuel and Johnston, "General Introduction," 188.
15 Albanese, "Metaphysical Asia," 351.
16 Olcott, "Introduction," i–iv, and Baier, "Mesmeric Yoga and the Development of Meditation within Theosophical Society," 151–61.
17 Prakash, "Science 'Gone Native' in Colonial India," 154.
18 Prakash, "Science 'Gone Native' in Colonial India," 158.
19 Prakash, "Science 'Gone Native' in Colonial India," 158.
20 Nanda, "Madame Blavatsky's Children," 329. Monier Monier-Williams information is from the same source and page.
21 Bramho Samaj is separate from Brahminism. It is a societal organization or a sect of a monotheistic reformist movement of the Hindu religion. The Brahmos campaigned against the Hindu caste system and promoted the emancipation of women and all penalized groups and classes.
22 Nanda, "Madame Blavatsky's Children," 330.
23 Narayan, "Refractions of the Field at Home," and Chakravorty, "Why the Adivasi Will Not Dance."
24 Narayan, "Refractions of the Field at Home," 493.
25 Meduri, ed. *Rukmini Devi Arundale 1904–1986,* 123–46.
26 Olcott, *Old Diary Leaves, Fourth Series,* 184.
27 Alter, *Yoga in Modern India,* 115.
28 Coomaraswamy, *The Dance of Shiva.*
29 Vatsyayan, "Two Keynote Addresses," 3.
30 The Bengal Renaissance is an artistic and cultural movement that took place from the late eighteenth to the early twentieth century; see Sarkar, *Writing Social History,* 104.
31 The debates surrounding Bharatanatyam have emerged surrounding the caste identity of Bharatanatyam practitioners and the Hindu caste politics in Tamil Nadu in South India, which is different than other classical forms from different regions in India. For

example, in Kathak, which covers many regions of north India, the debates include other complexities of religion and region such as the Hindu–Muslim dynamic.
32 Singleton, *Yoga Body.*
33 Singleton, *Yoga Body,* 81, and De Michelis, *A History of Modern Yoga,* 91–100.
34 Singleton, *Yoga Body.*
35 Alter, *Yoga in Modern India,* 73–108.
36 Goldberg, *The Goddess Pose,* 171–74.
37 Singleton, *Yoga Body,* 44.
38 Allen, "Rewriting the Script for South Indian Dance," 69.
39 Stebbins, *Dynamic Breathing and Harmonic Gymnastics,* 144–47; and Singleton, *Yoga Body,* 146.
40 Vatsyayan, *Classical Indian Dance in Literature and the Arts,* 272.
41 Royo, "Guru Surendranath Jena," and Chakravorty, "Review of *Performing Konarak, Performing Hirapur,*" 92–95.
42 Bharucha, "In Memorium."
43 Chandralekha, "Probe Deeper Into Your Own Art."
44 For an elaboration of this argument about contemporary Indian dance and universalism, see Chakravorty, "Why the Adivasi Will Not Dance."
45 Fraleigh, "Why Consciousness Matters," 5.
46 Alter, *Yoga in Modern India,* 69; and Haraway, *Simians, Cyborgs, and Women.*
47 Hindutvization is an extremist form of rightwing political ideology of Hindu nationalism that preaches Brahmin supremacy and Islamophobia.

Bibliography

Albanese Catherine L. "Metaphysical Asia." In *A Republic Mind of and Spirit: A Cultural History of American Metaphysical Religion,* 330–93. New Haven, CT: Yale University Press, 2007.

Allen, Mathew. "Rewriting the Script for South Indian Dance." *TDR: The Drama Review* 41, n. 3 (1997): 63–100.

Alter, Joseph S. *Yoga in Modern India: The Body between Science and Philosophy.* Princeton, NJ: Princeton University Press, 2004.

Baier, Karl. "Mesmeric Yoga and The Development of Meditation within Theosophical Society." *Theosophical History: A Quarterly Journal Research* XVI, n. 3/4 (2012): 151–61.

Bharucha, Ruston. "In Memoriam: Chandralekha in Retrospect 1928–2006." *India Seminar,* Feb. 2007. https://www.india-seminar.com/2007/570/570_in_memoriam.htm.

Chakravorty, Pallabi. "Artistry and the Moral Body: Yoga, Classical, and Contemporary Indian Dance." Presented at the Annual Conference on South Asia, University of Wisconsin-Madison, Oct. 2019.

Chakravorty, Pallabi. "Review of *Performing Konarak, Performing Hirapur: Documenting the Odissi of Guru Surendranath Jena* and *Interpreting and (Re)Constructing Indonesian Dance and Music Heritage* by Alessandra Lopez y Royo." *Dance Research Journal* 40, n. 1 (2008): 92–95.

Chakravorty, Pallabi. "Why the Adivasi Will Not Dance: Yoga, Bharatanatyam, Chhau and Processes of Expropriation." In *Oxford Handbook of Indian Dance,* edited by A. Sengupta and P. Purkayastha. New York: Oxford University Press, forthcoming, 2024.

Chandralekha. "Probe Deeper Into Your Own Art: Chandralekha on the Duty of the Indian Classical Dancer." *Scroll.in* (Oct. 17, 2017). https://scroll.in/magazine/855678/probe-deeper-into-your-art-chandralekha-on-the-duty-of-the-indian-classical-dancer. Accessed June 18, 2020.

Coomaraswamy, Ananda. *The Dance of Siva.* New York: The Sunwise Turn, 1924.

De Michelis, Elizabeth. *A History of Modern Yoga: Patanjali and Western Esoterism.* London: Continuum, 2004.

D'Silva, Patrick. "Bodies in Translation: Esoteric Conceptions of the Muslim Body in Early-Modern South Asia." In *Transformational Embodiment in Asian Religions,* edited by George Patiband and Katherine C. Zubko, 168–86. New York: Routledge, 2020.

Ernst, Carl W. "Being Careful with Goddess: Yoginis in Persian and Arabic Texts." In *Performing Ecstasy: Poetics and Politics of Religion in India,* edited by Pallabi Chakravorty and Scott Kugle, 189–203. New Delhi: Manohar, 2009.

Fraleigh, Sondra. "Why Consciousness Matters." In *Moving Consciously: Somatic Transformations through Dance, Yoga, and Touch,* edited by Sondra Fraleigh, 3–23. Chicago: University of Illinois Press, 2015.

Goldberg, Michelle. *The Goddess Pose: The Audacious Life of Indira Devi, the Woman Who Helped Bring Yoga to the West.* New York: Vintage Books, 2016.

Haraway, Donna. *Simians, Cyborgs, and Women: The Reinvention of Nature.* New York: Routledge, 1991.

Hess, Linda, and Sukhdev Singh. *The Bijak of Kabir.* New Delhi: Motilal Banarsidass, 2015.

Kapstein, Matthew T. "An Inexhaustible Treasury of Verse: The Literary Legacy of the Mahasiddhas." In *Divine Madness,* edited by Rob Linrothe, 23–35. Chicago: Serindia 2005.

Lorenzen, David N. and Adrián Muñoz. "Introduction." In *Yogi Heroes and Poets: Histories and Legends of the Naths,* edited by David N. Lorenzen and Adrián Muñoz, xv–xviii. Albany: State University of New York Press, 2011.

Mallik, K. *Siddha-Siddhanta-Paddhati and Other Works of the Natha Yogis.* Pune: Poona Oriental Book House, 1954.

Mallinson, James. "Yoga: Haṭha Yoga." In *Brill's Encyclopedia of Hinduism Online,* edited by Knut A. Jacobsen, Helene Basu, Angelika Malinar, and Vasudha Narayanan, 770–81. https://referenceworks.brillonline.com/entries/brill-s-encyclopedia-of-hinduism/yoga-hatha-yoga-COM_000354?s.num=0&s.f.s2_parent=s.f.book.brill-s-encyclopedia-of-hinduism&s.q=mallinson. Accessed Feb. 21, 2022.

Meduri, Avanthi, ed. *Rukmini Devi Arundale 1904–1986: A Visionary Architect of Indian Culture and Performing Arts.* New Delhi: Motilal Banarasidass, 2005.

Mill, James, *The History of British India.* London: Baldwin, Cradock and Joy, 1817.

Nanda, Meera. "Madame Blavatsky's Children: Modern Hinduism's Encounter with Darwinism." In *Handbook of Religion and the Authority of Science,* edited by James R. Lewis and Olaf Hammer, 279–344. Leiden: Brill, 2010.

Narayan, Kirin. "Refractions of the Field at Home: American Representations of Hindu Holy Men in the 19th and 20th Centuries." *Cultural Anthropology* 8, n. 4 (1993): 476–509.

Olcott, Henry Steel. "Introduction." In *The Yoga Philosophy.* Edited by Tukaram Tatia, i–iv. Bombay: Theosophical Society, 1882.

Olcott, Henry Steel. *Old Diary Leaves, Fourth Series,* 184–92. Adyar, Madras: The Theosophical Publishing Houses, 1974.

Ondracka, Lubomír. "Perfected Body, Divine Body and Other Bodies in the Natha-Siddha Sanskrit Texts." *Journal of Hindu Studies* 8, n. 2 (Aug. 2015): 210–32.

Prakash, Gyan. "Science 'Gone Native' in Colonial India." *Representations* 40, Special Issue: Seeing Science (Autumn 1992): 152–78.

Royo, Alessandra L. "Guru Surendranath Jena: Subverting the Reconstituted Odissi Canon." In *Dance Matters,* edited by Pallabi Chakravorty and Nilanjana Gupta, 264–78. New Delhi: Routledge, 2010.

Sarkar, Sumit. *Writing Social History.* New Delhi: Oxford University, 1997.

Samuel, Geoffrey, and Jay Johnston. "General Introduction." In *Religion and the Subtle Body in Asia and West: Between Mind and Body,* v. 8, edited by Geoffrey Samuel and Jay Johnston, 33–47. New York: Routledge, 2013.

Stebbins, Genevieve. *The Delsarte System of Expression.* New York: E. S. Werner, 1885.

Stebbins, Genevieve. *Dynamic Breathing and Harmonic Gymnastics: A Complete System of Physical Aesthetic and Physical Culture.* New York: E. S. Werner, 1892.

Singleton, Mark. *Yoga Body: The Origins of Modern Postural Practice.* New York: Oxford University Press, 2010.

Vatsyayan, Kapila. *Classical Indian Dance in Literature and the Arts.* Sangeet Natak Akademy, 1968.

Vatsyayan, Kapila. "Two Keynote Addresses." In *Dance as Cultural Heritage,* edited by Betty True Jones, xvii–216. New York: Congress on Research in Dance, 1985.

Vivekananda, *Yoga Philosophy: Lectures Delivered in New York, Winter of 1895–6 by the Swami on Raja Yoga or Conquering the Internal Nature: Also Patanjali's Yoga Aphorisms, with Commentaries.* London: Longmans & Green, 1896.

White, Jessica A. "From Liberation to Self-Realization: The Evolution of the Subtle Body in Modern Educational Yoga." BA honors thesis, University of Queensland, 2016.

9

Viewing Indian Dance across Time and Space

Response to "The Paradox of the 'Subtle Body': Dance, Tantra, and Science" by Pallabi Chakravorty

Tiziana Leucci

Pre-nineteenth-century accounts of native yogis by Europeans reveal the suspicion and scorn characteristic of imperial attitudes toward "others." This essay reviews selected French sources that laid the groundwork for Western perceptions of yoga and yogis, which continued to evolve through the nineteenth century.

In eighteenth-century texts, yoga practitioners were widely perceived as fanatical, hypocritical, and lustful, begging for food while indulging in eccentric bodily penances and promiscuous sexuality. Abraham Hyacinthe Anquetil-Duperron (1731–1805) reported on ascetics of different regions and sects at holy pilgrimage sites, relating that most were poor, wandering almost naked with few belongings, although others carried fearful weapons and, at times, joined to form imposing armies.[1] French naturalist Pierre Sonnerat (1748–1814) wrote that the yogis considered themselves living saints, as did their devotees.[2] With their meticulous sense of purity, they declined the approach or touch of others, particularly Europeans, for fear of pollution. Though yogis' vows of poverty engendered respect among Europeans, their nudity, begging, and penances were deprecated.

Adventurer Joseph Alexis Pallebot de Saint-Lubin, who visited India between 1757 and 1779, wrote a richly documented source, owing to his knowledge of local languages.[3] Like his predecessors, he was fascinated by yogis' painful penances, which he ascribed to their efforts to expiate past faults, whereas others undertook these painful acts to achieve supernatural powers. Nevertheless, he condemned some extreme ordeals as the harmful and foolish results of religious fanaticism, also comparing yogis to contemporary Christian mystics like the *convulsionnaires.*[4] A report published by Abbé Jean-Antoine Dubois (1766–1848), translated into English in 1816, became influential among British colonial officers and Christian missionaries, though scholar Sylvia Murr demonstrated that Dubois plagiarized the work of Jesuit father Gaston-Laurent Cœurdoux.[5] Despite its messy history, the text confirms Europeans' views of Indian yogis as cunning and superstitious. It also refers to the perceived degradation of the yoga exercises he described, suggesting an ancient golden age of the discipline.[6] Despite their contempt, these authors drew attention to aspects of Indian devotion, leading eventually to a more serious study of yoga in the West as a complex philosophical and psycho-physical discipline, paving the way for new perceptions of yoga that animated speculation regarding the "subtle body" among scholars, Theosophists, artists, and dancers, as Pallabi Chakravorty's essay explains.

The French sources cited above circulated among European travelers and thinkers. These views coalesced into prejudices concerning local ascetic practices through the intersection of Western science and widespread interest in esoteric and occult doctrines which, in the late nineteenth century, made India the perceived birthplace of the highest forms of human spirituality. This new perception percolated through such disciplines as traditional Indian dance and music. The theory of the "subtle body," sublimated and purified, became associated mainly with the Hindu Sanskritic-Brahminical upper castes, marginalizing Hindu and Muslim sectarian and mystical traditions.

The theory of the "subtle body," emphasized by Indian and Western spiritualists and reformers, became a crucial element in virulent campaigns by anti-nautch activists against hereditary female performers and courtesans. This moralistic crusade effected the criminalization of artists and culture in South Asia. Consequently, the "sensual body" of the professional female temple, court, and salon artists—sexually active outside the highly controlled domestic sphere—were regarded as dangerous by many intellectuals. These performers' artistic and literary knowledge, their pride, independence, and self-reliance clashed with the only acceptable role for women—marriage, with economic and emotional dependence on men (father, husband, son). That imposed model was reinforced between the late nineteenth and early twentieth centuries, despite (and sometimes with support of) campaigns for women's emancipation. Scholars Kumkum Sangari and Sudesh Vaid pinpoint the contradictions and troubling evolution of patriarchy forced on Indian women in colonial and postcolonial times, which gave "a new lease of life to patriarchal practices under 'religious' sanction."[7]

Anthropologist Amrit Srinivasan paved the way for later research, including my work, on Indian women as performing artists and courtesans.[8] Teresa Hubel, for example, analyzes how early twentieth-century women's rights activists, such as physician Muthulaksmi Reddi, attacked the only category of women, the courtesans, who had relative freedom and independence in profession, education, sexuality, matrilineage, and control of their incomes and properties.[9] In November 1947, just three months after India's independence, the Indian state outlawed the courtesans' artistic and socioreligious traditions in the so called Devadasi Act.[10] This blow deprived those hereditary artists of their livelihood, threatening to erase their deeply rooted repertory of poems, songs, music, and dance.[11]

Fortunately, recent studies uncover the voices of female performing artists and courtesans, along with those of their masters, voices that historically were silenced. In interviews and during music and dance classes—which I had the privilege to attend in India in the 1980s and 1990s with the last exponents of the temple and court hereditary arts—some understanding of their neglected histories can be gleaned as they speak about their professional and personal lives, values, and struggles, offering a far different narrative from the mainstream.

Under imposed critical regimes, the perceived "sensual body" of the courtesan-performer—too independent sexually and economically to be tamed by patriarchal socioreligious systems—was supplanted by the newly formulated "spiritual" or "subtle body." In appropriating artistic knowledge from the courtesans' deplored body to the "respectable body" of the newly conceived dance and music, female practitioners of the Indian upper castes and urban elites modified, even purged, the traditional repertoire. The mythical origins of both dance and yoga, symbolized by the iconic image of *Śiva Naṭarāja* as Supreme Lord of dancers and yogis, were emphasized. In this sanitizing process, all sensual aspects of the dance, of yogic practice, and of Śiva, were erased.[12] Famed Theosophist and dance reformer Rukmini Devi Arundale expounded the "spiritual" view of dance: "The art very nearly died as it had become a means for remembering the body rather than of forgetting it. . . . After training the body, one forgets it. . . . This is why the dance is called a Yoga."[13]

The "subtle body," idealized and purified, became universalized—disconnected from the problematic bodies of "sensational" yogis and "immoral" female artists/courtesans. Thus anybody, irrespective of ethnicity or status, could learn Indian yoga and dance without stigmatization. These reformed disciplines became commodities, sources of profit for a new set of teachers. As Chakravorty highlights, interconnections among Western and

Indian scientific, medical, spiritual, and artistic spheres—claiming a commitment to modernity—sustained conservative ideologies in the West and in South Asian socioreligious, political, and artistic fields.

Like Chakravorty, some French scholars investigated the yogic "subtle body" historically, philologically, and anthropologically, revealing the complexity, depth, and wealth of those psycho-physical disciplines and practices.[14] Consequently, the yogis are no longer perceived as mere eccentric and fanatical ascetics, nor are the dancers seen as dangerous and "immoral," but rather as the receptacles of sophisticated philosophical and artistic doctrines.

Notes

1 Anquetil-Duperron (1771), quoted in Deleury, *Les Indes Florissantes,* 864–65.
2 Sonnerat, *Voyage aux Indes Orientales et à la Chine* (1782), 262–63, 266.
3 Deleury, *Les Indes Florissantes,* 7–8; D'Souza, "Le régime des terres," 163–64; Vaghi, *La France et l'Inde,* note 149.
4 Pallebot de Saint-Lubin (c. 1784), quoted in Deleury, *Les Indes Florissantes,* 733–34.
5 Murr, "Nicolas Jacques Desvaulx" and *L'Inde philosophique*; Dubois, *Hindu Manners, Customs and Ceremonies.*
6 Dubois, *Hindu Manners, Customs and Ceremonies,* 537.
7 Sangari and Vaid, "Recasting Women," 1–2.
8 See bibliography for sample works by Srinivasan and Leucci.
9 Hubel, "The High Cost of Dancing."
10 Jordan, *From Sacred Servant to Profane Prostitute.*
11 Hardy, "Ideology and Cultural Contexts," 138, 150.
12 See Doniger O'Flaherty, *Asceticism and Eroticism.*
13 Devi Arundale, "The Spiritual Background of Indian Dance," 194, 196.
14 Boivin, *The Hindu Sufis of South Asia*; Bouillier and Tarabout, *Images du corps dans le monde hindou.*

Bibliography

Boivin, Michel. *The Hindu Sufis of South Asia. Partition, Shrine Culture and the Sindhis of India.* London: I. B. Tauris, 2019.

Bouillier, Véronique, and Gilles Tarabout, eds. *Images du corps dans le monde hindou.* Paris: CNRS Éditions, 2002.

Deleury, Guy, ed. *Les Indes Florissantes. Anthologie des voyageurs français (1750–1820).* Paris: Robert Laffont, 1991.

Devi Arundale, Rukmini. "The Spiritual Background of Indian Dance." In *Bharatanatyam: A Reader,* edited by Davesh Soneji, 192–96. New Delhi: Oxford University Press, 2010.

Doniger O'Flaherty, Wendy. *Asceticism and Eroticism in the Mythology of Śiva.* Oxford: Oxford University Press, 1973.

D'Souza, Florence. "Le régime des terres dans les aidées du XVIIIe siècle." *Revue française d'histoire d'outre-mer,* v. 81, n. 303 (1994): 161–69.

Dubois, Jean-Antoine. *Hindu Manners, Customs and Ceremonies.* Trans. Henry K. Beauchamp. New Delhi: Oxford University Press, 1983.

Hardy, Friedhelm. "Ideology and Cultural Contexts of the Srīvaisnava Temple." In *South Indian Temples: An Analytical Reconsideration,* edited by Burton Stein, 119–51. New Delhi: Vikas, 1978.

Hubel, Teresa. "The High Cost of Dancing. When the Indian Women's Movement Went after the Devadasis." In *Bharatanatyam: A Reader,* edited by Davesh Soneji, 160–81. New Delhi: Oxford University Press, 2010.

Jordan, Kay K. *From Sacred Servant to Profane Prostitute. A History of the Changing Legal Status of the Devadāsīs in India, 1857–1947.* New Delhi: Manohar, 2003.

Leucci, Tiziana. "Au Royaume de l'Amour les Courtisanes sont Reines. *Devadāsī*s et *Rājadāsī*s comme *Apsaras* terrestres et incarnations de la Fortune (*Śrī-Lak⊠mī*)." In *Rājamaṇḍala. Le Modèle Royale en Inde,* edited by Emmanuel Francis and Raphaël Rousseleau, 189–221. Paris: EHESS 2020.

Leucci, Tiziana. "La danse en Inde du sud, entre conflits générationnels, identitaires, de genre et de caste." In *Generational Frictions in Musical Ethnography of South Asia,* edited by Kaley Mason and Margaret E. Walker, *MUSICultures,* v. 44 n. 1 (2017): 134–62.

Leucci, Tiziana. "Royal and Local Patronage of Bhakta Cult: The Case of Temple and Court Dancers." In *The Archaeology of Bhakti II: Royal Bhakti, Local Bhakti,* edited by Emmanuel Francis and Charlotte Schmid, 257–302. Pondichéry: IFP/EFEO, 2016.

Leucci, Tiziana. "South Indian Temple Dancers: 'Donated' to the Deity and 'Donors' for the Deity. Two Tamil Inscriptions on Music and Dance in the Rājarājeśvaram Temple at Tanjāvūr (11th Century)." In *New Dimensions in Tamil Epigraphy: Historical Sources and Multidisciplinary Approaches,* edited by Appasamy Murugaiyan, 209–53. Chennai: CRE-A Pub., 2012.

Murr, Sylvia. *L'Inde philosophique entre Bossuet et Voltaire, vol. I: Mœurs et coutumes des Indiens (1777), un inédit du Père G.-L. Cœurdoux S.J. dans la version de N.-J,* edited by Sylvia Murr. Paris: École Française d'Extrême-Orient, 1987.

Murr, Sylvia. "Nicolas Jacques Desvaulx (1745–1823) véritable auteur de *'Mœurs, Institutions et Cérémonies des peuples de l'Inde'* de l'Abbé Dubois." *Puru⊠ārtha* v. 3 (1977): 245–58.

Sangari, Kumkum, and Sudesh Vaid. "Recasting Women: An Introduction." In *Recasting Women. Essays in Colonial History,* edited by Kumkum Sangari and Sudesh Vaid, 1–26. New Delhi: Kali for Women, 1989.

Sonnerat, Pierre. *Voyage aux Indes Orientales et à la Chine, fait par ordre du Roi, depuis 1774 jusqu'en 1781 . . . par M. Sonnerat,* tome 1. Paris: Froulé Libraire, 1782.

Srinivasan, Amrit. "The Hindu Temple-Dancer: Prostitute or Nun?" *Cambridge Anthropology* v. 8, n. 1, (1983): 73–99.

Srinivasan, Amrit. "Reform and Revival: The *Devadāsī* and Her Dance." *Economic and Political Weekly* Mumbai, v. 20, n. 44 (Nov. 2, 1985): 1869–1876.

Srinivasan, Amrit. "Reform or Conformity? Temple 'Prostitution' and the Community in the Madras Presidency." In *Structure of Patriarchy: State, Community and Household in Modernizing Asia,* edited by Bina Agarwal, 175–98. New Delhi: Kali for Women, 1988.

Vaghi, Massimiliano (2016). *La France et l'Inde: Commerces et politique imperial au XVIII siècle.* Sesto San Giovanni: éditions Mimésis, 2016.

10

Exhibiting (Scientific) Grace

American Delsartism and Black Citizenship in the New South

Carrie Streeter

> All art is founded upon nature. A student on entering one of our elocution classes is first led to study nature as manifested in his own body.
>
> —Adrienne McNeil Herndon, "Our Work in Elocution," *The Bulletin of Atlanta University,* 1897

> [The American Negro] simply wishes to make it possible for a man to be both a Negro and an American without being cursed and spit upon by his fellows, without losing the opportunity of self-development.
>
> —W. E. B. Du Bois, "Strivings of the Negro People," *The Atlantic,* 1897

When surveying the nineteenth century for choreographies that engaged scientific discourse, no discussion would be complete without considering the fin de siècle creation of American Delsartism. Recognized by historians as the cultural origins of modern dance, the movement's very name leveraged the ascendant authority of psychological studies of emotion, including those of the French voice and acting teacher for whom the movement was named, François Delsarte (1811–1871). Though he never published his theories nor visited the United States, American elocutionists and actors promoted Delsarte's mapping of body language as a scientific framework for exercises that claimed to "free the channels of expression."[1] Reform-minded women proved uniquely attuned to the liberatory potential of Delsarte's theories and the embodied practices they could legitimize. Indeed, their rejection of restrictive expressive norms proved an essential force in the rise of American Delsartism. By authoring numerous Delsarte-inspired manuals and teaching countless classes in schools, parlors, and community centers, US women not only made the Frenchman's name a household word, but they also demonstrated the cultural value of linking scientific rhetoric with social reform. This was especially apparent through the proliferation of the Grecian-styled posings, pantomimes, and dances that they popularized under the Delsarte moniker.[2]

In the past forty years, scholars have established myriad ways in which the aesthetic practices of American Delsartism promoted social acceptance of the "New Woman," a feminist ideal that championed independence and individuality.[3] More recently, greater attention has been given to evaluating the movement's influence on the intertwined cultural production of gender and race.[4] Because Delsartism proposed as universally true a set of principles for identifying body language, it is particularly relevant to examine how practitioners engaged visual taxonomies that upheld scientific racism—taxonomies that (among other things) had long ascribed to white women an exclusive capacity for exhibiting qualities like poise and gracefulness.[5] As current scholarship suggests, the cultural authority of American Delsartism offered tools for both strengthening and subverting these hierarchical aesthetics. Such outcomes are well illuminated in Colleen Kim Daniher's recent work. As she demonstrates, Delsartism's "rhetoric of bodily legibility" reinforced the biopolitical power of whiteness while also providing "racialized and gendered subjects" with tools for contesting their marginalized status.[6]

This essay examines the biopolitics of American Delsartism within the context of postbellum Atlanta, Georgia, a symbolic and economic center for the development of the New South.[7] By chronicling the arrival and promotion of Delsartism in the city's educational spaces, it demonstrates that Black and white displays of "scientific grace" constructed radically different visions of selfhood and citizenship—outcomes especially evident through public responses to Delsartean performances at Spelman Seminary (later Spelman College) and Atlanta University. While proponents of Black citizenship largely praised and defended Black students' aesthetic endeavors, antagonists mocked such demonstrations of self-confidence, campaigned for the elimination of Black student's access to expressive studies, and racialized Delsartean semiotics in ways that reinscribed assumptions about Black inferiority. Far from inconsequential, such actions contributed to the fabrication of cultural norms and "facts" justifying violence against Black citizens and legitimizing the rise of racial segregation.

As the records of Spelman Seminary and Atlanta University attest, these assaults did not vanquish the schools' promotion of grace-affirming pedagogy, but they did have a profound influence on how Black expressivity was seen. While educators in the city's Black schools regularly staged eloquent recitations and Grecian-styled dances, they also navigated widespread hostility to Black talent and achievement. To deflect threats and violence, educators often found it necessary to keep Black students' graceful exhibitions out of public

Figure 10.1. A demonstration by Spelman Seminary's students of elocution and physical culture, c. 1890s. According to Delsartean principles, this gesture was considered a scientific expression of poise and grace. Courtesy of the Spelman College Archives.

view—a dynamic that scholars Imani Perry defines as "Black formalism" and Jarvis Givens terms "fugitive pedagogy." That is, educators remained committed to quotidian practices that dignified Black Americans' "expression[s] of grace and identity"[8] at the same time that they "concealed and held close" the "physical and intellectual acts" that "explicitly critiqued and negated white supremacy."[9]

In addition to exploring how educators in Atlanta's Black colleges engaged American Delsartism, this essay's focus on the aesthetic politics of racial segregation also helps explain why Black exhibitions of gracefulness were excluded from mainstream records, the very sources that historians have long relied upon for assessing the movement's demographic reach. Placing Black Americans' Delsartean activities center stage not only changes this picture; it also encourages greater understanding about the ways in which these performances of "scientific" grace embodied one of the nineteenth century's most consequential projects: the reconstruction of the nation as a multiracial democracy.

Social Equality on Display: American Delsartism at Spelman Seminary

In the spring of 1893, the graduates of Spelman Seminary's normal school gave a newsworthy performance. As reported by the *Spelman Messenger,* these teachers-in-training elicited "high words of praise" after delivering "well thought out" speeches and donning "soft flowing robes of pink and white," "[keeping] time to the accompanying music," and rhythmically moving through a "physical culture drill."[10] In many ways, this display was nothing out of the ordinary, for Spelman made such activities a standard part of campus life. But in a southern state that had only recently legalized Black literacy, this emboldening of Black expression was also quite revolutionary—a fact that one student, Fannie Showers (1868–1977), emphasized in her graduation speech. In addition to helping Black women feel like "pictures of health" and "modern belles," Showers positioned the "study of Delsarte" as a great social equalizer; in her assessment, it placed the cultivation of grace not only in the hands of white social elites, but "within the reach of all."[11]

Fannie Showers's enthusiasm for "the study of Delsarte" occurred during an era of dramatic transformation for Black citizenship. In 1860, for example, Georgia counted the second largest number of enslaved individuals in any state. Soon after the Civil War ended in 1865, the state's capital boasted more Black colleges than any city in the nation. Largely funded by northern philanthropists and missionary societies, these institutions were especially instrumental in developing public schooling throughout the South.[12] By the 1880s, for example, most of Atlanta's Black youth were taught by graduates of the city's Black colleges, as were thousands of children across the region. This was just one reason why these schools could rightly characterize their alumni as "moulders of public sentiment."[13]

These new opportunities for civic leadership were also among the reasons why the city's Black colleges emphasized elocution, a study that imparted practical exercises for sustaining vocal health and cultivating confident self-expression.[14] The benefits of elocution training were especially evident at student recitals and commencement ceremonies. In addition to providing celebratory exchanges among family, friends, and community members, these events also generated positive press in Black and white publications. Throughout the 1880s, Black eloquence was a staple feature in college bulletins, like the *Spelman Messenger,* and in local newspapers such as the *Atlanta Constitution,* a prominent white paper. These media notices broadcast Black self-assurance in unprecedented ways.[15] On May 26, 1888, for example, readers of the *Atlanta Constitution* were informed that the speeches of Spelman students were nothing less than a "revelation to those who consider[ed] the colored race incapable of thinking deeply and accurately."[16] Notable for their praise of Black voices, such compliments also underscored the ways in which student performances were entangled with racialized ideas about aptitude.

As individuals tasked with orchestrating these reputation-forming events, elocution teachers in Atlanta's Black colleges played a significant role in shaping public perceptions about Black Americans' capabilities. During the 1880s, many of these teachers were northern white women whose missionary interest in Black education compelled them to move south. This was true of Mae B. Peckham (1860–1931), Spelman Seminary's teacher of elocution and physical culture from 1885 until 1910. The daughter of a Civil War veteran who supported his family by selling furniture, Peckham's middle-class upbringing in Boston, Massachusetts, provided access to an array of educational and entertainment experiences that deliberately affirmed women's health and expressive powers.[17]

These reform-minded interests were evident in Peckham's elocution classes at Spelman. During her twenty-five-year tenure, Peckham's students performed many of the aesthetic exercises that elocutionists across the United States strategically promoted in their efforts to broaden the range of acceptable "feminine" expressions. Such practices largely confronted understandings of "true womanhood," a gender ideal that (among other things) purported timidity and demureness as immutable features of women's nature. As scholars of rhetoric have established, this ideal purposefully constrained women's opportunities for public speaking and upheld a narrow range of acceptable feminine expressions.[18] Believing that facts about physiology and emotion could animate new conversations about women's "nature," postbellum elocution reformers increasingly turned to science as an aid for delegitimizing cultural assumptions about women's expressive abilities.[19]

In 1885, the year that Mae Peckham arrived at Spelman, the quest for scientific principles of expression gained a notable boost when Genevieve Stebbins (1857–1934), a New York-based actor and elocution teacher, published *Delsarte System of Expression*. Filled with theories and charts that revealed a "science of semiotics," Stebbins's book was the first manual that both explained François Delsarte's Laws of Expression and provided instructions about how to teach them.[20] With support of a burgeoning number of female educators and influential new women's clubs, this book quickly became a standard text in schools and parlors across the country.[21]

Especially effective was Stebbins's claim that Delsarte's theories were consistent with the aesthetic semiotics of ancient Greek statuary. Indeed, by 1890, the term "Delsarte" had become synonymous with an array of Grecian-styled exercises that claimed to cultivate grace, poise, and power. Establishing a scientific basis for statue-posing exercises also emboldened more cultural and choreographic risks. In many "Delsarte" classes, instructors led women through sequences that slowly transitioned from one pose to the next, often in rhythm with music. Though Stebbins and others began calling such movements "dance," that term remained heavily freighted with concerns about women's sexual impropriety or moral purity. To skirt such objections, women often referred to these interpretive dances as "drills" or "pantomimes."[22] Citing such practices as "Delsarte" became a popular way of claiming scientific soundness for these norm-defying activities.

While most of the publicity about these Delsarte-inspired activities celebrated white women's claims to expressive freedom, the movement's popularization of universal semiotics also offered tools for dispelling racialized assumptions about Black women's capacity for gracefulness.[23] Spelman's elocution recitals exemplified this type of contestatory work, and the school's newsletter did not miss the beat.[24] "Nothing could be finer," the *Spelman Messenger* proclaimed in 1890 after students donned "classic flowing robes of the Greeks," and staged a "Gesture Drill." In addition to appreciating the "perfection" of students' embodied emotions, the paper also praised the "charm" and "poetry of their movements."[25] In June 1898, the *Spelman Messenger* described a "very pretty entertainment," in which the "leading feature was the Scarf Drill," wherein students "went through various movements to the time of soft music and concluded with a succession of poses which elicited much applause."[26] In the spring of 1900, similar praise was afforded to performers of a "Ribbon Drill and Tableaux," a piece that featured rhythmic movement to celebratory songs about the prismatic colors of the rainbow.[27] The following year, students received accolades for embodying the spirit of Dionysius, a Grecian god of sensuous playfulness, and Niobe, a grieving mother of mythic proportion.[28]

Spelman's reporting about students' Grecian-styled performances also illuminates Peckham's behind-the-scenes instruction. In typical "Delsarte" fashion, her classes would have emphasized how this "poetic motion" demonstrated universal scientific principles. Two core theories were deemed especially applicable. Delsarte's Law of Correspondence established the immutable connections of mind, body, and soul. It stated, "To each spiritual function responds a function of the body. To each grand function of the body corresponds a spiritual act."[29] When applied to Grecian statuary, this law lent credibility to the claim that one could lighten one's spirits by posing like *Flying Mercury*, cultivate defiance by posing like *Fighting Gladiator*, and activate courage by posing like *Winged Victory*. Proponents also believed such exercises embodied Delsarte's Law of Control: "Strength at the center and freedom at the surface." In the words of the preeminent Black elocutionist Hallie Q. Brown, this Delsartean maxim revealed "the secret of grace in all bodily action."[30]

In addition to exploring these aesthetic theories in their own bodies, Spelman students also applauded Delsartean displays by prominent Black performers. In 1889, for example, Peckham hosted Frances and Lillie Preston, a mother-daughter team who lectured for the Woman's Christian Temperance Union (WCTU) and were lauded as talented "colored pantomimists."[31] Shortly before performing in Atlanta, Lillie Preston had reportedly dazzled Ohio audiences with pantomimes titled "Aesthetic Gestures" and "Lyre Movement." As one newspaper put it, "Her gestures expressing profound grief, anguish, supplication and remorse . . . were so natural as almost to cause a person to feel as if he were witnessing a dire disaster or calamity."[32] Her emotive display also made quite an impression on Fannie Showers and her classmates, for according to the *Spelman Messenger,* Preston's performance rendered the students "spell-bound."[33]

Spelman educators and students were not the only Georgians who admired these aesthetic exercises. In the early 1880s, white public schools and private seminaries had also begun delivering Grecian-styled Delsartean performances, a development that one educator cited as proof that Atlanta was "keep[ing] pace with the progress of the age."[34] Prominent New South boosters agreed, including Henry Grady, editor of the *Atlanta Constitution.* In the late 1880s, he ensured that white schoolteachers and society ladies could take Delsarte classes at his new Piedmont Chautauqua, an educational resort located a short twenty-minute train ride from the city. Modeled after the Chautauqua Institute in upstate New York, this facility represented a deliberate effort to demonstrate that white southerners were culturally on par with their northern peers.[35] When promoting Delsarte classes, Grady's influential newspaper defined their appeal clearly: "to teach grace and eloquence, properties very essential."[36]

These examples illustrate important points about Atlanta's 1880s Delsarte debuts. First, they establish that both Black and white educators were early and enthusiastic implementers of Delsarte-inspired pedagogy. Second, they demonstrate that civic leaders validated these activities as admirable exhibitions of feminine expressivity. Readers of the *Atlanta Constitution* could see this plainly, for its columns regularly acknowledged the graceful exhibitions of both Black and white students. For example, in May 1891, the paper noted that Spelman had staged another well-attended "Gesture drill in costume."[37]

Shortly thereafter, the paper praised a similarly styled performance at the Piedmont Chautauqua, where twenty-four white ladies in Grecian robes "expressed every emotion of the human soul."[38]

Though Black and white citizens shared admiration for these aesthetics, the cultural capital derived from these displays was far from equally distributed. For those who affirmed Black citizenship, this elevated visibility of Black poise signified an inspiring validation of Black selfhood. For those who believed Black citizenship was an affront to civilization, these performances conveyed something else entirely. In the 1880s, Henry Grady, editor of the *Atlanta Constitution*, bluntly articulated such thinking when he declared: "The white race is the superior race." Believing this "natural order" was under siege, he issued a rallying cry: "The domination of the negro race [must] be resisted at all points and at all hazards."[39]

Among other things, this effort to racialize the struggle for social and economic power targeted educational curricula that affirmed and strengthened Black selfhood. Thus, soon after Atlanta's Black and white schools began implementing "Delsarte," the city's newspapers also began racializing Delsartean theories and aesthetics in ways that affirmed entrenched beliefs about Black inferiority. For example, under the heading "Delsarte Index to Character," one story reinscribed the racial claims of earlier nineteenth-century physiognomic hierarchies. In this new "index to character" undesirable qualities like "viciousness" were tied to features associated with Blackness (for example, "heavy, sensuous lips") and qualities like "intellectual superiority" were tied to features associated with whiteness (for example, "full, high forehead").[40] The same newspaper also warned locals about the unsettling consequences of Delsartism's spread. In one humorous quip, a man reportedly pined for the "peace and comfort" of a simpler time when "poor people did not study Delsarte."[41]

In addition to distorting supposedly universal ideas about beauty and capability, such reactions demonstrated that racialized anguish about social equality shaped southerners' responses to Black and white Delsartean performances. This undermining of Black achievement also underscored why educators at Black colleges prioritized curricular activities that placed self-assurance "within the reach of all," for as long as assumptions of Black inferiority prevailed, performances of "well delivered" recitations and graceful poses would remain "revelation[s]" for "those who consider[ed] the colored race incapable of thinking deeply and accurately."[42] During the 1880s, these displays had helped convey Black Americans' fitness for citizenship. In the 1890s, as the nation moved toward codifying racial segregation, such "revelations" not only remained contestatory, but they also became increasingly fraught.

Dances, "Original and Graceful": American Delsartism at Atlanta University

The year 1895 was momentous for anyone living in Atlanta and teachers of elocution at the city's Black colleges were no exception. As the school year began, they helped students prepare for an opportunity to share their voices at the Cotton States and International Exposition, a fair that championed the recent industrialization of the New South. The year was also significant as the thirtieth anniversary of slavery's constitutional abolition. Though white fair organizers were uninterested in commemorating this achievement, Black leaders insisted on celebrating the fruits of emancipation. To counter the fair's preponderance of derogatory, plantation-themed sideshows, they successfully lobbied for a Negro Building for exhibiting evidence of Black Americans' progress.[43] In this space, educational achievements took center stage. That fall of 1895, when students spoke out their recitations, their well-rehearsed voices filled rooms that also displayed students' finely crafted wooden cabinets and elegantly styled hats.[44]

At Spelman Seminary, Peckham had prepared her students using the Delsarte-inspired methods that had won over local audiences in recent years. A new elocution teacher led this work at Atlanta University, and the story of her arrival to this position was itself a profound demonstration of the uplifting power of Black eloquence.

Raised in Savannah, Georgia, Adrienne McNeil Herndon (1869–1910) was the daughter of parents born enslaved. From such circumstances, she would live a life of many firsts. She was the first in her family to attend public school, where her vocal talents drew support from her teacher and helped secure a scholarship to Atlanta University. In 1891, she became the first in her family to graduate from college and teach in public schools. Two years later, her marriage to Alonzo Herndon (1858–1927), an enterprising barber, led to more fortuitous firsts. His support afforded her an opportunity to complete specialized expression training in Boston and New York. By the early 1900s, his successful businesses (which included investments in real estate and insurance) also distinguished her as the wife of Atlanta's first Black millionaire. While her husband was making those fortunes, Adrienne Herndon broke more records of her own. In 1895, when she began teaching elocution at her alma mater, she also became Atlanta University's first Black female faculty member.[45]

Within months of accepting this new position, Herndon helped raise awareness about declining commitments to Black citizenship when she recited "Exposition Ode" at the opening of the Negro Building. Though another educator had authored the words, the audience recognized Herndon's talents at embodying them. "The oration indicated that there had been much thought expended upon it," one reporter noted, adding that Herndon displayed "merit" and a "thorough appreciation and understanding of the situation."[46] "*The situation,*" as defined by the poem, referenced the nation's failure to uphold its recent legal commitments to equal protection under the law. "Exposition Ode" also subtly rebuked what would become known as Booker T. Washington's "Atlanta Compromise." Just a few weeks before Herndon's recitation, Washington's exposition speech endorsed racial segregation as an acceptable way of securing economic and political support from white Americans. In a mantra-like appeal to "cast down their buckets where they were," he encouraged Black Americans to forgo broader intellectual, artistic, and political pursuits, and to focus instead on proving their fitness for citizenship through industrial labor and conservative living.[47]

Adrienne Herndon's recitation assessed "*the situation*" rather differently. Decrying the nation's violation of Black Americans' constitutional rights, "Exposition Ode" employed the same phrase, "let down your buckets where they are," not as an appeal for labor and thrift; but as an acknowledgment of Black Americans' innate power for protesting "disenfranchisement, injustice, and prejudices."[48] As if on cue, a few weeks after reciting these words, Herndon gathered with other Black women to organize their collective power. The following year, they founded the first national alliance of Black women's clubs.[49] Recognizing the historic nature of their 1895 Atlanta meeting, one Black journalist thanked these "New Negro Women" for "lifting up" Black Americans. Praising their "intelligence, beauty, and virtue," and their ability "to speak for themselves," the journalist also endorsed Black women's expertise with a telling phrase: "She is in evidence on matters concerning her people."[50]

Through her involvement with women's clubs and her leadership at Atlanta University, Adrienne Herndon made it her life's work to affirm, strengthen, and broadcast the evidence of Black Americans' innate dignity. From 1906 until her death in 1910, she accomplished this through student productions of Shakespeare, work that scholar Patricia A. Cahill aptly characterizes as an "insurrectionary politics of embodiment."[51] As the following discussion demonstrates, Herndon also emboldened Black expressivity through pantomime and interpretive dance, a curriculum that was especially revolutionary given the times in which she taught. After the US Supreme Court's 1896 upholding of racial segregation, white philanthropists and state governments almost

Figure 10.2. May Day Dance at Atlanta University, c. 1905. Orchestrated by Adrienne Herndon, this performance was one of the university's many opportunities for celebrating students' gracefulness. Courtesy of Atlanta University Photographs Collection, Atlanta University Center Robert W. Woodruff Library.

exclusively backed schools that promoted Booker T. Washington's model of industrial and domestic training. Schools like Atlanta University that refused to narrow their curricular priorities found it increasingly difficult to secure funding or acclamation. In this context, Adrienne Herndon's classes became more than just spaces where she and her students explored elocution reforms; they were also a means for defying the broad cultural negation of Black selfhood and civil rights.

By unequivocally affirming student's individuality and their "nobleness and gracefulness" (her words), Herndon's pedagogical choices certainly met the moment.[52] To enliven these qualities, her classes prioritized tension-releasing breathing exercises, strengthening gymnastics, and pantomime explorations—practices that the rise of American Delsartism had helped normalize in schools across the country. These exercises were also the preferred methods of Herndon's teacher in Boston, Samuel S. Curry (1847–1921). In his estimation, pantomime was especially effective at synthesizing the distinctive powers of art and science. As he put it, this technique cultivated "imagination and feeling," while also requiring attentive "observation and development of reason."[53] Herndon concurred with this assessment. When describing her methods in the university's bulletin, she wrote that pantomime encouraged students to investigate "nature as manifest in [their] own bod[ies]," and to learn how to "imaginative[ly] grasp [any] situation." In an elegant summary of this practice, Herndon stated, "The student is taught to first see, then feel, and then tell."[54]

Like other elocution teachers of this era, Herndon's instruction of pantomime also facilitated experimentation with interpretive dance.[55] Though published records about these activities are spare, they are nevertheless illuminating. In 1902, for example, the university's bulletin noted that her students had performed such a "remarkably charming" pantomime that the "audience could not be satisfied without its repetition."[56] The following year, these dances brought "the friends of the graduates" out "in full force." In 1903, an audience was reportedly "carried away" by the students' "evening drill," and their "unique costumes and gestures" (likely Grecian in style).[57] Herndon's students also welcomed spring by personifying the spirit of the season and dancing around a maypole. In 1900, this dance was so popular that "by special request" several students gave an encore performance after supper at a campus lawn party.[58] A few years later, a visiting scholar was similarly awed by the students' aesthetic displays; as that year's commencement speaker, the esteemed sociologist (and radical abolitionist) Frank Sanborn (1831–1917) had the opportunity to watch Herndon conduct rehearsals.[59] Recalling this experience years later in an editorial for a Boston newspaper, Sanborn described their "dances, recitations and graduation exercises" as "original and graceful."[60]

As authors and readers of such reports knew, these acknowledgments of choreographic innovation performed important political work. For example, Sanborn recognized that Herndon's "original and graceful" dances directly refuted the aesthetic logics of social Darwinism, a scientific theory that assigned to Black Americans biological and hereditary inferiority.[61] Such claims were nourished by a steady diet of predictable imagery, stereotypes that art historian Francis Martin describes as depictions of "graceless Black Americans attempting to do what white Americans did with ease."[62] As was true during the antebellum rise of Jim Crow entertainments in the urban north, the postbellum proliferation of coon songs and caricatures traded on semiotics of awkwardness as a means of racializing socioeconomic hierarchies.[63] Indeed, in both eras, promoting racist aesthetics proved an effective tactic for justifying the disenfranchisement of Black Americans in order to extend symbolic privileges of whiteness to disadvantaged European immigrants.[64]

Depictions of American women in *Leslie's Weekly* offer a case in point. In 1897, this popular New York-based magazine regularly circulated images of middle- and upper-class white women enjoying a range of activities that personified New Womanhood. They stood in Grecian robes in "statuesque poise."[65] They rode bicycles. They dined in urban restaurants. They walked easefully on city streets. In Delsartean terms, such women epitomized the Law of Control: "strength at the center, freedom at the surface." *Leslie's Weekly* also sympathetically illustrated the plight and strivings of European immigrants. For example, one cover depicted a tenement girls' side-walk skirt dance as evidence of social progress. Remarking on the completion of this "terpsichorean feat," the magazine authenticated (albeit condescendingly) the girls' genuine exhibition of "grace" as a pleasant surprise.[66]

Such praise was noticeably absent in *Leslie's Weekly* depictions of Black women. Instead, the magazine ran mocking portrayals of their quotidian activities. For example, in 1897, several covers featured illustrations of an imaginary community named "Possumville," a place where Black citizens awkwardly attempted "American" activities like bicycle-riding and golfing. To mock Black women's civic acumen, the artist engaged the iconography of American Delsartism. Under the heading "The First Parade of the New Woman's Society in Possumville," a girl identified as "Venus" donned a tattered Grecian robe and led a disheveled assembly of illiterate suffragists.[67] Tellingly referencing postbellum and antebellum fashions associated with women's liberation, the caption read, "I isn't got no bloomer, but I'se pow'ful close to it."[68] Another illustration by the same artist continued this theme. In a book marketed for Christmas gift-giving, he featured a Black girl inelegantly performing a "skirt dance," a choreography described in newspapers, including the *Atlanta Constitution,* as "Delsarte posing" and a woman's "natural expression" of beauty.[69] Yet as the artist's caption made clear, there was nothing to admire in this Black girl's "natural" whirl. For gracefulness, it proclaimed, "ain' in it wif dis chile."[70]

With these racial semiotics ever present in national media and academic disciplines, affirming and exhibiting Black poise carried profound scientific and cultural significance. Indeed, this was one reason why W. E. B. Du Bois (1868–1963), esteemed sociologist and Adrienne Herndon's colleague at Atlanta University, also recognized the power of Black students' "original and graceful" dances. After joining the faculty in 1897 and taking charge of Atlanta University's new sociological laboratory, Du Bois organized several

Figure 10.3. Photographs of Atlanta University students featured in the "Exhibit of American Negroes" at the 1900 World's Fair. W. E. B Du Bois also included these images in his 1906 study *The Health and Physique of the Negro American,* where he described the student on the left as "of Amazonian build and carriage," and the student on the right as "tall and graceful," qualities that were notably affirmed and emboldened in Adrienne Herndon's elocution classes. Library of Congress LC-USZ62-124725 and LC-USZ62-124736. Published with permission.

public-facing projects aimed at dispelling scientific racism through honest representations of Black individuality and accurate data about Black life. The school's curricular commitments to Black expressivity proved vital to such work.

In 1900, for example, several of Herndon's students contributed photographs of themselves to an exhibition Du Bois curated for the World's Fair in Paris, France. There, on this global stage, their poised expressions irrefutably defied what Du Bois termed "conventional American ideas" about race.[71] Leveraging the utility of this visual evidence, Du Bois also included many of the student's photographs in a 1906 study about "Negro" health. In this publication, he added his expert assessments of the students' personalities. These "typical Negroes," he determined, were "graceful," "vivacious," "quiet," "serious," "kind," and "reliable." In a phrase that resonated with Delsartean references, Du Bois described one young woman as "Amazonian in build." Another student, he stated, was "full of fun."[72] Such descriptors largely conveyed what Du Bois and Herndon knew that any unprejudiced reading of contemporary science could confirm: proof of dignity and personhood.

Yet as both scholars also recognized, the thrust of America's cultural and political institutions were far more committed to upholding rather than dispelling scientific racism.[73] This was evident in *Werner's Magazine,* the professional journal for elocutionists. Though Black educators constituted a sizable demographic of subscribers, the publication by and large marginalized their accomplishments and interests. For example, in a feature story on southern teachers, *Werner's* briefly acknowledged Adrienne Herndon as Atlanta University's instructor, but not a single image of her nor of any other Black teacher was included among the article's numerous photographs.[74]

Werner's promotion of segregationist politics was especially evident in its February 1901 edition. Under the title "Expressional Power of the Colored Race," the journal filled several pages with photographs and reports of talented Black elocutionists, singers, and actors. It also coupled this praise with notably racist claims about Black lives and history. "Other peoples have risen from slavery by force of inherent qualities. Not so the negro," the article declared. By defining slavery as "a school for childlike virtues," and a time when Black people enjoyed carefree safety from "sword, pestilence, and famine," the story legitimized the continuation of social hierarchies premised on white paternalism and Black disenfranchisement.[75]

A few pages away from these assertions about slavery and emancipation, *Werner's* also printed another race-baiting article. Under the title "Unconscious Delsartism in Colored Women," a "southern physician" claimed that Black women were "naturally" poised when they carried buckets of water, beat clothes, and kneaded bread. His medical expertise also deemed Black women's expressions of restful postures (for example, hands on hips) as "lazy" demonstrations of "vulgar ease," adding that in such positions Black women appeared "sinuous and willowy [like] a panther in its native jungle."[76] Clearly trading on malicious stereotypes about Black women's sexuality, this racialization of Delsartean theories also implied that Black women were only capable of exhibiting "scientific" gracefulness when they performed domestic labor. These views also predominated in many of the "Negro dialect" entertainments advertised in *Werner's Magazine* and subsequently performed in schools and parlors across the United States. Indeed, during the 1890s and early 1900s, it became commonplace for white women and school children to perform "Delsarte" poses alongside pieces like "Pickaninny's Lullaby," and "Everyone Has a Country but the Coons."[77]

This national racialization of Delsartean aesthetics also helped Atlanta's white clubwomen expand their social power. During the early 1890s, their staging of Delsarte-styled plantation dramas facilitated strategic collaborations with esteemed northern teachers. In 1893, for example, the "society queens of Peachtree Street" hosted a prominent Philadelphia-based teacher (and colleague of Genevieve Stebbins) for a Grecian-styled "Delsarte matinee" at Atlanta's whites-only YMCA. According to the newspaper, the "most luminous and tender" part of the program occurred when she pantomimed a Black mammy rocking a child to sleep. Reportedly, her personification of a "real dark" was performed so "correctly" that the women longed for the restful summer days of their youth, when their own "Black mammy" sang to them "through [their] dreams."[78]

In addition to affirming the moral authority of white womanhood, performances such as this 1893 "Delsarte matinee" also signaled a consequential shift among reform-minded women's organizations. During the 1890s, influential northern clubs like the Woman's Christian Temperance Union largely abandoned previous commitments to multiracial membership as they welcomed southern white women into their organizations.[79] Since the late 1870s, these northern clubs had comprised a substantial clientele for Delsarte classes, and as their white members accepted or accommodated racial segregation, many teachers of expression followed suit. By the late 1890s, for example, Genevieve Stebbins, arguably the most influential American Delsartist, began teaching summer classes for segregationist white educators. In the early 1900s, she also performed at least one recital for the United Daughters of the Confederacy.[80]

These cultural alliances between white women's clubs emboldened strident assaults on Black citizenship. In 1899, for example, a prominent white women's club in Atlanta appealed to Boston philanthropists to fund their efforts to build and direct Black kindergartens. Under the guise of progressive benevolence, the clubwomen aimed to undermine the cultural and institutional power of Black teachers. "Let the teachers of the negroes throw the isms and ologies to the winds," the club's leader exclaimed. "To teach the negro children social equality would be ruinous." To further emphasize this point, she added, "To educate [Black students] in the classics . . . unfits them for work." Indeed, the clubwomen blamed the Black and white educators of the city's Black colleges for the "thousands of negro men and women," who were "roving the streets, looking impudent, and avoiding work."[81]

When these vicious words hit the press, Adrienne Herndon was a new mother and one of the teachers whose educational commitments the white clubwomen derided. She had the experience and expertise to see through such antics, as did her colleagues, fellow alumni, current students, and fellow women's club allies. They knew that most Black citizens were not roving the streets but walking with poise, not displaying impudence but self-assuredness, not avoiding work but investing in labor that dignified their talents and freedoms. They also knew that on balance, poverty and struggle were political problems, not individual failings. By speaking out in evidence of these facts, Black clubwomen successfully blocked the attempted takeover of Black kindergartens.[82]

This struggle for educational leadership underscored the critical relationship between curricular decisions and democratic vitality. It was no coincidence that white elites who endorsed racial segregation often sought to deny Black Americans' access to the very pedagogical "isms" and "ologies" that the "society queens of Peachtree Street" relied on for freeing their own "channels of expression." Such hostility to "higher education" curricula was also bluntly reactionary: by 1899, Black and white Atlantans had been implementing such programs, including Grecian-styled Delsartism, for well over a decade. In a New South interested in fortifying white supremacist "facts," expressions of Black selfhood could only be seen as "hazardous."

This thinking also guided educational decisions at the national level. In 1897, for example, an adviser in the federal Office of Education published the following assessment in *Popular Science Monthly*: "Scientific research affords proof of the fundamental unity of mind, but it gives no less decisive proof of differences due to ancestry and training. The negro child is psychologically different from the white child."[83] By tethering social Darwinism with supposedly universal truths about mind-body correspondence, this perspective fabricated a scientific justification for racist educational policies. Accordingly, the author of this report joined many such leaders whose undermining of a multiracial democracy was premised on their belief that Black Americans were "naturally" suited for subordinate educations and occupations.

These claims of Black inferiority, deeply normalized across American society, gave cover for a range of systemic inequalities and daily assaults. The attempted kindergarten takeover in Atlanta was just one of the city's many efforts to police Black education and expressivity. As historian Tera Hunter has documented, in an effort to control the leisure of Black laborers in Atlanta, municipal ordinances shuttered several working-class Black dance halls and heavily surveilled those they allowed to remain open.[84] When the white press did write about "respectable" Black dance halls, it refused to dignify their patrons. Instead of complimenting graceful displays, newspapers reported only "grotesque exaggerations of politeness" and "stilted and pompous ethics."[85]

In a city that endorsed racist policies and a nation that willfully looked away, Black exhibitions of self-possession became far more than contestatory acts: they were often dangerous. Local and national newspapers reported numerous stories of white Americans who believed a "wrong look" was a justifiable reason for assaulting or lynching Black Americans. In this climate, teachers of elocution at Atlanta's Black colleges were far more than promoters of students' talents; they also became vital protectors of spaces for free Black expression. Adrienne Herndon spoke to this necessity directly in the winter of 1907, just months after white mobs had terrorized the city and murdered over twenty-five Black citizens, including one of her husband's employees. Declining an offer to publicize her work, she wrote: "I should like to hide from the eyes of the white man, or at any rate the southern white man, the things, I, as a Negro woman hold most sacred for fear they pause & look to jeer and ridicule." To deflect an accusation of pessimism, she cited reality. "I was born and reared in Ga [*sic*], you know and I live in Atlanta."[86]

* * *

As this essay has demonstrated, the mainstreamed culture of American Delsartism did little to counteract such hostile disregard for Black expressivity. Instead, the most publicized performances and promotions of this "science of grace" consistently defined feminine liberation as exclusively and naturally white. By erasing Black women's expressive accomplishments and reinscribing visual taxonomies that had long legitimized racial discrimination, such actions contributed to the broader undermining of hard-won constitutional commitments to social equality and multiracial democracy.[87]

The racialization of American Delsartism as a whites-only pursuit also skewed historical assessments about the cultural significance of elocution and dance among reform-minded women. As the records of Spelman Seminary and Atlanta University attest, seeing Black women's Grecian-styled pantomimes and "original and graceful" dances offers compelling opportunities for revising historiographic assumptions about the era's choreographic landscape. In spaces such as these, teachers like Mae B. Peckham and Adrienne McNeil Herndon assuredly emboldened their generation's efforts to "free the channels of expression" by creating spaces where Black Americans claimed time "to first see, then feel, and then tell."[88] And, when so moved, to dance.

Notes

1 "Free the channels of expression" was a common phrase in the literature of American Delsartism; see Stebbins, *Delsarte System*, 402.

2 To clarify, most practices that comprised American Delsartism were never actually taught by Delsarte. Instead, referencing Delsarte's name was one way that American women claimed scientific credibility for exercises they popularized.

3 Daly, *Done into Dance;* Kimber, *The Elocutionists;* Lynch, "Aesthetic Dance"; Mullan, "Forgotten 'New' Dancer"; Perpener, *African-American Concert Dance;* Preston, *Modernism's Mythic Pose;* Ruyter, *Reformers and Visionaries* and *The Cultivation of Mind and Body;* Suter, "The Arguments They Wore"; Tomko, *Dancing Class.*

4 J. Brown, *Babylon Girls,* 177; Burns, "The Culture of Nobility," 218, Daniher, "Looking at Pauline Johnson"; Nur Amin, "African American Dance," Streeter, "Wings to Their Heels"; Walsh, *Eugenics and Physical Culture,* 19–56.
5 See Cain, "The Art and Politics"; Cobb, *Picture Freedom;* Schuller, *The Biopolitics of Feeling,* Smith, *Photography on the Color Line.*
6 Daniher, "Looking," 3. Walker's *Beauty and the Brain* makes a similar argument regarding phrenology.
7 On Atlanta's role in the development of the New South, see Hunter, *To 'Joy My Freedom.*
8 Perry, *May We Forever Stand,* 7.
9 Givens, *Fugitive Pedagogy,* 11.
10 *Spelman Messenger* 9, n. 8 (June 1893): 5.
11 Showers, "Physical Culture," 2.
12 See Anderson, *The Education of Blacks.*
13 *Catalogue of Atlanta University,* 1903–1904, 48.
14 Hull, "Introduction"; Logan, *We are Coming;* Royster, *Traces of a Stream.*
15 Like other southern newspapers, the *Atlanta Constitution* included a "Negro section," a choice reflecting white southerners' "progressive" ethos. Such columns provided spaces for citing Black accomplishments, but their separation from "mainstream" news also maintained a racialized social hierarchy.
16 "Commencement Exercises," 1888.
17 On Peckham and her family, see "May B. Peckham," 1880 United States Census; and "Miss Mae B. Peckham," *The Progress-Bulletin.* On northern reforms, see Blair, *The Clubwoman as Feminist;* Mattingly, *Appropriate[ing] Dress;* Verbrugge, *Able-Bodied Womanhood.*
18 Mattingly, *Appropriate[ing] Dress,* 1–6.
19 See Streeter, "Wings to Their Heels," 19–87.
20 Stebbins, *Delsarte System,* 139.
21 Mullan, "Forgotten 'New' Dancer"; Preston, *Modernism's Mythic Pose;* Ruyter, *The Cultivation of Mind and Body;* Streeter, "Wings to Their Heels"; Suter, "The Arguments They Wore."
22 Mullan, "Forgotten 'New' Dancer," 98; Lynch, "Aesthetic Dance."
23 See Stepan, "Appropriating the Idioms of Science," 185.
24 My use of "contestatory" derives from Smith, *Photography on the Color Line,* 6–9.
25 *Spelman Messenger* 6, n. 8 (June 1890): 5.
26 *Spelman Messenger* 14, n. 8 (June 1898): 5.
27 *Spelman Messenger* 16, n. 8 (June 1900): 5; Schell, "Ribbon Tableaux."
28 *Spelman Messenger* 17, n. 8 (June 1901): 6.
29 Stebbins, *Delsarte System,* 420.
30 Brown, *Elocution,* 7. On Hallie Brown, see Kates, *Activist Rhetorics,* 53–74.
31 "Sketch of Madame Preston," 1890.
32 *Republican* (Springfield, Ohio) quoted in Majors, *Noted Negro Women,* 100.
33 *Spelman Messenger* 5, n. 3 (June 1889): 5.
34 "Southern School of Elocution."

35 Lanier, "The Early Chautauqua," 9.

36 Lindley, "Physical Education." On racial politics of southern white progressivism, see Cox, *Dixie's Daughters;* Gilmore, *Gender and Jim Crow;* Gold and Hobbs, *Educating.*

37 "Annual Exercises."

38 "Delsarte Pictures."

39 Shurter, *Complete Orations,* 33.

40 "Delsarte Index." On phrenology, see Walker, *Beauty and the Brain.*

41 Nye, "Billy Nye."

42 "Commencement Exercises," 1888.

43 Wilson, *Negro Building,* 30–84.

44 "This is Negro Day."

45 On Herndon's biography, see Cahill, "Adrienne Herndon's Homeplaces"; Henderson, "The Work and Legacy"; Merritt, *The Herndons.*

46 "Free Thirty Years." "Exposition Ode" was written by Daniel Webster Davis, a prominent Black educator from Richmond, Virginia.

47 On Washington's influence, see Anderson, *Education of Blacks in the South,* 79–110.

48 "Free Thirty Years."

49 "Second Day's Session." On Black women's clubs, see Gilmore, *Gender and Jim Crow;* Jones, *Vanguard.*

50 Penn, "'New' Negro Woman."

51 Cahill, "Adrienne Herndon's Homeplaces," 53–54.

52 Herndon, "Our Work," 1.

53 Curry, "Expression," 1. On her studies with Curry, see Merritt, *The Herndons,* 23.

54 Herndon, "Our Work," 1. Though Herndon's classes drew upon developments of American Delsartism, she did not use the term "Delsarte" in her curriculum description. This was not uncommon for the late 1890s, when Genevieve Stebbins and other prominent teachers had moved away from using the "Delsarte" designation, favoring terms like "psycho-physical culture" and "aesthetic gymnastics," labels they believed better represented the various sources that inspired their pedagogy. See Streeter, "Wings to Their Heels," 139–98.

55 See Lynch, "Aesthetic Dance."

56 "Class Exercises."

57 "Commencement Exercises," 1903.

58 "On the Campus."

59 In the 1850s, Sanborn helped fund John Brown's militant efforts to end slavery. His support of Du Bois's sociological work in the Reconstruction Era exemplified his continued agitation against racism. See Morris, *The Scholar Denied,* 79–81.

60 Sanborn, "The 'Inferior' Race."

61 Sanborn, "The 'Inferior' Race."

62 Martin, "To Ignore Is to Deny," 675.

63 Though this essay does not address the phenomenon of the cakewalk dance, the biopolitics of that choreography are relevant to understanding the cultural world of American Delsartism. See Brown, *Babylon Girls;* Brooks, *Bodies in Dissent.*

64 Cox, *Dreaming of Dixie;* Hale, *Making Whiteness.*

65 "Greater New York."

66 Roush, "Side-walk Dancing."
67 See Lange, *Picturing Political Power,* 126–57.
68 Kemble, "First Parade." "Bloomers" were loose pants worn under knee-length dresses, named for women's-rights advocate Amelia Bloomer.
69 Martin, "To Ignore," 675; Cone, "Skirt Dancing."
70 Kemble, *Kemble's Coons,* 23.
71 Du Bois, "The American Negro at Paris," 577. On photography in this exhibition, see Smith, *Photography on the Color Line.*
72 Du Bois, *The Health and Physique,* 31–33.
73 On Du Bois's frustrations with academic racism, see Morris, *The Scholar Denied.*
74 "Expression in the South," 349. Mae Peckham was also listed on page 356.
75 "Expressional Power," 459.
76 "Unconscious Delsartism," 529.
77 "Our Card Basket"; "Musical Romances."
78 "Social News." The teacher was Minnie M. Jones, who studied with Stebbins and hosted "Delsarte matinees" in Philadelphia. On mammy stereotypes, see Hale, *Making Whiteness,* 85–88.
79 Smith, "The Fight to Protect," 482–85; Parker, "Frances Watkins Harper;" Streeter, "Wings to Their Heels."
80 "Are You a Progressive Teacher"; "Miss Stebbins Recital." On the UDC, see Cox, *Dixie's Daughters.* Stebbins's work with southern women remains an understudied feature of her career.
81 "Colored Kindergartens."
82 Smith, "The Fight to Protect."
83 Smith, "A Study in Race Psychology," 360.
84 Hunter, *To 'Joy My Freedom,* 170–76.
85 "Down in the Dance Halls." This derision of Black gracefulness shares similarities with antebellum assaults described in Brooks, "Race, Rank, and Reform."
86 Herndon to Washington letter, 1907.
87 On race, science, and beauty, see Cain, "The Art and Politics," 29.
88 Herndon, "Our Work," 1.

Bibliography

Primary Sources

"Annual Exercises," *Atlanta Constitution* (May 29, 1891).
"Are You a Progressive Teacher." *The Progressive Age* (June 5, 1902).
Brown, Hallie Q. *Elocution and Physical Culture.* Wilberforce: Homewood Cottage, 1910.
Catalogue of the Officers and Students of Atlanta University. Atlanta: Atlanta University Press, 1904.
"Class Exercises." *The Bulletin of Atlanta University* (June 1901): 1.
"Colored Kindergartens." *Atlanta Constitution* (May 25, 1899).
Cone, Ada. "Skirt Dancing." *Atlanta Constitution* (June 5, 1892).
"Commencement Exercises." *Atlanta Constitution* (May 26, 1888).

"Commencement Exercises." *The Bulletin of Atlanta University* (June 1903): 1.
Curry, S. S. "Expression." *Expression: A Quarterly Review* (June 1895): 1–7.
"Delsarte Index to Character." *The Sunny South* (Nov. 15, 1890).
"Delsarte Pictures at Chautauqua." *Atlanta Constitution,* (July 26, 1891).
"Down in the Dance Halls of Decatur Street." *Atlanta Constitution* (July 13, 1902).
Du Bois, W. E. B. "The American Negro at Paris." *The American Monthly Review of Reviews* 22, n. 5 (Nov. 1900): 575–77.
Du Bois, W. E. B. *The Health and Physique of the Negro American.* Atlanta: Atlanta University Press, 1906.
"Expression in the South." *Werner's Magazine* 25, n. 4 (June 1900): 325–67.
"Expressional Power of the Colored Race." *Werner's Magazine* 26, n. 6 (Feb. 1901): 459–78.
"Free Thirty Years." *Champaign Daily Gazette* (Oct. 29, 1895).
"Greater New York." *Leslie's Weekly* 85 (July 1, 1897): 391.
Herndon, Adrienne McNeil. "Our Work in Elocution." *Bulletin of Atlanta University* (May 1897): 1.
Herndon, Adrienne McNeil to Booker T. Washington, Feb. 12, 1907. In *The Booker T. Washington Papers,* edited by Louis R. Harlan and Raymond W. Smock, 216–17. Urbana: University of Illinois Press.
Kemble, E. W. "First Parade of the New Woman's Society of Possumville." *Leslie's Weekly* 84 (June 10, 1897): 375.
Kemble, E. W. *Kemble's Coons.* New York: R. H. Russell, 1896.
Lindley, Marguerite. "Physical Education." *Atlanta Constitution* (Sept. 2, 1888).
"May Peckham." United States Census (1880) Boston, Suffolk County, Massachusetts, Roll 561, p. 42c. Located using Ancestry.com. Accessed July 27, 2022.
Majors, M. A. *Noted Negro Women.* Chicago: Donohue & Hennebery, 1893.
"Miss Stebbins Recital." *Asheville Citizen-Times* (July 19, 1904).
"Miss Mae B. Peckham." *The Progress-Bulletin* (Dec. 24, 1931).
"Musical Romances." *Indianapolis News* (June 25, 1900).
Nye, Edgar W. "Billy Nye and Grover." *The Sunny South.* Apr. 27, 1895.
"On the Campus." *The Bulletin of Atlanta University* (June 1900): 1.
"Our Card Basket." *The Excelsior* [Omaha, Nebraska] (Mar. 18, 1899).
Penn, I. Garland. "New' Negro Woman." *Atlanta Constitution* (Dec. 22, 1895).
Roush, L. L. "Side-walk Dancing in New York." *Leslie's Weekly* 85 (July 8, 1897): 17.
Sanborn, Frank. "The 'Inferior' Race." *Boston Evening Transcript* (May 9, 1910).
Schell, Stanley. "Ribbon Tableaux and Drill." *Werner's Magazine* 24, n. 1 (Sept. 1899): 289–91.
"Second Day's Session." *Evening Star* (July 15, 1896).
Showers, Fannie. "Physical Culture." *Spelman Messenger* 10, n. 5 (Mar. 1894): 2.
Shurter, Edwin. *The Complete Orations and Speeches of Henry W. Grady.* Norwood, MA: South-West Publishing, 1910.
"Sketch of Madame Preston." *The Freeman* (Nov. 15, 1890).
Smith, Anna Tolman. "A Study in Race Psychology." *Popular Science Monthly* 50 (1897): 354–60.
"Social News." *Atlanta Evening Herald* (Apr. 5, 1893).
"Southern School of Elocution." *Atlanta Constitution* (Oct. 11, 1885).

Spelman Messenger, volumes cited from the 1880s to 1904.
Stebbins, Genevieve. *Delsarte System of Expression* 6th ed. New York: Werner Publishing, 1902.
"This is Negro Day." *Atlanta Constitution,* Dec. 26, 1895.
"Unconscious Delsartism in Colored Women." *Werner's Magazine* 26, n. 6 (Feb. 1901): 529.

Secondary Sources

Anderson, James D. *The Education of Blacks in the South, 1860–1935.* Chapel Hill: University of North Carolina Press, 1988.
Blair, Karen. *The Clubwoman as Feminist: The Womanhood Redefined, 1868–1914.* New York: Holmes & Meier, 1980.
Brooks, Daphne. *Bodies in Dissent: Spectacular Performances of Race and Freedom, 1850–1910.* Durham, NC: Duke University Press, 2006.
Brooks, Lynn Matluck. "Race, Rank, and Reform in Antebellum Philadelphia Social Dance." *Pennsylvania Magazine of History and Biography* 144, n. 2 (2020): 147–78.
Brown, Jayna. *Babylon Girls: Black Women Performers and the Shaping of the Modern.* Durham, NC: Duke University Press, 2008.
Burns, Judy. "The Culture of Nobility/The Nobility of Self-Cultivation." In *Moving Words: Re-Writing Dance,* edited by Gay Morris, 203–26. New York: Routledge, 1995.
Cahill, Patricia A. "Adrienne Herndon's Homeplaces: Shakespeare and Black Resistance in Atlanta, c. 1906." *Journal of American Studies* 54, n. 1 (2020): 51–58.
Cain, Mary Cathryn. "The Art and Politics of Looking White: Beauty Practice among White Women in Antebellum America." *Winterthur Portfolio* 42, n. 1 (2008): 27–50.
Cobb, Jasmine. *Picture Freedom: Remaking Black Visuality in the Early Nineteenth Century.* New York: New York University Press, 2015.
Cox, Karen L. *Dixie's Daughters: The United Daughters of the Confederacy and the Preservation of Confederate Culture.* Gainesville: University Press of Florida, 2003.
Cox, Karen L. *Dreaming of Dixie: How the South Was Created in American Popular Culture.* Chapel Hill: University of North Carolina Press, 2013.
Daly, Ann. *Done into Dance: Isadora Duncan in America.* Middletown, CT: Wesleyan University Press, 1995.
Daniher, Colleen Kim. "Looking at Pauline Johnson: Gender, Race, and Delsartism's Legible Body." *Theatre Journal* 72, n. 1 (Mar. 2020): 1–20.
Gilmore, Glenda Elizabeth. *Gender and Jim Crow: Women and the Politics of White Supremacy in North Carolina, 1896–1920.* Chapel Hill: University of North Carolina Press, 1996.
Givens, Jarvis R. *Fugitive Pedagogy: Carter G. Woodson and the Art of Black Teaching.* Cambridge, MA: Harvard University Press, 2021.
Gold, David, and Catherine L. Hobbs. *Educating the New Southern Woman: Speech, Writing, and Race at the Public Women's Colleges, 1884–1945.* Carbondale: Southern Illinois University Press, 2014.
Hale, Grace Elizabeth. *Making Whiteness: The Culture of Segregation in the South, 1890–1940.* New York: Pantheon Books, 1998.
Henderson, Alexa Benson. "The Work and Legacy of Adrienne Elizabeth McNeil Herndon at Atlanta University, 1895–1910." *Phylon* 53, n. 1 (2016): 80–101.

Hull, Akasha (Gloria). "Introduction." In *The Dunbar Speaker and Entertainer*, by Alice Moore-Dunbar, xv–xxv. New York: G. K. Hall, Co., 1996.

Hunter, Tera. *To 'Joy My Freedom: Southern Black Women's Lives and Labors after the Civil War*. Cambridge, MA: Harvard University Press, 1998.

Jones, Martha S. *Vanguard: How Black Women Broke Barriers, Won the Vote, and Insisted on Equality for All*. New York: Basic Books, 2020.

Kates, Susan. *Activist Rhetorics and American Higher Education, 1885–1937*. Carbondale: Southern Illinois University Press, 2001.

Kimber, Marian Wilson. *The Elocutionists: Women, Music, and the Spoken Word*. Champaign: University of Illinois Press, 2017.

Lange, Allison K. *Picturing Political Power: Images in the Women's Suffrage Movement*. Chicago: University of Chicago Press, 2020.

Lanier, Doris. "The Early Chautauqua in Georgia." *Journal of American Culture* 11, n. 3 (Fall 1998): 9–18.

Logan, Shirley Wilson. *We Are Coming: The Persuasive Discourse of Nineteenth Century Black Women*. Carbondale: Southern Illinois University Press, 1999.

Lynch, Kelly Jean. "Aesthetic Dance as Women's Culture in America at the Turn of the 20th Century: Genevieve Stebbins and the New York School of Expression." *Journal of Feminist/Modernist Studies* 5, n. 22 (2023): 247–60.

Martin, Francis. "To Ignore Is to Deny: E. W. Kemble's Racial Caricature as Popular Art." *Journal of Popular Culture* 40, n. 4 (2007): 655–82.

Mattingly, Carol. *Appropriat[ing] Dress: Women's Rhetorical Style in Nineteenth-Century America*. Carbondale: Southern Illinois University Press, 2002.

Merritt, Carole. *The Herndons: An Atlanta Family*. Athens: University of Georgia Press, 2002.

Morris, Aldon. *The Scholar Denied: W. E. B. Du Bois and the Birth of Modern Sociology*. Berkeley: University of California Press, 2015.

Mullan, Kelly. "Forgotten 'New' Dancer of New York City's Gilded Age: Genevieve Lee Stebbins and the Dance as Yet Undreamed." *Dance Research Journal* 52, n. 3 (Dec. 2020): 97–117.

Nur Amin, Takiyah. "African American Dance Revisited: Undoing Master Narratives in the Studying and Teaching of Dance History." In *Rethinking Dance History: Issues and Methodologies*, edited by Geraldine Morris and Larraine Nicholas, 44–55. New York: Routledge, 2019.

Parker, Allison M. "Frances Watkins Harper and the Search for Women's Interracial Alliances." In *Susan B. Anthony and the Struggle for Equal Rights*, edited by Christine L. Ridarsky and Mary M. Huth, 145–71. Rochester: University of Rochester Press, 2012.

Perpener, John O. *African-American Concert Dance: The Harlem Renaissance and Beyond*. Champaign: University of Illinois Press, 2001.

Perry, Imani. *May We Forever Stand: A History of the Black National Anthem*. Chapel Hill: University of North Carolina Press, 2018.

Preston, Carrie J. *Modernism's Mythic Pose: Gender, Genre, Solo Performance*. New York: Oxford University Press, 2011.

Royster, Jacqueline. *Traces of a Stream: Literacy and Social Change among African American Women*. Pittsburgh: University of Pittsburgh Press, 2000.

Ruyter, Nancy Lee Chalfa. *The Cultivation of Mind and Body in American Delsartism.* Westport, CT: Greenwood Press, 1999.

Ruyter, Nancy Lee Chalfa. *Reformers and Visionaries: The Americanization of the Art of Dance.* New York: Dance Horizons Press, 1979.

Schuller, Kyla. *The Biopolitics of Feeling: Race, Sex, and Science in the Nineteenth Century.* Durham, NC: Duke University Press, 2018.

Smith, Mary Jane. "The Fight to Protect Race and Regional Identity within the General Federation of Women's Clubs, 1895–1902." *Georgia Historical Quarterly* 94, n. 4 (2010): 479–513.

Smith, Shawn Michelle. *Photography on the Color Line: W. E. B. Du Bois, Race, and Visual Culture.* Durham, NC: Duke University Press, 2004.

Stepan, Nancy Leys, and Sander L. Gilman. "Appropriating the Idioms of Science: The Rejection of Scientific Racism." In *The 'Racial' Economy of Science: Toward a Democratic Future,* edited by Sandra Harding, 170–200. Bloomington: Indiana University Press, 1993.

Streeter, Carrie. "Wings to Their Heels: Self-Expression and Health and the Rise of the New Woman." PhD diss., University of California San Diego, 2023.

Suter, Lisa Kay. "The Arguments They Wore: The Role of the Neoclassical Toga in American Delsartism." In *Rhetoric, History, and Women's Oratorical Education,* edited by David Gold and Catherine L. Hobbs, 134–53. New York: Routledge, 2013.

Tomko, Linda. *Dancing Class: Gender, Ethnicity, and Social Divides in American Dance, 1890–1920.* Bloomington: Indiana University Press, 2000.

Verbrugge, Martha. *Able-Bodied Womanhood: Personal Health and Social Change in Nineteenth Century Boston.* New York: Oxford University Press, 1988.

Walker, Rachel E. *Beauty and the Brain: The Science of Human Nature in Early America.* Chicago: University of Chicago Press, 2023.

Walsh, Shannon L. *Eugenics and Physical Culture Performance in the Progressive Era: Watch Whiteness Workout.* Berlin: Springer International Publishing, 2020.

Wilson, Mabel O. *Negro Building: Black Americans in the World of Fairs and Museums.* Berkeley: University of California Press, 2012.

11

Interrupting Jim Crow

Response to "Exhibiting (Scientific) Grace: American Delsartism and Black Citizenship in the New South" by Carrie Streeter

Susan C. Cook

Black women singers, dancers, and choreographers, like Aida Overton (1880–1914), had begun to reshape entertainments as the practices Carrie Streeter explores emerged. Overton and others pursued professional careers in the only way possible, within shows still tethered to blackface minstrelsy. Wesley Morris, in *The 1619 Project*, provides a cogent reading of the minstrel show, whose influence was "too big for any American performer to escape,"[1] and which did monstrous work, as audiences at home and abroad consumed this American creation indelibly marked by its origin in chattel slavery. Post-emancipation, minstrelsy further normalized beliefs about a benign Old South while its dehumanizing caricatures "drew a comforting contrast with a white person's sense of honor and civility."[2] Minstrelsy's "residue"—the word used both by Morris and Thomas Riis in his essay on nineteenth-century Black musical culture—adhered to performing bodies, like Overton's, caught in the "upside-down, inside-out phenomenon" of Black portrayals of minstrelsy's racialized imitations.[3] As Morris argues, minstrelsy clung to Black psyches and poisoned "a race's collective self-esteem."[4] Freeing "the channels of expression," as Streeter narrates, was a tall order indeed.

Morris finds a resistant "anti-minstrelsy" in oration: "Black orators rejected whatever it was that minstrelsy had convinced the country Black people were, and they did so within institutions that were Black built and Black-maintained."[5] His exemplars, like Frederick Douglass and Frances Ellen Watkins Harper, drew on the tenets of racial uplift, shared by Streeter's subjects. Her focus on Frances and Lillie Preston and especially Adrienne McNeil Herndon, though, grounds Morris's anti-minstrelsy alternative further.

Streeter provides evidence for how localized and repeated acts of free self-expression made it possible for Douglass, Harper, and others to lead a larger chorus of defiant voices testifying for collective self-determination in the face of pseudo-"scientific" racism.

Minstrelsy depended on a fabricated Black dialect that offered comic proof that its speakers were unfit for citizenship. While Morris deems Black oration "aspirational" and "artful," Streeter demonstrates that performative acts of speech and attendant movement could be taught, practiced, and drawn upon in a host of situations.[6] She reveals the anti-minstrelsy of local practitioners who worked within Black-maintained systems of higher education that emboldened Black voices and inspired acts of eloquent resistance as students learned how to interrupt minstrelsy.

The instrument of oration and elocution—the voice—is also the instrument of the body. Because Black women's bodies were objectified, policed, and abused, embodied and envoiced choices are especially important for us to uncover in order to restore agency to women like Overton and Herndon who faced a doubly second-class status. However, the post-emancipation bodies of virtuosic professional Black women dancers (I am thinking of Jayna Brown's *Babylon Girls* and their "heterogeneous, sweating world"[7]) and Streeter's Black pantomimists and elocutionists (with their cool, bodily control) could appear to be on a collision course. As Terri Brinegar explores in *Voices of Black Folk*, Black vocal expression that retained too much of its emotional, "southern" roots was at odds with racial uplift.[8] Discourses of control can invoke the body-phobic ideologies of the mind/body binary. What might Herndon have thought of Overton, who described her virtuosic cakewalk choreography in terms of grace and sobriety?[9] She may well have found Overton "tainted"—to use Morris's word—by minstrelsy, regardless of Overton's resistant claims. How do we argue then for a both/and space of New Black Womanhood where "free expression" empowered a range of vocal and embodied alternatives across the shifting lines of class, race, and gender?

To close, I return to a fleeting image from Streeter's essay: a maypole dance. At my own public institution, cocurricular activities, including blackface minstrel shows, remained the purview of its overwhelmingly white male students. One of the few public spectacles that female students controlled was similarly the annual maypole dance. White women students in Wisconsin, like Streeter's students in Georgia, performed this Anglophile throwback to pre-Christian celebratory rites of female sexuality, as growing numbers of non–English-speaking emigrants fed Eugenic "scientific" theories about racial decline. Cakewalks, scarf drills, maypole dances: each of these female-centered acts of self-expression offered contradictory narratives of citizenship and national origin, of improvisational virtuosity and grace, of self-control and empowerment, of raced and gendered bodies, and of multiple ways to move, to speak, and to claim one's humanity. And for Streeter's students, to dissolve blackface residue still adhering to body, mind, and spirit. To riff on both the formidable Nina Simone and Adrienne McNeil Herndon, they wanted to know—as well as to model, to teach, and to learn—how it would feel to be free.

Notes

1 Morris, "Music," 371.
2 Morris, "Music," 370–71.
3 Morris, "Music," 372; Riis, "Defying Boundaries and Escaping Stereotypes," 216.
4 Morris, "Music," 374.
5 Morris, "Music," 373.
6 Morris, "Music," 373.
7 Brown, *Babylon Girls,* 4.
8 Brinegar, *Voices of Black Folk.*
9 Beerbohm, "The Cake-Walk and How to Dance It."

Bibliography

Beerbohm, Constance. "The Cake-Walk and How to Dance It." *London Tatler* (July 1, 1903): 13.

Brinegar, Terri. *Voices of Black Folk: The Sermons of Reverend A. W. Nix.* Jackson: University Press of Mississippi, 2022.

Brown, Jayna. *Babylon Girls.* Durham, NC: Duke University Press, 2001.

Morris, Wesley. "Music." In *The 1619 Project: A New Origin Story,* edited by Nikole Hannah-Jones, Caitlin Roper, Ilena Silverman, and Jake Silverstein, 359–79. New York: One World, 2021.

Riis, Thomas L. "Defying Boundaries and Escaping Stereotypes." In *Rethinking American Music,* edited by Tara Browner and Thomas L. Riis, 200–20. Urbana: University of Illinois, 2019.

III

Physical Cultures

Disciplining and Improving the Self

12

The "Muscular Sense" and Therapeutic Modernism in the Eurhythmics of Émile Jaques-Dalcroze

ANDREA HARRIS

In 1878, Stanley Hall wrote of a sensation that had recently preoccupied physiologists—the feeling of motion. This sensation was "absolutely unique" in that it did not occur in response to an external excitation, like the other senses, but rather to the "motion of the limb, the muscle, the nerve-end itself." Notably, Hall claimed the importance of the feeling of movement not only for the new science of experimental physiology, but further, as "one of the most important epochs in the history of *philosophy*."[1]

The discovery of the sense of movement—the "muscular sense," as it was known after 1830—was a breakthrough in neurophysiology. Hall's nod to its philosophical significance underscores the way in which the scientific study of the muscular sense altered understandings about the relation between body and mind. As scientists gained knowledge of the role of the nervous system in human movement, a parallel "physiology of the will" developed that attempted to scientifically understand, and often defend, human volition and the freedom of the will.[2] This essay examines the intersection of that science with aesthetics in the movement training method created by Émile Jaques-Dalcroze, the Swiss music pedagogue, body culture pioneer, and forerunner of early modern dance. The physiology of the muscular sense and the will found its way directly into Jaques-Dalcroze's method and its core concepts of bodily awareness and, especially, bodily mastery.

Jaques-Dalcroze's work is paradigmatic of what I call—building on the work of T. J. Jackson Lears—a "therapeutic modernism" that marked cultural practice at the turn of the twentieth century.[3] For fin de siècle writers, the fact that modern society sickened its citizens seemed indisputable. It was especially clear in the epidemic of nervousness, or neurasthenia, a term coined by physician George Beard to describe a new "paralysis of the will."[4] Neurasthenia supposedly marked the early symptoms of a larger "degeneration," widely known as a pathological and inheritable state of deterioration in the European race. Degeneration was a complex scientific, medical, and cultural discourse that encompassed racial- and gender-based anxieties exacerbated by colonialism and immigration, as well as fears that the chaotic experience of industrial society inevitably led to physical fatigue, moral confusion, and the erosion of individual autonomy.[5] Drawing on the science of the muscular sense, Jaques-Dalcroze asserted that his method of rhythmic movement training, known in English as "Eurhythmics," would heal the nervous subject. In so doing, he put forth new theories of the therapeutic impact that performative movement could have on the self, on society, and even on future generations.

Jaques-Dalcroze developed Eurhythmics as part of a larger approach to music pedagogy that addressed not only ear training, but also "the muscular and nervous response of the *whole organism*."[6] Jaques-Dalcroze's ideas about the relationship of rhythm to movement took root during his interactions with North African musicians in Algiers, where he worked as a theatrical music director from 1886 to 1887, and were reinforced by his exposure to Arabic and other non-Western music at the 1889 Paris Exhibition.[7] As the new professor of harmony at the Conservatoire of Geneva in 1892, he drew on those experiences, attempting to train his students "to react physically to the perception of musical rhythms" through exercises in rhythmic walking.[8] But soon the problem arose: not all students were able to execute physical responses to rhythmic variations in the same way. "Some responded too slowly, some too quickly to the word of command; some who were able to carry out an exercise with ease in a certain tempo were unable to change to another quickly . . . ; others again could begin an exercise perfectly but were not capable of sustained effort."[9] Jaques-Dalcroze concluded that the problem was physiological, or more precisely, a disorder in the functioning of the nervous system. Performing rhythmic movement required that "before all else, *communications* should be established, between the mind that conceives and analyzes, and the body that executes."[10] It was not long before he declared that a pervasive "lack of rhythm" plagued the whole of Western modern civilization, "almost like a disease." This condition pointed to widespread "nervous troubles" caused by the "insufficient co-ordination between the mental picture of a movement and its performance [by] the body."[11]

Historians are establishing Jaques-Dalcroze's formative influence on early modern dance. Several of his students at Hellerau went on to become important dancers, including Mary Wigman, Suzanne Perrottet, Michio Ito, and others who taught Eurhythmics across Europe, England, and North America.

His influence in Germany shaped the development of Weimar body culture and *Ausdruckstanz* and was carried into new expressive movement forms on both sides of the Atlantic.[12] In the US, his method was central to women's physical education and to the roots of dance in higher education in the 1910s and 1920s.[13] Eurhythmics remained part of American modern dance education well into the twentieth century.[14]

Further, Jaques-Dalcroze contributed to a new notion of bodily experience as efficacious, laying the conceptual groundwork for early modern dance. Ana Isabel Keilson argues that his "idea that bodily movement contained previously untapped information about human life, behavior, and organization" was influential for the work of Rudolf Laban and Wigman as well as other German dancers and critics.[15] And Hillel Schwartz locates Jaques-Dalcroze near the beginning of a larger cultural shift in which movement was viewed as not only expressive, but transformative.[16] As Jaques-Dalcroze's importance in modern dance history comes to light, Eurhythmics points to an emergent belief in the constructive potential of bodily movement. Modern dancers embraced this new way of thinking about movement, as examples at the end of this essay will highlight.

Michael Cowan stresses that rather than seeking to liberate the body, Eurhythmics aimed "to strengthen the dominance of the mind over the body . . . through bodily performance itself."[17] There is ample evidence in Jaques-Dalcroze's writings to support this claim. The project of retraining the body to obey the will characterized the therapeutic thrust of Jaques-Dalcroze's method. This hierarchical relationship, mind directing body, was what was meant by his repeated use of "harmony" to describe the coordination of intellectual and physical processes achieved through Eurhythmic practice.[18]

Yet what remains to be explained is how this was supposed to work. Exactly *how* would bodily movement strengthen the will and bring the body under its control? Jaques-Dalcroze conceived the body as a "mechanism" or an "instrument" capable of being perfected and operationalized though practice to serve the mind, or the soul—an idea that was completely dependent upon his knowledge of the nineteenth-century neurophysiology of the muscular sense and the physiology of volition.[19]

Jaques-Dalcroze's most direct scientific source was the Genevan neurophysiologist and psychologist Edouard Claparède. The two shared a lasting friendship, and Jaques-Dalcroze relied on Claparède for the scientific terminology and sources that supported his theories of movement and rhythm.[20] As he was preparing the first publication of his method, Jaques-Dalcroze sought Claparède's expertise:

> When I speak of the essence of rhythm, i.e., of energy . . . I establish several degrees of muscular innervation, and define innervation as follows: the elements and nervous tissue put in action by extension, by conscient contractibility of the muscles. Is this definition correct? I would like to establish a difference between muscular tension and the feeling of this tension.[21]

Jaques-Dalcroze's effort to distinguish the contraction of the muscles from the *feeling* of that action recalls Hall's fascination with the "absolutely unique" properties of the muscular sense, felt not because of any external stimuli, but through motion itself. Such a distinction points to the fact that although often taken for granted today, "self-consciousness about bodily posture and movement was . . . not always present: it has a history."[22] This is the history of the muscular sense.

The Muscular Sense

Interest in the physical feeling of movement emerged as empiricist philosophers examined the senses and their role in knowledge. Muscular sensation was distinguished as a separate sense, distinct from touch, by the early nineteenth century.[23] In the nineteenth century, scientists in the emergent field of experimental physiology, powerful new instruments in hand, made rapid advances in the understanding of the anatomical structures responsible for muscular sensation and motor action. Of particular importance was the discovery of sensory and motor nerves in the spinal cord, which showed that movement depended not only on signals sent by the brain to the muscles, but also on sensations that originated in the muscles themselves and guided the brain to bring about the desired action.

In 1811, Charles Bell found that when he cut across the posterior nerve root at the spinal cord, it produced no movement in the muscles of the back, but when he touched the anterior nerve root with his knife, the muscles immediately convulsed.[24] He concluded that sensory and motor nerves were different entities, with different functions. Tracing the path of the sensory nerve, Bell found that it was distributed profusely in the muscles. Muscles thus possessed a sensibility that indicated that they had another purpose than simply to contract when stimulated.[25] He determined that the job of the sensory nerves was to communicate the condition of the muscles to the brain, and that this information was necessary for movement. "In standing, walking, and running, every effort of the voluntary power, which gives motion to the body, is directed by a sense of the condition of the muscles, and without this sense we could not regulate their actions," he argued.[26] In 1854, Bell named the muscular sense the "sixth sense," on the same footing as the other five senses.[27] The physician H. Charlton Bastian later argued for replacing the term "muscular sense" with "kinesthesis" to indicate that it was not only the muscles that were responsible for the impressions of movement and position that guided the performance of a motor action, but a host of bodily organs including fascia, tendons, joints, and skin.[28]

The discovery of the motor and sensory nerves allowed for greater understanding of reflex action, or the way in which a sensory nerve and a motor nerve, linked together by the spinal cord, acted together to produce a muscular movement in response to a stimulus.[29] A reflex movement could be "evoked and guided to its completion by afferent impressions which [were] wholly unfelt."[30] In other words, reflex movements, which represented the most basic organization of nervous function, were fully a bodily activity, occurring automatically through correspondences in the nervous system. As research on the reflex progressed, scientists realized that much of the coordination of the muscles took place through nervous structures in the spinal cord and lower brain—not in the higher brain, which was viewed as the seat of consciousness and volition.[31]

The investigation of reflex action highlights the way in which nineteenth-century neurophysiology "so obviously and colorfully challenged conventional representations of mind, spirit, will, and self."[32] Some scientists refused to separate the reflex from consciousness, claiming that the purposeful nature of reflex action evidenced the operation of an immaterial principle at work.[33] In particular, the experience of effort in movement preoccupied physiologists: Was not one's awareness of the degree of effort put into the execution of a movement indicative of the existence of an active will and a person's capacity for subjective action? The investigation of the muscular sense remained inextricably linked with questions about the mind-body relation and the will throughout the nineteenth century.

Writing in 1897, Claparède divided the work of physiologists and psychologists on the muscular sense into "two camps" with opposing views on the characteristics and location of the muscular sense. In the first camp were those who took the perception of movement and effort expended in movement as proof of the active exercise of the will upon the body. These scientists identified the muscular sense with particular "feelings of innervation" that accompanied the efferent flow of nervous energy from the brain to the motor nerves in the muscles.[34] In this theory, the muscular sense did not originate in the body or the "periphery," but rather was centrally located in the brain itself and was a sensation accompanying the exertion of the will.[35]

The scientists in Claparède's second camp argued that the muscular sense was located not in the brain, but in the muscles (and related bodily organs). For these physiologists, the muscular sense was not the sensation of the will discharging nervous action from the brain to the body, but precisely the opposite—it was the result of "impressions emanating from the moving organs themselves during the actual accomplishment of movements."[36] The ability to execute a desired movement was dependent upon muscular sensations sent from the body to the brain.

Claparède was in this second camp.[37] He concurred with physiologists who argued that the execution of movement was guided by impressions sent by kinesthetic organs in the body to the brain. In this account, any movement that was performed was preceded by a mental idea of a movement, or motor image. And those motor images arose not from mental processes, like volition, but rather from afferent impressions produced by movements themselves, particularly ones that had been previously executed.

With knowledge of sensory and motor nerves, physiologists organized bodily movement into different classes. At one end of the spectrum was the reflex, which was executed entirely by afferent impressions and without any conscious awareness. At the other end was volitional movement, which was conventionally considered to be opposite to the reflex because it involved consciousness.[38] However, while voluntary movements took a great deal of conscious attention when they were new or unfamiliar—as in learning to dance or to play an instrument—that need for concentration gradually diminished as the movement was practiced. Through repetition, the once-difficult movement became "a movement so easy of execution as to recur independently of conscious attention, with machine-like regularity."[39] This could only be explained by the fact that in the process of learning a movement, new nervous connections were laid down between the sensory centers and the related motor mechanisms.[40] Once those nerve tracts had been established, the movement that had once required full attention changed rank in the schema of movement types to become more like an automatic movement that could be performed "quite independently of, and certainly without any need for, Consciousness."[41] All voluntary movements had this potential—or rather, this fate—of becoming automatic movements through repetition.

The muscular sense played a central role in this process. Learning a movement initially required both the visual and the muscular sense. But as new nervous connections were established, reliance on the visual sense dimin-

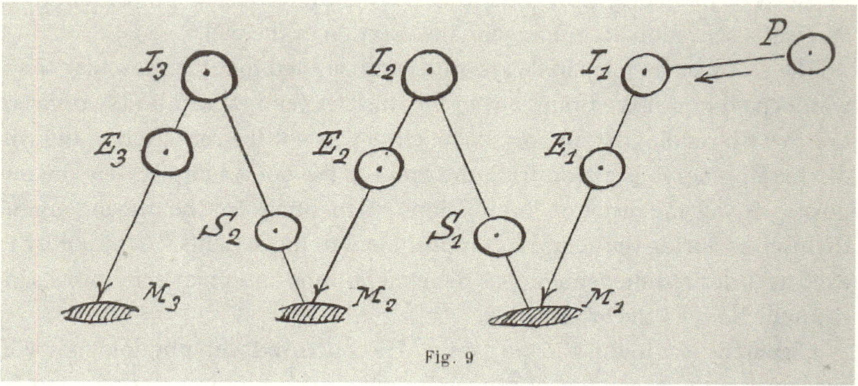

Figure 12.1. Diagram of successive voluntary movement by Edouard Claparède, in *Du Sens Musculaire: à Propos de Quelques Cas d'Hémiataxie Posthémiplégique* (Geneva: Eggimann, 1897), 70. Public domain.

ished and muscular, or kinesthetic, impressions, became "the supreme guiding influence" in voluntary movements.[42] The muscular sense educated the brain on how to continue the movement being executed so as to reach the desired end.[43] And, as kinesthetic impressions were sent to the brain via the process of movement, they were stored there "for future use in the guidance of all kinds of voluntary movements."[44] The brain thus possessed only memories, appearing as mental images, of the sensory effects of movements that had been previously performed. These motor images could be aroused by the feeling of a reflex movement or by an external sensation. These "ideal revivals of kinesthetic impressions" formed the basis for the "execution of all the voluntary movements we perform."[45]

If voluntary acts were incited and guided by a mental image of movement that was facilitated by past muscular sensations, then what of the will? As Bastian noted, many critical issues in "psychological doctrine" turned upon this question.[46] But even physiologists who rejected the idea that the muscular sense represented the felt action of the will on the body did not abandon the belief in a personal will altogether. Many physiologists, Bastian and Claparède among them, "saw themselves as enriching the explanatory potential of mind-body dualism, not abolishing it in the name of materialism."[47] As Smith notes, what was at stake was nothing less than "the ideological burden . . . to represent the human subject *as active*, indeed knowingly active, and hence to represent a person as a moral and responsible, creative and civilised, subject."[48] An effort to understand the will scientifically—a "physiology of the will"—emerged alongside the experimental study of the nervous system, in both scientific and popular literature.[49]

Claparède's diagram of successive voluntary movement is one such attempt to explain the will physiologically (see figure 12.1). It depicts the combination of simple movements into a compound, coordinated series of movements in the course of a voluntary action. The process begins when a perception, "P," arouses a desire to act—for instance, seeing a piece of fruit and wishing to grab it.[50] That perception evokes a mental image of a movement, "I," which Claparède describes consists of a feeling of a movement done in the past (like raising the arm). Motor image "I" arouses the motor cell "E1" and causes the first simple movement in the series, "M1." This first step in the voluntary coordinated movement illustrates the physiological law we have seen above, that "a voluntary act is always preceded by an idea or conception of the movement we desire to execute; and that this idea or conception is . . . compounded of two kinds of past impressions, namely, those of the visual sense and those of the kinesthetic sense."[51] The initiation of the voluntary movement takes place at the sensorimotor level of the nervous system.

It is immediately after that first simple movement, "M1," that we want to focus our attention. The act of movement awakens "S1," a combined muscular/visual sensation that delivers the information that the cerebrum needs to direct the next movement in the series. Or, as Claparède put it, once the muscular sensations evoked by the first movement have been delivered to consciousness, "the will will be able to continue the movement and the existing 'S1' sensations will evoke the image of the movements that must still be made in order to seize the fruit."[52] The muscular sense, produced by the first movement, awakens the motor images in the brain that the will can use to guide the next movements. Claparède explained, "Pieces of information from the muscle sense only become useful when, the first simple movement having been completed, consciousness needs to be informed of it so that the will may execute the following movement." The initial simple movement, Claparède concluded, could be done in an automatic manner, without consciousness. But to combine simple movements into complex movements in a way that would reach the desired goal, "the cooperation, that is to say the association of the consciousness and will, is indispensable."[53]

Thus, the muscular sense was intimately connected to, even provocative of, the operation of the will in the performance of voluntary movements. But what, Claparède asked, was the origin of the movement image "I1" that incited the first movement in the chain? How had our ability to perform voluntary movements come about? Science did not totally understand the origin of

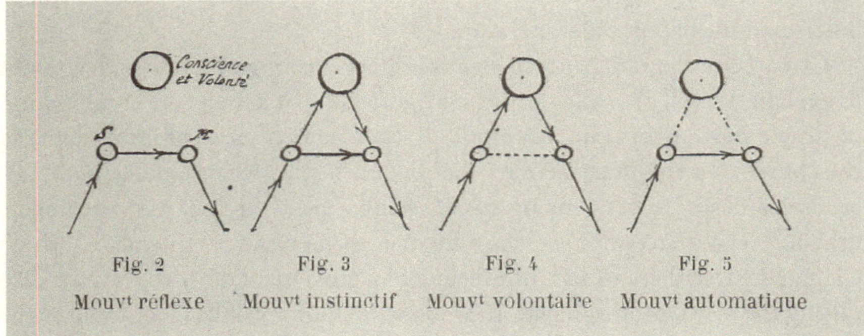

Figure 12.2. Diagram of the evolution of movement by Edouard Claparède, in *Du Sens Musculaire: à Propos de Quelques Cas d'Hémiataxie Posthémiplégique* (Geneva: Eggimann, 1897), 61. Public domain.

voluntary movement, he acknowledged. But the reflex movement alone could not explain it, as the reflex, by definition, was not accompanied by consciousness and therefore could not convey the kinesthetic impressions needed for voluntary movement to the brain. There thus existed an intermediate movement that was "mechanical, like the reflex, but conscious, like a voluntary act."[54] This was the instinctive movement.

In physiology, the instinctive movement was an evolution of a reflex action. It emerged after a species had been habitually subjected to a recurring sensation, resulting in the formation of new nervous pathways that were then inherited by offspring. In Claparède's diagram of the evolution of movement (see figure 12.2), instinctive movement is unique in that the bodily phenomena of sensation and movement ("S" and "M") are in full cooperation with the mental aspects of "consciousness and volition." Because it bridged body and mind in this way, instinctive movement provided the images to the brain that would facilitate future voluntary movements.[55] Claparède's diagram shows a developmental process in which reflex movements become instinctive movements, instinctive movements facilitate voluntary movements, and voluntary movements eventually become automatic—their "last stage of evolution."[56]

The instinctive act was organized—as a structured pattern of nervous action, it developed a "rhythmical manner," becoming, after multiple repetitions, "rapid and regular."[57] It was also malleable, becoming more complex when the nervous system met new stimuli in the environment. As new nervous pathways acquired by the individual were passed to subsequent generations, more complex instincts developed, and these eventually gave rise to higher mental powers in animals with more developed nervous systems.[58]

"Reason, Imagination, and Volition" were each "higher developments arising out of previous processes" that emerged when animals were exposed to new stimuli and experiences and their nervous systems adapted accordingly.[59] The reorganization of nervous action was the foundation for the evolution of higher mental states.

Physiologists' discovery of the plasticity of the nervous system—its ability to form new structures through repetition and in response to changes in environmental conditions—dovetailed with Neo-Lamarckianism, the dominant theory of evolution at the time. Neo-Lamarckians fused Jean-Baptiste Lamarck's theory of acquired characteristics—in which the use and disuse of organs over time could lead to adaptations that were passed to subsequent generations—to Darwin's theories of natural selection and extinction. This allowed them to argue that "while degeneration could lead to the extinction of a race, conversely regeneration could lead to their evolution, particularly through the inheritance of positive traits or beneficial characteristics."[60] Neurophysiologists saw regenerative potential in the nervous system's inherent "proclivities to the development of new nervous mechanisms."[61] If this neuroplasticity was the basis for all volitional movements, both those that developed in infancy because of inherited nervous structures, and those acquired by the individual, like a new dance step, then all volitional acts were "merely Automatic acts in process of formation, first of all for the Individual and subsequently . . . for the Race."[62] The reformed nervous structures that accompanied new movement automatisms were transmitted by heredity to become the basis for future volitional movements.

Bastian drew on a thermodynamic conception of the human body to describe the mental progress achieved when motor processes evolved to an automatic state. As the "the machine-like" workings of the motor side of the nervous system took over the performance of movements, the individual became "free to follow up the threads of our conscious life unhindered by the multitudinous details pertaining to the varying states of innumerable muscles acting in ever changing combinations."[63] This concept drew on an understanding of the mind-body as a closed system of energy conservation and conversion in which increased work on one side of the system, the bodily mechanism, would free up energy on the other side of the system, the mind, for higher mental acts.[64]

This is the science behind Jaques-Dalcroze's assertion "that experiments in rhythm, and the complete study of movements simple and combined, ought to create a fresh mentality."[65] Under Claparède's tutelage, Jaques-Dalcroze endeavored to understand and apply the physiology of the muscular sense and its relation to mind to his theory and system of rhythmic training. In so doing, he forged a manifesto for the transformative effects of creative bodily performance, laying historical roots for early modern dance. Jaques-Dalcroze proposed that rhythmic movement was the key, as captured in Claparède's diagram, to activating the will and acquiring new nervous tracts that would eventually become automatic. This process would create new regenerative possibilities not only for the individual, but also for a future human being better adapted to environmental demands.

Eurhythmics

Jaques-Dalcroze published his full method in 1906, devoting Volume I to Eurhythmics. The introduction framed the main problem: modern education had failed to develop the physical life of the body. "The living muscular forces subjected to the influences of the nervous centers, vibrate with neither control nor goal," Jaques-Dalcroze claimed. "The will is not exercised to dominate and direct them, and intellectual desires grow parallel to muscular capacities without educators seeking to put the latter at the service of the former."[66] Eurhythmics aimed to "educate" the will so that instinctive bodily movements would be better "regulated by a reasoned and conscious intelligence." By practicing how to apply varying degrees of muscular exertion in movement, the student would gain greater control over a more powerful nervous system, and a "complete harmony" would be established between the "cerebral intelligence" and the physical "instinctive impulses," the former gaining independence from the latter, and not "allowing itself to be dominated by it."[67]

These "instinctive impulses" of the body manifested in rhythm. Jaques-Dalcroze's definition of rhythm was rooted in physiology: it was "the reflex of instinctive corporeal movements."[68] Recall that physiologists took instinct as an organized and inherited pattern of nervous action in which mental and physiological events were simultaneously engaged. Jaques-Dalcroze's linking of rhythm to instinct helps to explain his varied and flexible use of the term. On the one hand, "rhythm" pointed to the unique aspects of a person's temperament, "the very image of the soul" that emerged in individual variations in movement.[69] On the other hand, it referred to the inherited and shared expression of a group of people; the observable way in which "the influence of climate, customs, and historical and economic circumstances must have produced certain differences in the rhythmic sense of each people, which are reproduced and perpetuated in such a way as to imprint a peculiar character on the dynamic and nervous manifestations at the root of every original corporal rhythm."[70] And in still another usage, rhythm was an instinctive capacity possessed by all humans that linked bodily movement and consciousness, through the muscular sense.[71] As such, it was a powerful access point for the nervous connections between body and mind and thus for "the training of the physical will, or the disciplining of the nerve-centres."

Each of the thirty lessons in the first volume of *Méthode Jaques-Dalcroze* begins with general exercises in breathing, balance, strength, and flexibility. The volume includes a detailed explanation of the anatomy and physiology of breathing, including the movements of the diaphragm, intercostal muscles, ribs, and clavicles. It gives exercises for increasing awareness of these muscular and bony actions and for manipulating them by altering the tempo and force of the breath.[72] Odom aligns these warm-up exercises with Swedish gymnastics, a contemporaneous method that focused on functional movement.[73]

The walking exercises in the *Méthode* divide the step into "strong" beats (when the foot hits the ground) and "weak" beats (when the foot lifts). The foundational rule of "walking in time" in Eurhythmics is to give each beat the same duration. As the exercises develop, quicker and slower tempos are combined with smaller and larger step-lengths, and patterns of breathing and beating time with the arms are integrated.[74] Through these exercises "the pupil is trained to adapt different muscular processes for short and long durations respectively, to estimate durations according to the sensations of tension and extension of muscles . . . to coordinate the different dynamic forces of the body, and to apply the measure of space to the control of the duration and intensity of muscular contractions."[75] The Eurhythmics student practiced regulating the rhythm of the breath, relaxing and contracting different muscle groups, and increasing and decreasing muscular force—lessons that taught students, first, how to be conscious of the body and its movement, and second, how to control it.

It is noteworthy that Eurhythmics was focused just as much, if not more, on the *repression* of movement as on its execution.[76] This is especially evident in the exercises that involved the command "*hop*," which required students to respond suddenly to the instructor's order by changing what they were doing—shifting directions, alternating which arm was beating the bar, replacing a step with a hop, or some combination of these.[77] As the student progressed, the command signaled more challenging modifications: replacing steps with one or more bars of stillness, altering the phrasing or the note value of the steps, or adding changes of level such as kneeling and rising.[78] The "hop" exercises were followed by "*exercises d'arrêt*" which required the student to stop the walk mid-step, weight on the forward foot, back foot "slightly on the tip," and stay still for a measure while "measuring time by intensifying the counting *mentally:* 'one,' 'two.'"[79] Such stopping exercises forced the student "to contrive to prevent his arm from contracting, or his foot from stamping."[80]

The objective of the "hop" lessons was to develop the student's "Spontaneous Will-power" and "Faculties of Inhibition."[81] This places Eurhythmics within a larger history of "inhibition" in the late nineteenth century. As Roger Smith details, the term appeared across multiple disciplines to represent regulation and control in a world where disorder threatened to break out at all levels. In physiology, "inhibition," with its antonym, "excitation," became known as the process by which the cerebrum controlled and organized bodily movement in a purposive way.[82] Moreover, these accounts often rested on a hierarchical organization of the nervous system, conflating the cerebrum, as the "highest" brain structure, and the soul in ways that ultimately defended "Cartesian dualism and a Christian idea of the human essence."[83] Smith writes,

> There was a gradual reformation of religious, moral, and political "higher" powers of control in terms, initially, psychological, of the mind's control over the body and, later physiological, of the brain's control over the nervous system. Language changed: . . . spirit became mind, and mind became the upper brain; flesh became body, and body became the spinal nervous system.[84]

Referencing the power of the mind over the body, "inhibition" became a key term in studies of the will from a physiological standpoint.[85]

Nineteenth-century science's understanding of the nervous system as the central controlling mechanism in the body, and inhibition as the power of the will, is the larger context for the strong emphasis on order and control that characterizes not only the "hop" exercises, but also the entire system of Eurhythmics. By practicing how to deliberately repress movement, the student would learn to engage varying degrees of "muscular innervation," or nervous inhibition and excitation.[86] Retraining the nervous system in this way would create the proper "co-ordination between the *mind* which conceives, the *brain* which orders, the nerve which transmits and the muscle which executes" clearly and with little delay.[87] The image Jaques-Dalcroze crafts in this order of events—the mind acting downward, through the brain and the nervous system to execute bodily movement—rests on the same hierarchical organization of the nervous system we have just seen, including its connection to Christian beliefs in the soul's control over the flesh. Indeed, establishing the power of the soul over the body was at the heart of Eurhythmics.

An "education 'by rhythm'" strove to place "the individual's motor powers [in] immediate contact with the cerebral and the emotional faculties, for soul and body to be in mutual and intimate communication."[88] And while Eurhythmic training was designed "to harmonise mind and body," this was not an interactive relationship, but rather one that was definitively hierarchical.[89] As Jaques-Dalcroze reiterated many, many times, "The aim of all exercises in eurhythmics is . . . to accustom the body to hold itself, as it were, at high pressure in readiness to execute orders from the brain."[90] Facilitating such pressure was precisely the purpose of the "hop" and "*d'arrêt*" exercises, which forced the mover to "produce either movements or sudden halts . . . to enable the mind to choose from all the muscles the one most necessary for the action demanded and to keep the other muscles motionless; to train the nervous system in such a way that the commands transmitted by the mind may be immediately and completely performed."[91] Eurhythmics drew on the association of nervous inhibition with the power of the will, as the agent of the soul, over the body.

"Our mental powers are developed by conscious control of the body," Jaques-Dalcroze advised.[92] Claparède's diagram of successive voluntary movement (figure 12.1) illuminates the physiology supporting that claim. Physiological research had demonstrated that many bodily actions took place solely between sensory and motor nerves, without any need for the intervention of mind. Even voluntary actions eventually became automatic through repetition. What power, then, did the will have? The answer lay in the inhibitory role of the higher nervous system. A complex voluntary movement involved two different classes of nervous action: "the cerebral incitation to the movement—the *volition*" and "the immediate execution of the movement by means of properly co-ordinated muscular action."[93] The latter was accomplished entirely by the nervous system; once nervous connections in the spinal cord had been established through repeated movements, "the will interferes no more with them." It remained the responsibility of the will, however, to determine *how* the movement was done. In "voluntary acts we have not merely to account for the production of a movement," wrote Bastian:

> but of a movement of a certain kind, in which the action of each set of muscles brought into play is duly regulated so as to lead to the exact result desired. And, seeing that the volition or desire to bring about such and such movements is purely cerebral in its origin, so we are bound to admit that such qualities of movement as are contained under this head must also depend upon cerebral influence. Hence the strength, the con-

> tinuance, the rapidity, and the direction of movements, are variable according to the precise nature of the cerebral incitation or volition.[94]

The fact that movements occurred was due to nervous operations at the bodily level. But how they occurred—their force, progression, speed, and direction—was dependent upon higher cerebral processes. Bastian called these the "volitional qualities."[95] The work of the will was *regulatory:* it used the information from the muscular sense to decide whether to increase or decrease the force, to continue or bind the progression, to accelerate or decelerate the tempo, and how to direct the movements in space. In Claparède's diagram, these qualities are activated at the critical point "S" when the muscular sensation reaches consciousness and brings the will into the process. This is where Eurhythmic training attempted to intervene—the point at which the muscular sensation aroused the will to assert its control of the body through the movement qualities of time, resistance, continuation, and direction. This physiology, with its close relation to Cartesian dualism, underlies Jaques-Dalcroze's theory of Eurhythmics as a means of "educating" the "will of the individual, transmitted to the muscles by the nerves, [in order to] make them the rapid and precise executors of his desires and aspirations."[96]

Establishing such "clear and rapid communication between the two poles of our being"—mind and body—was only one of the goals of Eurhythmics. The second was to "to create automatisms and to assure the effective working of the muscular system."[97] The regular practice of Eurhythmics would create new nervous tracts and accelerate the evolution of movement to the final stage when volitional actions became automatic (figure 12.2). Communication from body to brain and back to body would become more efficient, cutting "the time lost between the conception and realization of the movement."[98] New automatisms would give rise to powerful new physical and mental potentialities, resulting in a state of heightened concentration in which the student was "executing a rhythm already heard, [while listening] to a second: one automatism being in operation while another is preparing; the body in the past, the mind intent on the future."[99]

Taking a page out of physiology, Jaques-Dalcroze drew on the thermodynamic conception of the body:

> If we can teach our bodies to work automatically, our minds will have more time and freedom for higher things. If we are obliged always to be thinking of our bodies we must perforce lose some of our liberty of mind. Without doubt the majority of mankind are the slaves of their bodies, prisoners in matter, and, contrary to what is generally believed, the over-cultivation of intellectualism, of analytical studies and special-

> ized psychology, tends rather to trouble and disturb the mind than to render it lucid and calm.[100]

Eurhythmic practice would capitalize on the potential of the nervous system to form new habits and reflexes, thereby redistributing the forces in the psycho-physical economy. As the output needed to operate the bodily mechanism was minimized, the total energy available to the mind would be maximized. This would give rise to a new level of spiritual agency. "The more automatism possessed by our body, the more our soul will rise above material things," Jaques-Dalcroze insisted.[101]

Eurhythmics prepared the body to become the active instrument of the mind in a world full of forces that threatened a loss of control. This was its therapeutic promise, which spoke to contemporaneous anxieties about human degeneration under the conditions of industrial society. In this argument, the ever-proliferating new technologies of communication, media, commerce, and transportation produced a barrage of stimuli that overwhelmed the nervous system, resulting in a physiological state of fatigue that manifested in a host of disorders including alcoholism, hysteria, and nervous diseases.[102] Medical and psychiatric studies of degeneration took it as a biological and hereditary process of pathological decline. The fatigued body suffered a debilitating loss of willpower, "a progressively intensifying tyranny of the body over the spirit or soul."[103]

Indeed, Jaques-Dalcroze saw the lack of control over the body as the root cause of nervous disorders. "Neurasthenia is often nothing else than intellectual confusion produced by the inability of the nervous system to obtain from the muscular system regular obedience to the order from the brain," he wrote. "Training the nerve centres, establishing order in the organism, is the only remedy for intellectual perversion produced by lack of will power and by the incomplete subjection of body to mind."[104] The "music cure" could restore the mind's dominance over the unruly body and thus provide "a new vital impulse" that would strengthen not only the individual, but all of humanity.[105]

Fusing neurophysiology with neo-Lamarckian evolutionary theory, Jaques-Dalcroze ascribed regenerative and inheritable benefits to Eurhythmics. Through rhythmic training, the individual would

> learn that we are masters of our fate, that heredity is powerless if we realize that we can conquer it, that our future depends upon the victory which we gain over ourselves. However weak the individual may be, his help is required to prepare a way for a better future. Life and growth are one and the same, and it is our duty by the example of our lives to develop those who come after us. Let us therefore assume the

> responsibility which Nature puts upon us, and consider it our duty to regenerate ourselves; thus shall we help the growth of a more beautiful humanity.[106]

Jaques-Dalcroze translated the physiology of the muscular sense and its relation to the will directly into Eurhythmics to make new claims for the therapeutic impact of performative movement on the individual, society, and even the future of humanity.

Jaques-Dalcroze and the Dance

As Jaques-Dalcroze's place in dance history comes into view, I will spotlight four traces of his work in modern dance. First, his formulation of movement training as learning how to manipulate degrees of muscular tension in combination with different time durations and spatial range was taken up and expanded by later dancers, perhaps most obviously in Laban's theory of space harmony. As Laban conceived his movement scales, the moving body not only maps spatial forms, but also generates them through the rhythmic pathways traced by the mover's limbs. Laban's idea that movement and space are mutually constructive—that "there is neither space without movement nor movement without space"—built on Jaques-Dalcroze's claims that "all movement needs space and time [and] space and time are connected by the matter which crosses them in an eternal rhythm."[107]

Second, Jaques-Dalcroze's deliberate application of principles of physiology to physical training helped establish the focus on bodily awareness and regulation that drove emerging forms of somatics, body culture, and modern dance. Dalcrozian ideas echo, for instance, in Mabel Elsworth Todd's approach to "structural hygiene," which employs principles from physics, mechanics, and physiology to develop the kinesthetic sense and achieve an economy of effort in posture and movement that will help prevent fatigue, tension, anxiety, and other maladies of life and work in the modern era.[108]

Third, the emphasis on "mastering" the body to bring it under control of the mind in Eurhythmics resonates in *Ausdruckstanz* and beyond. Take, for instance, Expressionist poet and dance critic Ernst Blass's 1921 claim that the new German modern dance had tamed the body so that "rendered obedient, it allows itself to be guided by the will and the spirit."[109] Related to this idea is Jaques-Dalcroze's elaboration of the body as an "instrument," an image that enjoyed a long lineage in modern dance after his work. Laban put the relationship succinctly: the body was an "instrument of our mind and soul."[110] Such Dalcrozian imagery also courses through Margaret H'Doubler's early theories of dance education, which emphasized the importance of strengthening the kinesthetic sense so that the "physical being [would be] perfected but primarily to serve as a well-ordered instrument correctly tuned, and sensitive to the impressions of the mind. It must . . . be able to respond instantaneously without hindrance."[111] As an "instrument," the body could be operationalized for deliberate ends.

Finally, through his prolific writings on rhythmic gymnastics and its therapeutic effects, Jaques-Dalcroze put forth a modernist manifesto that advocated for movement practice as a unique means of transforming the self and addressing social problems—an effort that Dee Reynolds places at the heart of early modern dance.[112] Former Wigman dancer and pioneering dance therapist Liljan Espenak credited Jaques-Dalcroze for creating a method that sought "to develop the mind as a result of the physical experience," which drove not only early modern dance, but also the development in the US of dance therapy in the late 1930s and 1940s.[113] With strong echoes of Jaques-Dalcroze, dance therapist Blanche Evan ascribed to her early work the need to repair the loss of "instinctive rhythm" caused by life in a mechanized urban setting.[114]

Jaques-Dalcroze's integration of physiology and aesthetics was vital to early modern dance. In his preoccupation with the physiology of the muscular sense and his endeavors to mold that science into a method of working on the mind through bodily movement, we find a therapeutic impulse in the very origins of the form that continued well into—and perhaps beyond—the twentieth century.

Notes

1 Hall, "Muscular Perception," 435, emphasis added.
2 Smith, "The Physiology of the Will," 81.
3 Lears, *No Place of Grace.*
4 Lears, *No Place of Grace,* 50.
5 Pick, *Faces of Degeneration,* 20–21.
6 Jaques-Dalcroze, *Rhythmic Movement and Education,* vi, emphasis in original.
7 Odom, "Delsartean Traces," 139; see also Keilson, "Making Dance Modern," 47–51.
8 Jaques-Dalcroze, *Rhythm, Music and Education,* vi.
9 Jaques-Dalcroze, *Rhythmic Movement,* iv.
10 Jaques-Dalcroze, "Rhythmic Movement, Solfège, and Improvisation," 116, emphasis in original.
11 Jaques-Dalcroze, "From the Lectures" [1912], 30.
12 Odom, "Jaques-Dalcroze;" Vertinsky, "Transatlantic Traffic," 154.
13 Vertinsky, "Transatlantic Traffic," 151–52.

14 For instance, Dalcroze Eurhythmics was part of the training offered by the New Dance League School in the mid-1930s, as evidenced in several references in *New Theatre*, published by the group; for example, "A Recent Activity . . . ," 29.
15 Keilson, "Making Dance Modern," 42. Also see Huxley and Burt, "Concerning the Spiritual," 252, 254.
16 Schwartz, "Torque," 77.
17 Cowan, *Cult of the Will*, 188.
18 Jaques-Dalcroze, *Méthode Jaques-Dalcroze*, viii. All translations from this volume are my own.
19 Jaques-Dalcroze, "Initiation into Rhythm," 83; "Rhythm as a Factor," 18. Jaques-Dalcroze used both "*esprit*" and "*âme*" in his writings, which are commonly translated into English as "mind" and "soul," but also sometimes as "spirit" and "soul" (see Habron and van der Merwe, "A Conceptual Study," 179). I believe Jaques-Dalcroze drew on a Cartesian framework, prevalent in mid- to late nineteenth-century physiology and psychology, in which "mind" and "soul" can be used interchangeably, and the will is understood as "an essential force of the soul" (Smith, "The Physiology of the Will," 94, 90–98).
20 Spector, *Rhythm and Life*, 66; Ingham, "Jaques-Dalcroze Method," 38–39.
21 Jacque-Jaques-Dalcroze, letter to Claparède, June 12, 1906, quoted in Spector, *Rhythm and Life*, 144.
22 Smith, "The 'Sixth Sense,'" 220. On bodily experience as a means of knowing in Eurhythmics, see Habron, "'Through Music,'" 94; Juntunen and Westerlund, "Digging Jaques-Dalcroze," 206–209. I follow these authors in their conclusions about embodiment and therapeutics in Eurhythmics but disagree with their claims that Jaques-Dalcroze rejected mind-body dualism.
23 Smith, "Sixth Sense,'" 222.
24 Bell, "Idea of a New Anatomy," 161.
25 Bell, "On the Nervous Circle," 166–67.
26 Bell, "On the Nervous Circle," 167.
27 Bell, *The Hand*, 235.
28 Bastian, *The Brain as an Organ*, 543.
29 Smith, *Inhibition*, 68.
30 Bastian, "The 'Muscular Sense,'" 2.
31 Smith, "The 'Sixth Sense,'" 251.
32 Smith, "The Physiology of the Will," 89.
33 Clark and Jacyna, *Nineteenth-Century Origins*, 129–32.
34 Claparède, *Du Sens Musculaire*, 29–30. All translations from this volume are my own, with the assistance of Louis Betty.
35 Claparède, *Du Sens Musculaire*, 42
36 Bastian, "The 'Muscular Sense,'" 45.
37 Claparède, *Du Sens Musculaire*, 43–44.
38 Bastian, "The 'Muscular Sense,'" 2.
39 Bastian, "The 'Muscular Sense,'" 3–4.
40 Bastian, "The 'Muscular Sense,'" 53.
41 Bastian, "The 'Muscular Sense,'" 4.

42 Bastian, "The 'Muscular Sense,'" 58.
43 Bastian, "The 'Muscular Sense,'" 56.
44 Bastian, "The 'Muscular Sense,'" 7.
45 Bastian, "The 'Muscular Sense,'" 37.
46 Bastian, "The 'Muscular Sense,'" 50.
47 Smith, *Inhibition,* 79.
48 Smith "The Sixth Sense,'" 232, emphasis in original.
49 In "The Physiology of the Will," Smith argues that the understanding of the will as both psychological and physiological in nineteenth-century physiology became the discursive origins of psychology.
50 Claparède, *Du Sens Musculaire,* 70–71.
51 Bastian, "The 'Muscular Sense,'" 59.
52 Claparède, *Du Sens Musculaire,* 71. Subsequent quotation on same page.
53 Claparède, *Du Sens Musculaire,* 71. Claparède is citing the physiologist Fulgence Raymond, but it is unclear which text he is quoting.
54 Claparède, *Du Sens Musculaire,* 60.
55 Claparède, *Du Sens Musculaire,* 61.
56 Claparède, *Du Sens Musculaire,* 71, 61.
57 Bastian, *The Brain as an Organ,* 220, 221.
58 Bastian, *The Brain as an Organ,* 246.
59 Bastian, *The Brain as an Organ,* 252–53.
60 Brauer, "The Janus Face of Evolution," xxiii.
61 Bastian, *The Brain as an Organ,* 563.
62 Bastian, *The Brain as an Organ,* 563.
63 Bastian, "The 'Muscular Sense,'" 53, 59.
64 Rabinbach, *The Human Motor,* especially 64–68.
65 Jaques-Dalcroze, "Rhythm as a Factor," 19.
66 Jaques-Dalcroze, *Méthode Jaques-Dalcroze,* vii. Subsequent quotation on the same page.
67 Dalcroze, *Méthode Jaques-Dalcroze,* viii.
68 Jaques-Dalcroze, "An Essay in the Reform of Music," 51.
69 Jaques-Dalcroze, "An Essay in the Reform of Music," 51. Jaques-Dalcroze is quoting Diderot here.
70 Jaques-Dalcroze, "Rhythm, Time and Temperament," 320.
71 Jaques-Dalcroze, "The Initiation into Rhythm," 86–87. Subsequent quotation, 88.
72 Jaques-Dalcroze, *Méthode Jaques-Dalcroze,* 11–14.
73 Odom, "Delsartian Traces," 143–44.
74 Jaques-Dalcroze, *Méthode Jaques-Dalcroze,* 25–29.
75 Jaques-Dalcroze, "Rhythmic Movement, Solfège, and Improvisation," 123–24.
76 Jaques-Dalcroze, "Rhythmic Movement, Solfège, and Improvisation," 124.
77 Jaques-Dalcroze, *Méthode Jaques-Dalcroze,* 30. The word in the French text is "hop"; English translations use "hopp." Boyarsky reports that "hopp" meant "change," and was accompanied at some point by "heep" and "hupp" which could mean "change back" or "go on to the next series" (Boyarsky, "Dalcroze Eurhythmics," 16).
78 Jaques-Dalcroze *Rhythmic Movement,* 29.

79 Jaques-Dalcroze, *Méthode Jaques-Dalcroze,* 31, emphasis in original.
80 Jaques-Dalcroze, "Rhythmic Movement, Solfège, and Improvisation," 122.
81 Jaques-Dalcroze, "Rhythmic Movement, Solfège, and Improvisation," 124.
82 Smith, *Inhibition,* 71–72.
83 Smith, *Inhibition,* 75.
84 Smith, *Inhibition,* 2–3.
85 Smith, *Inhibition,* 17.
86 Jaques-Dalcroze, "The Initiation into Rhythm," 88.
87 Jaques-Dalcroze, "Rhythm as a Factor," 15, my emphasis.
88 Jaques-Dalcroze, preface to *Eurhythmics, Art, and Education,* vi, vii.
89 Jaques-Dalcroze, foreword to *Rhythm, Music and Education,* vii.
90 Jaques-Dalcroze, "Rhythmic Movement, Solfège, and Improvisation," 118.
91 Jaques-Dalcroze, "The Nature and Value of Rhythmic Movement," 6.
92 Jaques-Dalcroze, *Rhythmic Movement,* v.
93 Bastian, "Remarks on the 'Muscular Sense,'" 462. Subsequent quotation on the same page.
94 Bastian, "Remarks on the 'Muscular Sense,'" 462.
95 Bastian, "Remarks on the 'Muscular Sense,'" 462.
96 Jaques-Dalcroze, *Méthode Jaques-Dalcroze,* vii.
97 Jaques-Dalcroze, "Rhythmic Movement, Solfège, and Improvisation," 117.
98 Jaques-Dalcroze, "Rhythmic Movement, Solfège, and Improvisation," 125.
99 Jaques-Dalcroze, "Rhythmic Movement, Solfège, and Improvisation," 128.
100 Jaques-Dalcroze, *Rhythmic Movement,* iv.
101 Jaques-Dalcroze, "Rhythmic Movement, Solfège, and Improvisation," 116.
102 As a classic example of this construction, see Nordau, *Degeneration,* 37–42.
103 Pick, *Faces of Degeneration,* 51.
104 Jaques-Dalcroze, "Rhythm as a Factor," 17.
105 Jaques-Dalcroze, "Music and the Child," 111.
106 Jaques-Dalcroze, "From the Lectures," 32.
107 Laban, "Choreutics," 192; Jaques-Dalcroze, *Méthode Jaques-Dalcroze,* viii.
108 Todd, *The Thinking Body,* 41–44.
109 Blass, *Das Wesen des neuen Tanzkunst,* cited in Cowan, *Cult of the Will,* 222.
110 Laban, "Gymnastics and Dance," 91.
111 H'Doubler, *A Manual of Dancing,* 9.
112 Reynolds, *Rhythmic Subjects,* 19.
113 Espenak, *Dance Therapy,* 62.
114 Evan, "The Child's World," 67.

Bibliography

"A Recent Activity of Importance . . . ," *New Theatre,* May 1935, 29.

Bastian, H. Charlton. *The Brain as an Organ of Mind.* New York: D. Appleton, 1880.

Bastian, H. Charlton. "The 'Muscular Sense': Its Nature and Cortical Localisation." *Brain* 10 (1887): 1–89.

Bastian, H. Charlton. "Remarks on the 'Muscular Sense' and on the Physiology of Thinking." *British Medical Journal* (May 22, 1869): 461–63.

Bell, Charles. *The Hand: Its Mechanism and Vital Endowments, as Evincing Design.* London: John Murray, 1854.

Bell, Charles. "Idea of a New Anatomy of the Brain; Submitted for the Observations of his Friends." 1811, private printing of 100 copies; reprinted in *Journal of Anatomical Physiology* 3 (Nov. 1868): 147–82.

Bell, Charles. "On the Nervous Circle Which Connects the Voluntary Muscles with the Brain." *Philosophical Transactions of the Royal Society of London* 116, n. 1/3 (1826): 163–73.

Boyarsky, Terry. "Dalcroze Eurhythmics and the Quick Reaction Exercises." *The Orff Echo* (Winter 2009): 15–18.

Brauer, Fae. "The Janus Face of Evolution: Degeneration, Devolution, and Extinction in the Anthropocene." Introduction to *Picturing Evolution and Extinction: Regeneration and Degeneration in Modern Visual Culture,* edited by Fae Brauer and Serena Keshavjee, xv–xlii. Newcastle upon Tyne: Cambridge Scholars Publishing, 2015.

Claparède, Edouard. *Du Sens Musculaire à Propos de Quelques Cas d'Hémiataxie Posthémiplégique.* Geneva: Eggimann, 1897.

Clark, Edwin, and L. S. Jacyna. *Nineteenth-Century Origins of Neuroscientific Concepts.* Berkeley: University of California Press, 1987.

Cowan, Michael. *Cult of the Will: Nervousness and German Modernity.* University Park: Pennsylvania University State Press, 2008.

Espenak, Liljan. *Dance Therapy: Theory and Application.* Springfield, IL: Charles C. Thomas, 1981.

Evan, Blanche. "The Child's World, It's Relation to Dance Pedagogy: Article V, Technique for the City." In *Collected Works by and about Blanche Evan,* compiled by Ruth Gordon Benov, 66–70. San Francisco: Blanche Evan Dance Foundation, 1991.

Habron, John. "'Through Music and Into Music,' Through Music and Into Well-Being: Dalcroze Eurhythmics as Music Therapy." *TD: The Journal for Transdisciplinary Research in Southern Africa* 10, n. 2 (Nov. 2014): 90–110.

Habron, John, and Liesl van der Merwe. "A Conceptual Study of Spirituality in Selected Writings of Émile Jaques-Dalcroze." *International Journal of Music Education* 35, n. 2 (2017): 175–88.

Hall, G. Stanley. "The Muscular Perception of Space." *Mind* 3, n. 12 (Oct. 1878): 433–50.

H'Doubler, Margaret. *A Manual of Dancing: Suggestions and Bibliography for the Teacher of Dancing.* Madison, WI: Tracy & Kilgore, 1921.

Huxley, Michael, and Ramsay Burt. "Concerning the Spiritual in Early Modern Dance: Émile Jaques-Dalcroze and Wassily Kandinsky Advancing Side by Side." *Dance, Movement, and Spiritualities* 1, n. 2 (2104): 251–69.

Ingham, Percy. "The Jaques-Dalcroze Method." In *The Eurhythmics of Jaques-Dalcroze,* translated by Percy and Ethel Ingham, 35–53. Boston: Small, Maynard & Company, 1918.

Jaques-Dalcroze, Émile. "An Essay in the Reform of Music Teaching in Schools." In *Rhythm, Music, and Education,* translated by Harold F. Rubinstein, 13–57. New York: G. P. Putnam and Sons, 1921.

Jaques-Dalcroze, Émile. "From the Lectures of Émile Jaques-Dalcroze." In *The Eurhythmics of Jaques-Dalcroze*. Translated by Percy and Ethel Ingham, 29–34. Boston: Small, Maynard & Company, 1918.

Jaques-Dalcroze, Émile. "The Initiation into Rhythm." In *Rhythm, Music, and Education,* translated by Harold F. Rubinstein, 79–92. New York: G. P. Putnam and Sons, 1921.

Jaques-Dalcroze, Émile. *Méthode Jaques-Dalcroze, Vol I: Gymnastique Rythmique.* Paris: Sandoz, Jobin & Cie, 1906.

Jaques-Dalcroze, Émile. "Music and the Child." *Rhythm, Music, and Education,* translated by Harold F. Rubinstein, 95–112. New York: G. P. Putnam and Sons, 1921.

Jaques-Dalcroze, Émile. "The Nature and Value of Rhythmic Movement." In *Eurhythmics, Art, and Education,* edited by Cynthia Cox, translated by Frederick Rothwell, 3–13. New York: Arno Press, 1976.

Jaques-Dalcroze, Émile. "Rhythm as a Factor in Education." In *The Eurhythmics of Jaques-Dalcroze,* translated by Percy and Ethel Ingham, 12–22. Boston: Small, Maynard & Company, 1918.

Jaques-Dalcroze, Émile. "Rhythmic Movement, Solfège, and Improvisation." In *Rhythm, Music, and Education,* translated by Harold F. Rubinstein, 115–42. New York: G. P. Putnam and Sons, 1921.

Jaques-Dalcroze, Émile. *Rhythmic Movement, Vol. 1 of The Jaques-Dalcroze Method of Eurhythmics.* London: Novello and Company, 1920.

Jaques-Dalcroze, Émile. *Rhythm, Music and Education.* Translated by Harold F. Rubinstein. New York: G. P. Putnam and Sons, 1921.

Jaques-Dalcroze, Émile. "Rhythm, Time, and Temperament." In *Rhythm, Music, and Education,* translated by Harold F. Rubinstein, 309–34. New York: G. P. Putnam and Sons, 1921.

Juntunen, Marja-Leena, and Heidi Westerlund. "Digging Dalcroze, or, Dissolving the Mind-Body Dualism: Philosophical and Practical Remarks on the Musical Body in Action." *Music Education Research* 3, n. 2 (2001): 203–14.

Keilson, Ana Isabel. "Making Dance Modern: Knowledge, Politics, and German Modern Dance, 1880–1927." PhD diss., Columbia University, 2017.

Laban, Rudolf. "Choreutics" (excerpts). In *The Laban Sourcebook,* edited by Dick McCaw, 175–96. Translated by Stefanie Sachsenmaier and Dick McCaw. New York: Routledge, 2011.

Laban, Rudolf. *Gymnastics and Dance* (excerpts). In *The Laban Sourcebook,* edited by Dick McCaw, 83–96. Translated by Stefanie Sachsenmaier and Dick McCaw. New York: Routledge, 2011.

Lears, T. J. Jackson. *No Place of Grace: Antimodernism and the Transformation of American Culture, 1880–1920.* Chicago: University of Chicago Press, 1994.

Nordau, Max. *Degeneration.* Lincoln: University of Nebraska Press, 1993. Translated from the second edition of the German work. Reprint, New York: D. Appleton, 1895.

Odom, Selma Landen. "Delsartian Traces in Dalcroze Eurhythmics." In "Essays on François Delsarte," ed. Nancy Lee Chalfa Ruyter and Thomas Leabhart, special issue, *Mime Journal* (2004–2005): 137–51.

Odom, Selma Landen. "Jaques-Dalcroze, Emile." In *The International Encyclopedia of Dance* (online), edited by Selma Jeanne Cohen and Dance Perspectives Foundation.

New York: Oxford University Press, 2005. https://www.oxfordreference.com/display/10.1093/acref/9780195173697.001.0001/acref-9780195173697-e-0858?rskey=ZGlknX&result=1.

Pick, Daniel. *Faces of Degeneration: A European Disorder, c. 1848–c. 1918.* New York: Cambridge University Press, 1989.

Rabinbach, Anson. *The Human Motor: Energy, Fatigue, and the Origins of Modernity.* Berkeley: University of California Press, 1992.

Reynolds, Dee. *Rhythmic Subjects: Uses of Energy in the Dances of Mary Wigman, Martha Graham and Merce Cunningham.* Hampshire, UK: Dance Books, 2007.

Schwartz, Hillel. "Torque: The New Kinaesthetic of the Twentieth Century." In *Zone 6: Incorporations,* edited by Jonathan Crary and Sanford Kwinter, 70–127. New York: Zone Books, 1992.

Smith, Roger. *Inhibition: History and Meaning in the Sciences of Mind and Brain.* Berkeley: University of California Press, 1992.

Smith, Roger. "The Physiology of the Will: Mind, Body, and Psychology in the Periodical Literature, 1855–1875." In *Science Serialized: Representations of the Sciences in Nineteenth-Century Periodicals,* edited by Geoffrey Cantor and Sally Shuttleworth, 81–110. Cambridge, MA: MIT Press, 2004.

Smith, Roger. "The 'Sixth Sense': Towards a History of Muscular Sensation." *Gesnerus* 68, n. 1 (2011): 218–71.

Spector, Irwin. *Rhythm and Life: The Work of Émile Jaques-Dalcroze.* Stuyvesant, NY: Pendragon Press, 1990.

Todd, Mabel Elsworth. *The Thinking Body: A Study of the Balancing Forces of Dynamic Man.* New York: Dance Horizons, 1979. First published 1937 by Paul B. Hoeber (New York).

Vertinsky, Patricia. "Transatlantic Traffic in Expressive Movement: From Delsarte to Margaret H'Doubler and Rudolf Laban." In *Gymnastics, a Transatlantic Movement from Europe to America,* edited by Gertrud Pfister, 143–63. New York: Routledge, 2014.

13

Dualism in Jaques-Dalcroze's Theory of Movement

Response to "The 'Muscular Sense' and Therapeutic Modernism in the Eurhythmics of Émile Jaques-Dalcroze" by Andrea Harris

Dick McCaw

For the past twenty or so years, I have been researching the connection between neuroscience and actor training and have spent the same amount of working time in the Brotherton Library, Leeds, England, cataloguing an archive relating to movement pioneer Rudolf Laban. This work informs my response to Andrea Harris's essay on the work of Émile Jaques-Dalcroze.

Jaques-Dalcroze's "therapeutic modernism" was aimed at "neurasthenia," which he tellingly understood as a "disease of the will." It is precisely the concept of "Will" that is at the heart of his dualistic understanding of movement. This is evident in how he defined concepts like "harmony" and "rhythm." For Jaques-Dalcroze, the former was the "coordination of intellectual and physical processes achieved through Eurhythmic practice." Harris explains that nineteenth-century physiologists understood instinct as "an organized and inherited pattern of nervous action in which mental and physiological events were simultaneously engaged." Rhythm was the organization of these instincts. Thus, a rhythm could be considered as both onto- and phylogenetic, about both an individual character and a particular group; it was a way of describing their characteristic behavior, the manner in which a person controlled movement instincts. Will is the agent that, through rhythmic exercise, exerts control over inherited patterns of movement.

What is the connection between Laban and Jaques-Dalcroze? Two of the teachers working in Jaques-Dalcroze's school, Mary Wigman and Suzanne Perrottet, left his center at Hellerau to work with Laban. Both were attracted by Laban's belief in the autonomy of dance as an art form. In Laban's correspondence with Perrottet,[1] there is an inferred dialogue with Jaques-Dalcroze about the nature of rhythm and the challenge of creating dance without music—what Laban championed as "Free Dance." It was precisely this notion of dance without music (and by extension a rhythm without music) that placed Jaques-Dalcroze and Laban in opposite camps. I also find two points of close connection. In *The World of the Dancer* (1920) Laban wrote, "Action is the expression of the will and feeling through dance, integrating desire, feeling and knowledge into a unit."[2] As with Jaques-Dalcroze, action is seen as being led by will. The second similarity is how they both describe movement behavior. When discussing the brain's control of movement, Jaques-Dalcroze notes that although movements originate in the muscles, "their force, progression, speed, and direction" are "dependent upon higher cerebral processes," leading to decisions about the uses of force, progression, tempo, and spatial direction. These point to Laban's four movement factors—Weight, Flow, Time, and Space—which he was already using in his *Choreographie* (1926). Of course, at bottom, these elements are pure physics and not specifically Laban's thinking, but in terms of culture and influence, it is hard not to see here a connection between Laban and Jaques-Dalcroze. I would add that the later Laban was a dualist in his explicit distinction between everyday working movement (as in his book *Effort*) and the spiritual dimension of dance (as in his later book, *Choreutics*).

Jaques-Dalcroze's lifelong dialogue with the physiologist Eduard Claparède influenced, even shaped, his ideas about mind and body, particularly "the explanatory potential of mind-body dualism."[3] Central to this dualistic conception is an instrumental approach to the body whereby the bodily instrument "serves" mind or soul. In his *Intelligence in the Flesh* (2016), Guy Claxton argues for a "new materialism" that is explicitly opposed to such a dualistic conception. While books such as J. A. Scott Kelso's *Dynamic Patterns* (1997) offer a materialist theory of how the brain's complexity is self-organized into intelligible patterns of behavior, we intuitively still hold onto the dualistic notion that there is a *me* inside the brain that wills and makes decisions. Philosopher Mark Johnson argues that such dualism is so "deeply embedded in our Western ways of thinking that we find it almost impossible to avoid framing our understanding of mind and thought dualistically."[4]

Underpinning Jaques-Dalcroze's notion of the brain controlling the bodily instrument was the idea that the sense of movement begins in the muscle, not in the brain. Forty years later, Soviet neurophysiologist Nikolai Bernstein would argue that precision and dexterity are not achieved through motor engrams (memorized or instinctive movement patterns) but are controlled by the brain through sensory-motor loops (where the motor idea is constantly transformed through sensory feedback).[5] Bernstein and Johnson notwithstanding, many teachers in performer training still talk about "muscle memory" and the body as the actor's "instrument"—examples of dualism at work in how we think about ourselves and operationalize these thoughts in professional stage training.[6]

Even if our contemporary understanding of movement control is now quite different to that of Claparède and Jaques-Dalcroze, what Harris calls the "thermodynamic conception" of brain regulation still holds. In this model, the mind-body is a "closed system of energy conservation and conversion." When work increases in the brain, it draws energy from the body and vice versa. The aim of Dalcrozian training was to reduce mental effort through practice, repetition, and assignment of movement responses, thereby freeing the brain's energy for "higher mental acts." This account holds good in current neurophysiology theory, the difference being that this regulation is understood to happen entirely *within* the brain and its nervous systems and not *between* brain and body.

One of the most valuable aspects of Harris's chapter is that it lays bare the mechanics of dualistic thinking and thereby allows space for a consideration of a materialist theory of movement. This is important because, as Raymond Gibbs observes, mind-body dualism gives rise to a whole series of other dualisms: "subjective as opposed to objective knowledge, knowledge as opposed to experience, reason as opposed to feeling, theory as opposed to practice, and verbal as opposed to nonverbal."[7] The human body is not a fleshly instrument controlled by our brain or Will; it is, rather, a hugely complex organism that learns through doing.

Notes

1 See McCaw, *The Laban Sourcebook,* 25–26, and Chapter 2 of *The Art of Movement.*
2 McCaw, *The Laban Sourcebook,* 17.
3 Smith, *Inhibition,* 79.
4 Johnson, *The Meaning of the Body,* 7
5 Bernstein's most accessible account of this process is *The Development of Dexterity,* translated from the Russian by Mark Latash, in *Dexterity and its Development.*
6 See, for example, Kubik, "Biomechanics," 6.
7 Gibbs, *Embodiment and Cognitive Science,* 4.

Bibliography

Bernstein, Nicholai. *On Dexterity and its Development,* Part I of *Dexterity and its Development,* edited by Mark Latash and Michael Turvey. New York: Psychology Press, 1996.

Claxton, Guy. *Intelligence in the Flesh: Why Your Mind Needs your Body Much More than It Thinks.* New Haven, CT: Yale University Press, 2016.

Gibbs, Raymond. *Embodiment and Cognitive Science.* Cambridge: Cambridge University Press, 2005.

Johnson, Mark. *The Meaning of the Body.* Chicago: University of Chicago Press, 2007.

Kelso, J. A. Scott. *Dynamic Patterns: The Self-Organization of Brain and Behavior.* Cambridge, MA: MIT Press, 1997.

Kubik, Marianne. "Biomechanics: Understanding Meyerhold's System of Actor Training." In *Movement for Actors,* edited by Nicole Potter, 3–15. New York: Allworth Communications, 2002.

McCaw, Dick. *The Art of Movement: Unpublished Writings of Rudolf Laban.* Abingdon: Routledge, 2024.

McCaw, Dick. *The Laban Sourcebook.* Abingdon: Routledge, 2011.

Smith, Roger. *Inhibition: History and Meaning in the Sciences of Mind and Brain.* Berkeley: University of California Press, 1992.

14

From Animal Magnetism to Materialist Transcendentalism

Margaret Fuller on Fanny Elssler

Johanna Pitetti-Heil

> Everything tends in the civilized world to a reinstatement of the body in the rights of which it has been defrauded, as an object of care and the vehicle of expression.
>
> —Margaret Fuller, "Entertainments of the Past Winter"

> When the soul trembles, it does indeed tremble, just as much as water does when it approaches its boiling point. What we currently call the "soul" is no different from arousal and receptiveness to motion and emotion. The soul is the body that is touched, vibrating, receptive and responding. Its response is the sharing of touch, its awakening to it.
>
> —Jean-Luc Nancy, "Rühren, Berühren, Aufruhr"

Introducing Margaret Fuller's Magnetic Aesthetics

It was the fall season of 1841 when the theatergoers and transcendentalist intellectuals of New England were moved—stirred emotionally, aesthetically, and morally—by the performances of the Austrian ballerina Fanny Elssler, who toured the United States in the early 1840s. While Elssler was not the first European ballerina to perform in the United States, her performances on New England stages are of particular interest because she was the only concert dancer whose performance affected transcendental writers (Sam Ward, Henry David Longfellow, Ralph Waldo Emerson, and Margaret Fuller) and their idealist approach to the body to such an extent that they commented on them in private correspondence and published reviews.[1] Their treatment of Elssler's corporeality and movement reinforced conservative and moralist as well as idealist reservations against women's expressive bodies;[2] Elssler also elicited a particular response from the transcendentalist essayist and women's rights advocate Margaret Fuller, who was experimenting with mesmerist treatments, through which she was becoming aware of sensual, corporeal dimensions of her own well-being around the time she saw Elssler's performances. She came to understand the body as magnetically charged—and therefore materially significant—through her (self-)practice of mesmerism / animal magnetism, a therapeutic technique that sought to realign the electromagnetic fields within human bodies when they had been disturbed and were causing physical, emotional, or mental discomfort.[3] This practice and Fuller's experience of being able to manipulate the vital flows of her electric and magnetic body, I argue, correlated with (and, perhaps, influenced) her positive assessment of the ballet and Elssler's performances in New England to such a degree that it impacted her aesthetic theory in general.

Fuller's review of Elssler brings to the fore the ways in which her thinking intertwined several traditions of thought, which often appear to be contradictory. Fuller was an active member of the transcendentalist club, a group of thinkers that loosely based their positions on European idealism in that they emphasized the significance of nature, the soul, and the spirit over the corporeal realm.[4] Idealist approaches in philosophy often conflicted with materialist and socialist reform movements, which foregrounded the very material basis on which consciousness is built (property, physical well-being, working and living conditions), but groups of transcendentalist reformers, inspired by the socialist Charles Fourier, turned their attention to projects of communal living and shared farm labor.[5] Fuller was incited to think about the relationship of body and soul in altogether different ways: through her experiences with animal magnetism, she considered the material body itself as soulful, vibrant, and spiritually meaningful. Her transcendentalism thereby became inflected by a specific materialism—a materialism that was clearly distinguished from the sociopolitical form of Fourier's materialism, and which I call "materialist transcendentalism." This materialist transcendentalism was ontological, epistemological, and aesthetic—it addressed the questions of what are the body and the soul, how can the soul know the body and vice versa, and how do these aspects translate into artistic production and experience. My reading of Fuller's review of Elssler alongside Fuller's approach to animal magnetism carves out these configurations of Fuller's materialist transcendentalism, which is best described as a feminist-aesthetic approach to the formation of a woman's subject position: feminist because it explores practices through which women could acknowledge their situated body as a site of agency and knowledge; aesthetic because Fuller came to understand the body as source of selfhood through philosophical contemplations of the art of the dance. My reading shows that mesmerism allowed Fuller to find ways to move her philosophical and activist practices beyond those of transcendentalist idealism, to emancipate herself from the authority of men, and to feel her body in its vital corporeality. As a practice of heightening and deepening her senses, mesmerism prepared Fuller to experience dance performances such as Elssler's as corporeal and aesthetic enactments of self-care and self-culture—that is, as exercises through which she could tend to her corporeal and spiritual needs (self-care) and develop and cultivate her intellectual and spiritual faculties (self-culture). Fuller's review of Elssler thus needs to be acknowledged as an important stepping stone in the development of a form of aesthetics that understands dancing as a corporeal, cultural, and intellectual practice of the lived and living body, which is instrumental in developing a sense of self. Ultimately, I argue, feeling the body as trembling soul and the soul as trembling body, Fuller prepared the ground for more radical corporeal approaches in philosophy and for movement explorations of dancers subsequent to this period.

Fuller's Evaluation of Fanny Elssler: Dance as a Way of Forming the Subject

Elssler's arrival on American shores was preceded by her international fame, comparable only to the widely beloved Marie Taglioni. Taglioni was often referred to in language like that of Johanne Heiberg, who saw her as possessing "the ideal of Beauty that radiated from the depths of the soul into this body, animated it, lifted it with such power that something marvelous took place before our eyes as we saw the invisible made visible."[6] In contrast, Elssler's reputation was that of being both graceful and sensual. Heiberg vividly captured Elssler's effect on others: Elssler was

> a bacchante: delightful, intoxicating, suffused with graceful sensuality, with a beautiful earthly body, full of life, and a mime whose clarity and firmness one would find it hard to equal. She spoke without words, roared with soundless laughter, cried with pain though her lips remained sealed. She captivated spectators so that they flew into a kind of frenzy. . . . Yet, with her everything remained within the bounds of refinement. Coarse, vulgar coquetry was still remote. It was spirit, all-conquering life, that held the audience in its thrall.[7]

Because of and despite Elssler's "intoxicating" performances, her debut in New York in the spring of 1840 was very well received. The emerging penny presses directed an "exceptional amount of publicity" toward Elssler's shows,[8] and by 1841, some of these popular presses in New York City began to react less benevolently to the city's "Fannyelsslermaniaphobia."[9] Spurred by what one scholar has called an evangelical "humbug,"[10] several papers published satirical stories and mocking cartoons of Elssler, drawing an even wider audience to her performances, and every show at New York City's Park Theatre sold out.

Elssler's life off stage sparked immense public, commercial, and journalistic interest; her appearance and her movements challenged the viewing habits of theatergoers because her dancing and her costumes "emphasized the physicality of the body's movements and freed the dancers to move in an unconfined manner."[11] Although her dancing body stood at the center of all discussions, the assessment of her body diverged significantly. Maureen Needham Costonis's analysis of reviews confirms Heiberg's impression that Elssler was often evaluated as an "exhibitionist," a "coquette," and/or as "lascivious, seductive, and voluptuous." She was noted for the "'intoxicating' sensual effect of her presence," or regarded as an enchantress—and, thus, as the opposite of proper American women, who were viewed as "innately more chaste and moral."[12] At the same time, Elssler's dancing "charmed, fascinated . . . and electrified the audience" due to her "irresistible effect of her magic."[13] Costonis interprets the audience's reactions as mechanisms for coping with desires that the spectators themselves often perceived to be morally wrong. She suggests that using the term "enchantress" assuaged audience members' guilt at spectating Elssler's shows, for "if they had been magically enchanted, their secret or repressed desires could not be considered immoral or improper."[14] This coping mechanism, Costonis proposes, functioned as a "form of displacement" that elevated Elssler above the "ranks of ordinary women."[15]

For those who conceptualized Elssler as such, she no longer was "a real woman" but "instead a goddess, particularly when she danced on the tips of her toes or when her partner lifted her high into the air."[16] Under such circumstances, spectators were able to view Elssler's dancing body as effectually unsexed and spiritualized, allowing those who strove for transcendence to participate in meaningful experiences while watching her perform. Mark Knowles, for instance, reports that Emerson "called her dancing a religion,"[17] acknowledging the spiritual dimension of her corporeal practice. Emerson's response was more nuanced than Knowles presents it, however, and this nuance underlines the tension between the general transcendentalists' idealism and Fuller's bourgeoning materialism, between aesthetics and morals that the transcendentalist had to navigate. Carlos Baker notes that in contrast to many of his contemporary New England intellectuals, "energetic athleticism was of less consequence in [Emerson's] eyes."[18] What Emerson did emphasize about

Elssler was "the extreme grace of her movement," "the freedom and determination," and "the air of perfect sympathy" that her movement created.[19] Baker thus argues that for Emerson, Elssler's performance was "all at work in perfect concord,"[20] while the moralist in Emerson also understood that "immorality the immoral will see" (by which he meant "college boys, like those from Harvard") while "the pure will not heed it" and "perhaps will not see it at all." Although he debated the morals of dancing and the effect it would have on young men, Emerson also compared Elssler's dancing to a spiritual practice; yet elevating Elssler to the divine realm automatically removed her from the embodied and corporeal realities of dancing women (and men).

It is Fuller's journalistic reviews of music concerts and ballet performances in *The Dial* in the early 1840s that explicitly and wittingly interrogate the moralist anti-dance sentiments of her time, exposing the hypocrisy that many theatergoers exhibited after seeing Elssler's 1841 performances in New York and Boston.[21] Fuller, in fact, wrote with transcendentalist enthusiasm about all theater performances she saw. Her review of piano concerts in the 1840–1841 winter season, for instance, already hints at Fuller's appreciation of the *physical* rapture of the ephemerality of sounds that the music transmitted: "They are rich, brilliant, wild, astonishing. They revel in insatiable rapture and rage of all fantastic motions. They are the heaving of the billowy deep, now dark, now lit by gleams of lightning; they are the sweeping breeze of the forest; they are the flickering aurora; they are the cool flow of the summer evening zephyr; they are the dance of the elves by moonlight; they are everything marvelous and exquisite."[22] Describing the music's movement, Fuller vividly captures the satisfaction of the physical exhaustion that she experienced while listening. She stresses "the aesthetic satisfaction to be derived from music" but—and this is important—she adds, "without regard to its moral function or spiritual message."[23] She thereby emphasizes the lived experience of her corporeal reactions to the music and the elevation of the soul that such a cathartic experience yielded.

In most of her reviews, Fuller concentrated on concert and opera music, but in her 1842 review, "Entertainments of the Past Winter," she published a notable discussion of Elssler's opera dancing: the full and sensuous richness that Fuller experienced in music, she also found in Elssler's dance. But, while Fuller's reviews of music concerts and her other art criticism stand by themselves as articles, her review of Elssler's dance concerts responds to the moral and aesthetic disputes ignited by Elssler's performances. Pushing back against established moral concepts, Fuller proposed a careful corporeal approach to transcendentalist philosophy. In an appeasing notion, Fuller agreed with Emerson that "opera dancing" "must have a demoralizing effect where it is looked upon in any way but as an art," but she did not blame the artist for performing her art. Rather, she pointed to those who abuse the art form and asked whether "any one look[s] on beauty with the bodily eye alone? that degrades" when "it is the lust of the eye [that] brings sin and death."[24] If someone understands so little of what they see, they "should not witness the ballet" because "to him who looks with the eye of the soul also, every form in which beauty appears is religious, and casts some flower upon the alter of intelligence."[25] Thus elevating the art to serve only its own purpose, she formulated one of her most radical positions, which opened avenues for a vibrant, feminist, and corporeal materialism: "Everything," she writes "tends in the civilized world to a reinstatement of the body in the rights of which it has been defrauded, as an object of care and the vehicle of expression."[26] Here, in her appreciative review of Elssler's dancing, Fuller formulated one of the first—if tentative—American aesthetic theories of the body.

In transcendentalist fashion, however, Fuller's acknowledgment of the corporeal materialism of the dancing body was wrapped back into the transcendentalist notion that it is the spirit that moves the body and thus appears primary to the body. The body of Elssler "seems but thickened soul, and the subtlest emotion is seen at the fingers' ends"; her body and her movements were extraordinary, so that one's body "if not thus transparent, is no better than a soul case, or rude hut in which he lives, this is the lesser half."[27] But when the body has been thus educated and disciplined by the soul, "the motions of the ballet give it an advantage, on its side, perhaps commensurate with those derived by the drama from the beauty of poetic rhythm, and the elaborate and detailed expression of thoughts by means of words."[28] In this vein, Fuller understands that it takes a "genius to make a dancer." Implicit in Fuller's formulation of a body that is "thickened soul" is a sense of self that is agential, that makes the body become a version of itself that transcends untrained matter.

Fuller's Mesmerist Self-Culture: From Idealist Self-Culture to Materialist Technology of Self

Central to Fuller's materialist transcendentalism and to her specific way of approaching dance in her review of Elssler was her approach to self-culture. Self-culture, for the transcendentalists, was a spiritual practice of nurturing the soul, which was pitted against the doctrines of Calvinist predestination, in which individuals could not take their spiritual fate into their own hands because God had already decided upon it: self-culture was conceived of as an idealist and spiritual practice of cultivating the self. As Unitarian minister William Ellery Channing argued, it is within one's own power to develop oneself and to "cultivate any thing, be it a plant, an animal," or, what he was mainly concerned with, "a mind."[29] Self-culture was "moral," "religious," "intellectual," "social," and "practical."[30] And while he argued that "no man receives the true culture of a man in whom the sensibility to the beautiful is not cherished," he also warned of the need to maintain "control of the animal appetites" because, in order "to raise the moral and intellectual nature, we must put down the animal. Sensuality is the abyss in which very many souls are plunged and lost."[31] Fuller, for whom "the only object in life was to grow,"[32] departed from Channing's strictly idealist path of self-culture in several ways. First, Fuller connected the individual and spiritual aspects of the "growth of the soul" to her feminist cause of "diagnos[ing], and prescrib[ing] a remedy for, the condition of women."[33] And second, I argue, she turned the self-cultivation of the soul into a reciprocal relationship with the self-cultivation of her body, which, in her case, was afflicted by physical ailments that she came to treat by means of animal magnetism.[34]

Animal magnetism was introduced as a therapeutic practice by the German-Austrian physician Franz Anton Mesmer in the late eighteenth century. It only became widely known in the United States in the middle of the nineteenth century when it was already losing popularity in Europe. Mesmer's theory of animal magnetism was distinguished from the mineral magnetism found in magnets,[35] although it was clearly inspired by it to the extent that it was imbued with a sense of the gravitational pull that planets exercise upon one another and that can be observed in the tidal movements of the sea. Mesmer's assumption, formulated in proposition 2 in his *Abhandlung*, was that a fine fluid runs through human bodies, as it runs through other matter.[36] What followed for him was that the fluids within the human, and any animal body, flow back and forth in similar and cyclical ways, as he illustrates with the example of a woman's menstruation.[37] These movements have the power to influence and disturb the equilibrium of the human nervous system. Mesmer came to the conclusion, which he found validated by his own observations and his successful practice, that the fluids within people exhibited different polarities with varying intensities (formulated in propositions 7–12).[38] What followed was that a magnetizer, like Mesmer himself, was able to realign the flows of the afflicted by "'polarizing' magnetic fluids, bringing patients to 'crisis' through magnets and mesmeric trances designed to remove 'obstacles' in bodily channels and revive circulation of the vital magnetic fluids."[39] While mineral magnets might assist a magnetizer in achieving the magnetic effects in his patients, it was the magnetizer's own magnetism, according to Mesmer, that effected the desired pull within a patient's nervous system. Traditionally, then, a magnetizer (typically a man) treated a "sensitive" or "nervous" woman, putting her in a trance and aligning her magnetic fluids by his non-touch, leaving the woman's body and mind vulnerable to the practicing physician; but self-magnetism was also possible and this was what Fuller eventually practiced.

Mesmerism was conceived out of an Enlightenment concern with understanding the human body and its inner workings and devising a remedy for its ailments, but from early on, the concept of animal magnetism was received with much contention. Inquiries into scientific method in general and mesmerism in particular aimed at dismantling it as nonscientific.[40] In addition, nineteenth-century practices of animal magnetism turned increasingly toward public performances, often scandalous and sexualized, which spanned lay medicine, science, and stage entertainment. Thus, in Victorian Britain, animal magnetism became increasingly popular and commercial as mesmerists "drew on rich traditions of scientific lecturing, lay healing, and popular entertainment."[41] Profiting from the success of phrenology performances, mesmerism combined the practical benefit of promising healing with the sensational hope of spectators that the mesmerized and hypnotized subject "should say or even do something indecorous."[42] For the New England guardians of morality, mesmerist stage performance practices were thus doubly problematic because they not only highlighted the vulnerable corporeality of the mesmerized subject's body, but they also scandalously exhibited it publicly: the mesmerized woman and the ballet dancer were thus dangerously similar to one another in the eyes of moralist beholders.[43]

As predominantly practiced, mesmerism reiterated a gendered dualism of the active, knowledgeable, and potent man, on the one hand, and the passive and ailing woman in need of help, on the other hand. But as feminist critics have successfully argued, Fuller found in animal magnetism not only temporary relief from her physical pain (she suffered from severe headaches), but also a way of writing the body into her feminist and materialist transcendentalism: she "enjoyed the attention that her body received" during the treatment and she was pleased by the way in which she developed "self-confidence and empowerment" by means of the passions that the treatment aroused.[44] In the early 1840s, animal magnetism became a significant theme by which Fuller emancipated her thoughts and work from more traditional transcendentalist positions like Emerson's. Her short essay "Leila," published in *The Dial* in 1841, most clearly links the magnetic body to movement: Leila is a mystical and energized female figure who "seem[s] a key to all nature" and who leaves most men "angry and well-nigh baffled" because they are "bound in sense, time, and thought."[45] She appears to Fuller at night and offers an alternative way of being as a woman. But Leila is not only a womanly apparition, she is also "vital fluid" itself—the stuff that in mesmerism runs through our bodies and needs to be aligned: she, "with wild hair scattered to the wind, bare and often bleeding feet, opiates and divining rods in each over-full hand, walked amid the habitations of mortals as a Genius, visited their consciences as a Demon."[46] "At her touch," Fuller writes, "all became fluid" and Leila emerges as "the moving principle." Leila serves Fuller as inspiration—as an idea of a woman who embodies electricity, freedom, fluidity, and spirituality: she is the "fluid that harmonizes body and spirit: the trance-inducing power of mesmerism, the feminine goddess, the regulator of magnetic fluid. She allows Fuller to operate through her body to enter the spiritual realm, where Leila can then guide Fuller's spiritual journey."[47] I understand this body through which Leila allows Fuller to enter a spiritualized realm as both Leila's phantasmagorical and Fuller's very corporeal body: both bodies collapse in Fuller's writing because Leila also impersonates a part of Fuller herself, that is, the part of her that accomplishes her own trance-like magnetic states through which she achieves relief from pain and finds in her body the "object of care and the vehicle of expression" that she recognized in Elssler's dance performance.[48]

In this context, it is significant to consider that Elssler's performance served Fuller to make a number of materialist, transcendentalist, and aesthetic observations about human embodiment. What springs to the eye, when reading Fuller's description of Elssler against her rendering of Leila, is that Leila is depicted as an energetic, wild, and demonic ur-woman. While this suggests comparison to the way that Heiberg and Costonis describe Elssler's performance on stage, the description deviates markedly from Fuller's own impression of Elssler: it was Elssler's "sweetly childlike" yet alluring movements that Fuller emphasized in her review of Elssler's interpretation of *La Sylphide* (in contrast to the "more refined and poetic" movement qualities that, she read, Taglioni exhibited).[49] In Fuller's account, Elssler is characterized as a "young girl," "more than half conscious of her captivations," and exhibiting "that perfect innocence of gesture" that one associates with "a young child."[50] Fuller recognized in Elssler a primacy and fluidity of movement that seemed forceful yet unmeditated by larger societal conventions and enculturation. Fuller, of course, knew that this was an illusion created on stage, but the effect of Elssler's performance was similar to the ways Fuller imagined Leila to move through the world: fully owning and enjoying the splendor of her body and movements.

Fuller appreciated Elssler's style of dancing for her "coquettish play," that is, she largely enjoyed it for the very reasons that the moralists disdained it.[51] But to Fuller, Elssler's playfulness was not morally corrupt; instead, it allowed her to be "sparkling with life and joy, new to all the varied pulses of the heart."[52] Her interpretation of Elssler's Spanish dances, for instance, showcased her "impassioned feeling of life" that was accompanied by the sound of the castanets, which Fuller experienced as "count[ing] the pulses of a life of ecstasy, to keep time with the movements of an existence incapable of a dull or heavy moment." Overall, Elssler's dancing combined several aspects that appeared to Fuller as nothing less than "undoubtedly . . . the true state of man": a moving body that is so full of life and "thickened soul," so perceptive of the air and space surrounding it, so expressive and articulate of emotion, and so fully in touch with itself in/as nature that Fuller finds no other way than to liken Elssler's dancing to the "impersonation of spring" and the "free loveliness of Nature."[53] Reading these comparisons against Fuller's transcendentalist background illustrates her more general conclusions on the significance of dance. Through Elssler's vitality and her body that was "thickened soul," Fuller witnessed the ways in which dancing united body and soul: in Elssler's dancing, the inextricable unity of body and spirit became visible to the eye.[54]

Animal magnetism had taught Fuller similar corporeal lessons. Because of her mesmerist experiences, Fuller "felt that her body and spirit were in harmony," and "she had discovered a personal, autonomous spirituality."[55] In *Summer on the Lakes, in 1843,* Fuller thus weaves her magnetic experiences into a fictitious dialogue between the characters Free Hope (representing herself), Good Sense (representing the American common sense), Old Church (representing established religious authorities), and Self-Poise (representing an Emersonian voice). In her guise as Free Hope, Fuller is able to convey her own (and, to her eyes, differentiated) position on animal magnetism. The "vital principle," this magnetic force behind all living matter, she argues, has always preoccupied men of thought as a "principle of flux and influx, dynamic of our mental mechanics, human phase of electricity."[56] Although she readily admits that magnetism was "tampering unless done in a patient spirit and with severe truth," she underlines that those "who work in the true temper patient and accurate in trial" and "feeling there is a mystery . . . may learn . . . and teach." Free Hope's experience of the world is such that "all [her] days are touched by the supernatural, for [she feels] the pressure of hidden causes, and the presence, sometimes the communion, of unseen powers," which she must acknowledge and explore. By 1845, when Fuller published *Woman in the Nineteenth Century,* she "seems to have found empowerment in revelatory, ecstatic trances that did not employ a magnetizer."[57] She was thus fully self-reliant in her unity of body and soul: "The especial genius of woman, I believe to be electrical movement, intuitive function, spiritual in tendency. She excels not so easily in classification, or re-creation, as in an instinctive seizure of causes, and a simple breathing out of what she receives that has the singleness of life, rather than the selecting and energizing art."[58]

Fuller's Materialist Transcendentalism and the Body Magnetic

I have thus far argued that Fuller's review of Elssler marks the beginnings of an aesthetic theory based on what I call her materialist transcendentalism. But her materialism has to be carefully distinguished from the nineteenth-century socialist materialism against which Fuller expressed her explicit criticism. In her feminine-poetic-corporeal conception of the arts, Fuller pitted a form of vital and magnetic materialism (akin to the new materialism proposed by feminist thinkers in the late twentieth and early twenty-first centuries) against the functional and technocratic materialism of the nineteenth century that she recognized in Charles Fourier, despite her general appreciation of his reformatory endeavors and his "grand and clear" mind (she was especially pleased about his support for women's liberation and emancipation).[59] As her praise in *Woman in the Nineteenth Century* shows, she appreciated Fourier's endeavors for structural and institutional change in principle, but she also found them wanting in aesthetic depth.[60] One of the problems for Fuller might have been that Fourier's classist aesthetics were inspired by aristocratic traditions and thus lacked the qualities of simplicity and appreciation of nature that transcendentalism at large cherished.[61] But despite aesthetic and socialist-materialist disagreements between Fourier and Fuller, Fuller was interested in Fourier's adaptation of Mesmer's "concept of a universal fluid, which he called 'aroma,' to culture itself. In Fourier's vision, the production of a 'crisis' or an 'aromal state' [within a group or society] could lead to social harmony."[62] In this, Fourier develops the clear anti-Emersonian vision that not self-reliance but "'association' among men would form the only basis for a legitimate social critique."[63] Fuller's treatment of Fourier is thus instrumental for establishing her complex relationship to materialist approaches in the nineteenth century, and it is equally significant for demonstrating that her critique of Fourier's materialist backdrop in no way inhibits her materialist transcendentalism—a materialism informed by her aesthetics and her sexed and gendered situatedness by acknowledging the vibrancy of the material body.

Thus, taking opera dancing seriously in its own right, Fuller's appreciation of Elssler's dancing body was based on Fuller's aesthetic sense, which she developed systematically in her critical reviews in *The Dial*. At the same time, I argue that her appreciation of Elssler was also grounded in Fuller's own corporeal situatedness as an intellectual woman in the nineteenth century who suffered from severe headaches and back pains, which stemmed from the curvature of her spine. Fuller's body was visibly gendered and both visibly and invisibly marred, which for a long time inhibited her from fully participating in intellectual life in the ways that her male and more able-bodied contemporaries did. Just as the figure of Leila served Fuller in her dreams as a model of how to inhabit one's body, to take up space, and to sweep through (male-dominated) environments, my reading of Fuller's review establishes Elssler as an example of a woman in the real world who moves without the restrictions that Fuller felt in her own body. Thus, I suggest, Fuller began to develop a sense of self and a practice of self-culture that was aware of and incorporated corporeal aspects in addition to the transcendentalist idealist, intellectual, and spiritual practice of bettering the soul. Fuller's gendered body and her experience of sexual difference became the cornerstone of her feminist intellectual output, but her aching body also helped Fuller to grasp the corporeal aspect of the arts and philosophy because she could not ignore the fact that she was body as much as she was soul. In her search for relief from pain, the only treatment she could find for her ailments was offered by mesmerism, which demanded her physical and mental cognizance of her body as an active and vital agent. Thus, it was her aching body and her practice of mesmerism that shifted Fuller's transcendentalism from the idealist positions of Emerson toward her own feminist materialist point of view. This shift happened at the same time as she saw Elssler—a significant concurrence that I argue contributed to her corporeal understanding of the dance as a form of art: her use of mesmerist language in her review of Elssler suggests that for Fuller, the fluidity of Elssler's movement and the vital fluidity of animal magnetism overlapped conceptually (Fuller's style of thinking and writing followed associative chains rather than logical deduction). I see in Fuller's appreciation of dance and her turn toward the materiality of her body a first step toward conceptualizing the knowledge we gain through dancing as a Foucauldian technology of self, that is, as a practice through which we become philosophically aware of our corporeality as part of our subject position.

Magnetism and the Soul That Touches: Theoretical Implications

Evaluating Fuller's self-magnetism along with her review of Elssler, the idea of the dancer's movements as "thickened soul" resonates with the image of

Leila as the "vital fluid" and "moving principle." Through her experience with mesmerism, "Fuller so thoroughly intertwines the physical and mystical, the material and transcendental, that they are almost impossible to parse"[64]—the body is thickened soul and the soul is fluid body. Such a conception was only possible, first, through Fuller's investigation and practice of mesmerism and, second, because of the nineteenth-century imagery that linked aesthetic and "sensuous experience" to "scientific metaphors of electricity."[65] Mesmerism allowed Fuller to find ways through which she was able to transcend transcendentalist idealism and to emancipate herself from the authority of men and feel her body in its corporeality. I propose that mesmerism, as a practice of heightening and deepening her senses, prepared Fuller to experience dance performances, such as Elssler's, as what Foucault terms a technology of self, as corporeal and aesthetic practices of self-care.

Fuller's materialist transcendentalism smoothed the way for two interrelated developments. The first one pertains to the cultural and artistic history of the United States, even if only implicitly: Fuller's significant reconceptualization of the body and of movement was an early exploration and presentation of dancing as offering epistemic as well as aesthetic experiences. Her particular way of deploying animal magnetism for her feminist empowerment is instructional for her appreciation of opera dancing in general and for her response to Elssler in particular: because mesmerism allowed Fuller to develop a materialist form of transcendentalism, she recognized in Elssler's performance a level of body consciousness that led to the first (if tentative) American theory of corporeal aesthetics from which later dance practitioners such as Isadora Duncan[66] would implicitly benefit.

The second aspect that Fuller's mesmerist dance review heralded pertains to an aspect of dance philosophy: I evaluate Fuller's material transcendentalism and her mesmerist approach to Elssler as an early contribution to the discussion of the complex relationship of "touching and being touched." This "fundamental element of dance" includes the aesthetic and technical grounds of expression and movement production but also encompasses questions of subjectivity, agency, and self-knowledge.[67] Looking at touch from the perspective of the nineteenth-century transcendentalist Margaret Fuller looking at dance, it becomes clear that Fuller's kinesthetic empathy for the movements performed by Elssler touched Fuller's body and soul—magnetized her, so to speak—in ways that have been described by philosopher Jean-Luc Nancy, who draws attention to the "mobile, moving and dynamic value of touch."[68] For Nancy, "touch begins when two bodies move apart and distinguish themselves from one another," that is, when they understand their identity to be separate from one another yet intertwined through touch. "Only a separate body is capable of touching," he writes; "It alone can entirely separate its touch from its other senses—that is, attribute to one autonomous sense that which nevertheless traverses all the senses, as though differentiating in them while distinguishing itself as a kind of common reason. Reason or passion, drive, motion." The physical touch that Nancy describes and the mesmerist non-touch that Fuller cherished share the quality that Nancy delineates in almost mesmerist language as being "first and foremost this rocking, this floating and rubbing . . . renewing and replaying the desire to feel touched and to touch, the desire to be affected by the contact of the outside."[69] Our "entire being is contact," and our entire "being is touched/touches" because, as he proclaims, "we are contact itself." Nancy therefore resumes that "the sensory act creates the current effectivity" so that the "soul that feels is itself feeling, and hence feels itself feel . . . in the sense of touch." Nancy's theory of touch, Elssler's movement, and Elssler's touching of and being touched by her dance partner James Sylvain touched Fuller's soul and intellect and made her soul tremble. For Nancy, "when the soul trembles, it does indeed tremble, just as much as water does when it approaches its boiling point"—that is, the trembling soul is indicative of its corporeal materiality: "What we currently call the 'soul' is no different from arousal and receptiveness to motion and emotion. The soul is the body that is touched, vibrating, receptive and responding. Its response is the sharing of touch, its awakening to it."[70] What Nancy theorizes is what Fuller experienced in her body and understood in her soul when she watched Elssler. She described Elssler's dancing body as being "but thickened soul, and the subtlest emotion is seen at the fingers' ends" because it appeared to Fuller to be more than a mere "soul case."[71] Fanny Elssler demonstrated for Fuller in her dancing what Fuller was feeling during her animal magnetism sessions: the fluids that move when mesmerized, just as Elssler's fluid movements in dance enabled the touch and touching of the soul. Fuller's careful yet radical forays into a materialist transcendentalism in her review of Fanny Elssler are ground-breaking and instructive because they offer an idealist-cum-materialist approach to bodily techniques and practices that connect body and soul in a way that highlights the vibrant (or magnetic) dimension of touching and almost touching, of sensing the self and others as if the waves of movement were electrically, magnetically, or otherwise charged with natural energy.

Notes

1 Ralph Waldo Emerson and Margaret Fuller went together to see Elssler perform in Boston on October 13, 1841. See Fuller, *The Letters of Margaret Fuller,* 251 n. 3.

2 Conservative and moralist adversaries of the dance highlighted the ways in which dancing would corrupt especially young women's chastity; idealist positions value the metaphysical realm of the mind/spirit/soul over embodied approaches to ontology and epistemology.
3 I will use "mesmerism" and "animal magnetism" interchangeably.
4 The transcendentalists were not in any way systematic, and they did not strictly follow any one idealist framework; they were rather inspired by philosophical writing that was concerned with matters of the spirit and the soul.
5 George and Sophia Ripley founded Brook Farm; Amos Bronson Alcott and Charles Lane founded Fruitlands. Brook Farm and Fruitlands were communes in which transcendentalist principles were set into agrarian practice. Both projects were short-lived.
6 Heiberg, "Memories of Taglioni and Elssler," 15.
7 Heiberg, "Memories of Taglioni and Elssler," 16.
8 McCoy, "Fanny Elssler's Reception," 77.
9 Clapp Jr. quoted in "Fanny Elssler's Reception," 77.
10 Delarue, "Historical Background to America's Elsslermania," 6–7.
11 Costonis, "The Personification of Desire," 56–7.
12 Costonis, "The Personification of Desire," 55, 52, 53, 53.
13 Costonis, "The Personification of Desire," 53.
14 Costonis, "The Personification of Desire," 56.
15 Costonis, "The Personification of Desire," 58.
16 Costonis, "The Personification of Desire," 59.
17 Knowles, *The Wicked Waltz and Other Scandalous Dances,* 10. Paul Gilmore's footnote on Emerson's alleged praise, which goes back to an "apocryphal account of Fuller and Emerson's dialogue" on Elssler, shows that it is unclear whether it was Fuller or Emerson who referred to Elssler's dancing as poetry or religion respectively (84, n. 21). The general sentiment captured in this alleged conversation, however, was that "by giving full expression to a particularly feminine beauty and by restoring a more natural relationship between body and soul, ballet grants us access to a realm where the very distinctions between body and soul, male and female, are transcended" (Gilmore, "'The Poetical Side of Existence,'" 69).
18 Baker, "Moralist and Hedonist," 29.
19 Quoted in Baker, "Moralist and Hedonist," 29.
20 Baker, "Moralist and Hedonist," 29. Subsequent quotation, 30.
21 *The Dial,* edited by Ralph Waldo Emerson, Margaret Fuller, and George Ripley, was published in Boston by E. P. Peabody.
22 Fuller, "Concerts of the Past Winter," *The Dial,* 130.
23 Quoted in Saloman, "Margaret Fuller on Musical Life in Boston and New York, 1841–1846," 429.
24 Fuller, "Entertainments of the Past Winter." *The Dial,* 70.
25 Fuller, "Entertainments of the Past Winter," 65, 70.
26 Fuller, "Entertainments of the Past Winter," 65.
27 Fuller, "Entertainments of the Past Winter," 65. In "The Modern Drama," she writes that in ballet, the "body [is] made pliant to the inspirations of spirit"; quoted in Gilmore, "'The Poetical Side of Existence,'" 69.

28 Fuller, "Entertainments of the Past Winter," 65. Subsequent quotation, 67.
29 Channing, *Self-Culture,* 12.
30 Channing, *Self-Culture,* 12, 13, 14, 17, 18.
31 Channing, *Self-Culture,* 19, 26, 26.
32 Fuller, quoted in Emerson et al., *Memoirs of Margaret Fuller Ossoli,* 133. See also Robinson, "Margaret Fuller and the Transcendental Ethos," 85.
33 Robinson, "Margaret Fuller and the Transcendental Ethos," 84; see also 86.
34 See Davis, "Margaret Fuller, Body and Soul," 31–56.
35 Mesmer, *Abhandlung über die Entdeckung des thierischen Magnetismus,* 28.
36 Mesmer, *Abhandlung über die Entdeckung des thierischen Magnetismus,* 47.
37 Mesmer, *Abhandlung über die Entdeckung des thierischen Magnetismus,* 9.
38 Mesmer, *Abhandlung über die Entdeckung des thierischen Magnetismus,* 48.
39 Zwarg, "Footnoting the Sublime," 631.
40 As early as 1784, a commission of scientists (including Benjamin Franklin) furthered the scientific method by investigating Mesmer's animal magnetism. They demonstrated that mesmerism exhibited no effect that could be traced back to magnetization as such. Instead, the commission found that "subjects developed the characteristic mesmeric crises if and only if they *expected* to be magnetized, regardless of whether they were actually magnetized" (Lanska and Lanska, "Franz Anton Mesmer and the Rise and Fall of Animal Magnetism," 317). See also Franklin et al., "Report of the Commissioners," 332–63. For Mesmer's influence on the development of dynamic psychiatry, see Ellenberger, *The Discovery of the Unconscious,* esp. ch. 2. For a discussion of the scientific and metaphysical debates around mesmerism, see Kaplan, "'The Mesmeric Mania,'" 691–702. On the debates about scientific methods, experiments, and explanatory discourses in eighteenth- and early nineteenth-century medicine, see Sutton, "Electric Medicine and Mesmerism." 375–92. On the medical and psycho-social dimension of mesmerism, see Fulford, "Conducting the Vital Fluid," 57–78. For a nuanced account of different forms of science in Enlightenment discourse, see Brand, "Aufgeklärte Geisterseher," 128–39.
41 Parssinen, "Mesmeric Performers," 89. Nathanial Hawthorne's *The Blithedale Romance,* 6 and 138–40, shows a similar practice of animal magnetism as stage spectacle.
42 Parssinen, "Mesmeric Performers," 102, and on phrenology, 91; see also McCandless, "Mesmerism and Phrenology in Antebellum Charleston," 199–230.
43 See also McCarren, "The 'Symptomatic Act' Circa 1900," 748–74. McCarren reads Loie Fuller's dancing vis-à-vis theories of hypnosis and hysteria in psychoanalytical treatments of the late nineteenth century, especially in Charcot's clinic. McCarren fathoms dance "as metonymic self-elaboration" (774), "as connected to the unconscious, and as a privileged site of its expression" (752). Central to this reading is seeing L. Fuller as both the "hypnotized and hypnotist" when she dances (755), and I argue that M. Fuller's experience of self-mesmerism enabled her to recognize similar mechanisms of self-understanding and self-expression in Elssler's dancing.
44 Manson, "'The Trance of the Ecstatica,'" 307, 308. See also Blumenthal, "Margaret Fuller's Medical Transcendentalism," 553–95.
45 Fuller, "Leila," 53.
46 Fuller, "Leila," 55, 56–57.

47 Manson, "'The Trance of the Ecstatica,'" 314.
48 Fuller, "Entertainments of the Past Winter," 64.
49 Fuller, "Entertainments of the Past Winter," 66.
50 Fuller, "Entertainments of the Past Winter," 65, 66.
51 Fuller, "Entertainments of the Past Winter," 66. Elssler was not able to convince Fuller in her performance of *La Sylphide,* however: "The light hovering motions of the piece, however, suggest an order of grace more refined and poetic than hers, such as is ascribed to Taglioni" (66). Fuller never saw Taglioni perform and had to rely on her imagination and on reviews she had access to.
52 Fuller, "Entertainments of the Past Winter," 65.
53 Fuller, "Entertainments of the Past Winter," 67.
54 Paul Gilmore concludes that "by giving full expression to a particularly feminine beauty by restoring a more natural relationship between body and soul, ballet grants us access to a realm where the very distinctions between body and soul, male and female, are transcended" ("'The Poetical Side of Existence,'" 69).
55 Manson, "'The Trance of the Ecstatica,'" 312.
56 Fuller, *Summer on the Lakes, in 1843,* 79.
57 Manson, "'The Trance of the Ecstatica,'" 314.
58 Fuller, *Woman in the Nineteenth Century,* 68.
59 Fuller, *Woman in the Nineteenth Century,* 73. For Fourier's socialist reformatory ideas, see Beecher and Bienvenu, eds., *The Utopian Vision of Charles Fourier;* and Missé, "Fourier, Marx, and Social Reproduction," esp. 11–12.
60 Fuller, *Woman in the Nineteenth Century,* 73.
61 Beecher and Bienvenu, eds., *The Utopian Vision of Charles Fourier,* 234.
62 Zwarg, "Footnoting the Sublime," 631. Fourier's social critique surfaces in Chapter 5 of Fuller's *Summer on the Lakes, 1843* before Fuller discusses mesmerist treatments in reference to the case of Friederike Hauffe, known as the Seeress of Prevorst (cf. Zwarg, 631).
63 Zwarg, "Footnoting the Sublime," 631.
64 Blumenthal, "Margaret Fuller's Medical Transcendentalism," 586.
65 Blumenthal, "Margaret Fuller's Medical Transcendentalism," 554; Gilmore, quoted in Blumenthal, "Margaret Fuller's Medical Transcendentalism," 554.
66 Duncan was an avid reader of Walt Whitman's poetic renderings of the "body electric," which was itself a reference to animal magnetism (Duncan, *My Life,* 22).
67 Brandstetter, Egert, and Zubarik, "Touching and Being Touched," 3.
68 Nancy, "Rühren, Berühren, Aufruhr," 10. Subsequent quotation, 11.
69 Nancy, "Rühren, Berühren, Aufruhr," 13.
70 Nancy, "Rühren, Berühren, Aufruhr," 16.
71 Fuller, "Entertainments of the Past Winter," 65.

Bibliography

Baker, Carlos. "Moralist and Hedonist: Emerson, Henry Adams, and the Dance." *New England Quarterly* 52, n. 1 (1979): 27–37.

Beecher, Jonathan, and Richard Bienvenu, eds. *The Utopian Vision of Charles Fourier: Selected Texts on Work, Love and Passion.* Boston: Beacon Press, 1971.

Blumenthal, Rachel A. "Margaret Fuller's Medical Transcendentalism." *ESQ: A Journal of the American Renaissance* 61, n. 4 (2015): 553–95.

Brand, Klaus. "Aufgeklärte Geisterseher: Wissenschaft als Religion im frühen 19. Jahrhundert am Beispiel des Mesmerismus." *Zeitschrift für Religions- und Geistesgeschichte* 66, n. 2 (2014): 128–39.

Brandstetter, Gabriele, Gerko Egert, and Sabine Zubarik, "Touching and Being Touched: Motion, Emotion, and Modes of Contact," trans. Christine Henschel. In *Touching and Being Touched: Kinesthesia and Empathy in Dance and Movement,* edited by Gabriele Brandstetter, Gerko Egert, and Sabine Zubarik, 3–10. Berlin: de Gruyter, 2013.

Channing, William Ellery. *Self-Culture: An Address Introductory to the Franklin Lectures, delivered at Boston, September 1838.* Boston: James Munroe & Co., 1839.

Costonis, Maureen Needham. "The Personification of Desire: Fanny Elssler and American Audiences." *Dance Chronicle* 13, n. 1 (1990): 47–67.

Davis, Cynthia J. "Margaret Fuller, Body and Soul." *American Literature* 71, n. 1 (1999): 31–56.

Delarue, Allison. "Historical Background to America's Elsslermania." In *Fanny Elssler in America,* edited by Allison Delarue, 2–8. New York: Dance Horizon, 1976.

Duncan, Isadora. *My Life.* 1927. New York: Liveright, 2013.

Ellenberger, Henri F. *The Discovery of the Unconscious: The History and Evolution of Dynamic Psychiatry.* New York: Basic Books, 1970.

Emerson, Ralph Waldo, William Ellery Channing, and James Freeman Clarke, eds. *Memoirs of Margaret Fuller Ossoli.* 1884. New York: Burt Franklin, 1972.

Foucault, Michel. "Technologies of the Self." In *Technologies of the Self: A Seminar with Michel Foucault,* edited by Luther H. Martin, Huck Gutman, and Patrick H. Hutton, 16–49. Amherst: University of Massachusetts Press, 1988.

Franklin, Benjamin, Jean-Sylvain Bailly, Gabriel de Bory de Saint-Vincent, Jean d'Arcet, Joseph-Ignace Guillotin, Antoine Lavoisier, Jean-Baptiste Le Roy, Michel-Joseph Majault, and Charles Louis Sallin. "Report of the Commissioners Charged by the King with the Examination of Animal Magnetism." 1784. *International Journal of Clinical and Experimental Hypnosis* 50, n. 4 (2002): 332–63.

Fulford, Tim. "Conducting the Vital Fluid: The Politics and Poetics of Mesmerism in the 1790s." *Studies in Romanticism* 43, n. 1, Romanticism and the Sciences of Life (2004): 57–78.

Fuller, Margaret. "Concerts of the Past Winter." *The Dial: A Magazine for Nature, Philosophy, and Religion* 1 (July 1840), 124–33.

Fuller, Margaret. "Entertainments of the Past Winter." *The Dial: A Magazine for Nature, Philosophy, and Religion* 3 (July 1842): 46–72.

Fuller, Margaret. "Leila." In *The Essential Margaret Fuller,* edited by Jeffrey Steele, 53–58. New Brunswick, NJ: Rutgers University Press, 1992.

Fuller, Margaret. *The Letters of Margaret Fuller.* Volume II: 1839–1841, edited by Robert N. Hudspeth. Ithaca, NY: Cornell University Press, 1983.

Fuller, Margaret. *Summer on the Lakes, in 1843.* Urbana: University of Illinois Press, 1991.

Fuller, Margaret. *Woman in the Nineteenth Century.* New York: Norton, 1998.

Gilmore, Paul. "'The Poetical Side of Existence': Margaret Fuller, Early Mass Culture, and Aesthetic Transcendence." *ESQ: A Journal of the American Renaissance* 47, n. 1 (2001): 59–87.

Hawthorne, Nathaniel. *The Blithedale Romance,* edited by Richard H. Millington. New York: Norton, 2011.

Heiberg, Johanne Luise. "Memories of Taglioni and Elssler," trans. Patricia McAndrew. *Dance Chronicle* 4, n. 1 (1981): 14–18.

Kaplan, Fred. "'The Mesmeric Mania': The Early Victorians and Animal Magnetism." *Journal of the History of Ideas* 35, n. 4 (1974): 691–702.

Knowles, Mark. *The Wicked Waltz and Other Scandalous Dances: Outrage at Couple Dancing in the 19th and Early 20th Centuries.* Jefferson, NC: McFarland, 2009.

Lanska, Douglas J., and Joseph T. Lanska. "Franz Anton Mesmer and the Rise and Fall of Animal Magnetism: Dramatic Cures, Controversy, and Ultimately a Triumph for the Scientific Method." In *Brain, Mind and Medicine: Essays in Eighteenth-Century Neuroscience,* edited by Harry Whitaker, C. U. M. Smith, and Stanley Finger, 301–320. New York: Springer, 2007.

Manson, Deborah. "'The Trance of the Ecstatica': Margaret Fuller, Animal Magnetism, and the Transcendent Female Body." *Literature and Medicine* 25, n. 2 (2006): 298–324.

McCandless, Peter. "Mesmerism and Phrenology in Antebellum Charleston: 'Enough of the Marvellous.'" *Journal of Southern History* 58, n. 2 (1992): 199–230.

McCarren, Felicia. "The 'Symptomatic Act' Circa 1900: Hysteria, Hypnosis, Electricity, Dance." *Critical Inquiry* 21, n. 4 (1995): 748–74.

McCoy, Genevieve. "Fanny Elssler's Reception: Gender, Class, and Republicanism in the United States, 1840–42." In *Proceedings of the 25th Annual Conference of the Society of Dance History Scholars, Temple University, 2002,* compiled by Stephanie Rieke, 77–82. Philadelphia: Temple University, 2002.

Mesmer, Franz Anton. *Abhandlung über die Entdeckung des thierischen Magnetismus.* Tübingen: edition diskord, 1985.

Missé, Blanca. "Fourier, Marx, and Social Reproduction." *CLCWeb: Comparative Literature and Culture* 22, n. 2 (2020): 1–14.

Nancy, Jean-Luc. "Rühren, Berühren, Aufruhr," trans. Roxanne Lapidus. *SubStance* 40, n. 3, issue 126, Plus d'un toucher: Touching Worlds (2011): 10–17.

Parssinen, Terry M. "Mesmeric Performers." *Victorian Studies* 21, n. 1, Victorian Leisure (1977): 87–104.

Robinson, David M. "Margaret Fuller and the Transcendental Ethos: Woman in the Nineteenth Century." *PMLA* 97, n. 1 (Jan. 1982): 83–98.

Saloman, Ora Frishberg. "Margaret Fuller on Musical Life in Boston and New York, 1841–1846." *American Music* 6, n. 4 (1988): 428–41.

Sutton, Geoffrey. "Electric Medicine and Mesmerism." *Isis* 72, n. 3 (1981): 375–92.

Zwarg, Christina. "Footnoting the Sublime: Margaret Fuller on Black Hawk's Trail." *American Literary History* 5, n. 4 (1993): 616–42.

15

Labor and Laboratory of Kinesic Interplay in Nineteenth-Century Dance Theory

Response to "From Animal Magnetism to Materialist Transcendentalism: Margaret Fuller on Fanny Elssler" by Johanna Pitetti-Heil

Claudia Jeschke

Johanna Pitetti-Heil explores the personal and professional voyage of the American journalist and women's rights advocate, Margaret Fuller, from her embrace of a contemplative "idealist self-culture" to the more vigorous "materialist technology of self." Instrumental for this development is the physio-sensual evaluation of Fanny Elssler, the famous Austrian performer, whose dynamic dancing seems to have disclosed key experiences and parameters for Fuller's agential philosophical and scientific research on the forming of her own feminist self-concept.

My perspective, in this response, moves from the personal to the general, addressing issues of dancers' agency during the first half of the nineteenth century. I discover an exemplary, then-contemporary praxeological form of mental and physical education that preconditioned (West-)European dance artists toward a specific responsiveness in the works of Paulo Bruno Bartholomay. I ask how dancing bodies have been theatrically (and socially) trained in order to generate such interactive occurrences as the energetic scenic vibrancy that Margaret Fuller experienced when watching Elssler perform. I turn to the intricate physiological and motoric as well as philosophical, scientific, social, and aesthetic aspects of dance that underpinned its experiments in communication and kinesic interplay, all shaping nineteenth-century European dance theories.

As critical phraseologies and praxeological instructions, dance theories reflect a web of cultural contexts and borrowings from nondance disciplines for the framing of concepts of dancing. Especially during the first half of the nineteenth century, the increased publication of dance treatises in the many national languages of Europe is notable.[1] However, the vigorous nomadic activities of dance artists in Europe make it difficult to grasp distinctions revealing why, how, and where dancing appeared. The theories moved not only across geographic borders, but also among the realms of aesthetics, anthropology, and medicine, creating a common or at least comparable throughline: the discussions of the physiological dimension of dancing. They gave way to praxeological considerations of inner perception, or proprioception, in response to external stimuli, increasingly exploring the body's irritable and excitable capacities. Thus, nineteenth-century theories explicated the physiological conditions for energy transfers—that is, for interactions inside the dancers' bodies as well as those affecting and resonating in the bodies of recipients, of the audience. Fuller, clearly, was alive to such energy.

Here I will present excerpts of Bartholomay's 1838 treatise, *Die Tanzkunst in Beziehung auf die Lehre und Bildung des wahren Anstandes und des gefälligen Äußeren* (The Art of Dancing in Relation to the Teaching and Education of True Decency and Pleasing Appearance). I have chosen this work as an example of the then-multifold and vivid sociocultural negotiations undertaken at the time to address the materialities of doing dancing. Bartholomay's rhetoric exemplifies both the exploration of energetic textures of movement production and the reception of the physical effects of (e)motion.

Constructing Vital Postures

A dancing and fencing master at the University of Giessen, Bartholomay quotes, without naming the source, the guidelines established by the influential theorist and pedagogue Carlo Blasis, whose works had been translated into many European languages.[2] Bartholomay adapts and expands the Italian master's mostly dance-theatrical conceptualization by integrating material from medicine and early anthropology. Like Blasis, Bartholomay creates a cartogram of the dancer's body in which each limb is experienced and described in terms of its functional and (e)motional capacity.[3] The map lists first the sole of the foot, which forms the "true base" of the body and carries the whole weight of the "body architecture." The importance of the pelvis lies in the energetic connection between step and stance in order to maintain physical equilibrium. The effective interplay of foot, instep, legs, and hip is motorically supported by a free chest, an unconstrained placement of the head and both shoulders, and a lengthened neck.[4] The assertions by Bartholomay and other dance theorists of the time refer to their understanding of anatomy and correspond to its logic. Besides the initially motoric claims and their kinesic effects, Bartholomay also attributes symbolic value to the correct posture of the upright body with special regard to the head and face, underlining their importance as signs of the dancer's movement control, self-confidence, and ability to communicate. This bodily cartography, which unreservedly connects physical zones with their metaphysical and communicative capacities via energy, might be thought of as analogous to what Pitetti-Heil calls Margaret Fuller's materialist transcendentalism.

Sensing Inner and Outer Spheres

When Bartholomay leaves the traditional codifications of theater dance, which draw, for example, on Blasis's explanations, he expands his system through descriptions of corporeal behavior according to the (experientially felt) inside and the (socially stipulated) outside, perhaps akin to Fuller's reciprocal relationship between self-culture of the soul and self-cultivation of the body. Bartholomay describes arms and hands as "antenna [Fühlhörner] of the human body," which, like the tactile senses and receptors of insects, are responsible for the perception of the environment.[5] As outer extremities and through their mobility, they function as extensions of the body's movement, which they share or lead. The elbows are to be held with awareness and give volume to the posture; Bartholomay emphasizes, above all, the visibility of the arms, wrists, and hands reaching into the space around the body, establishing a mobile and thus communicative sphere.

(In)Visibilities—Oscillating Perspectives

Bartholomay's detailed movement descriptions discuss issues of inwardly and outwardly directed activities. His para-medical observations concerning the arteries, understood as concealed transport routes of the blood and, as such, a path of feelings, are loosely connected to remarks regarding the hands: the palms should also be hidden, because their visibility stands for a rejecting gesture. Bartholomay thus switches between examples of visceral (hidden, vegetative, involuntarily motivated) motions and examples of somatic (visible, skeletal, muscular, voluntarily motivated) movements, making statements about physiological functions as well as about meaningful features of body language. The flexibility of his maneuvers echoes the dancers' personal choices.

Equilibrium

Alongside the issues of body posture, head and gaze alignments, and arm-hand activities in their ability to project spatiality and to facilitate communication, Bartholomay discusses practical dance topics, mainly the conscious maintenance of equilibrium which is, "however, not a matter of instinct, but of practice and consideration; for with the manifold parsing of our body, the art of standing and walking is not learned in earliest youth without effort or without painfully felt apprenticeship."[6] Yet it is not the technique alone but rather the spiritual and inner mastery of the dancer that constitutes his or her art. Here, the concept of will takes on a significant role in shaping the dancer's identity: "All the muscles of our body, with the exception of the heart (for this continues to beat even if the will forbids it), depend on the influence of the will. It is therefore the will that determines that any muscle performs a bending or tension, an inward or outward turning."[7] According to Pitetti-Heil, Fuller shared a similar belief that the equilibrium of body and soul was an individual practice of self-discipline, which notably departed from the popular magnetism industry of her time.

Bartholomay's concept reveals a differentiated cartogram of the human body, just as it displays the functional capacity of physical and psychic motions. Equilibrium, the logic of verticality, and the handling of weight according to the line of aplomb (gravitational pull) ask the performer to be continuously aware of the interplay of different energetic qualities that are producing intra- and extracorporeal kinesic responses. Anatomical-physiological processes interact with deeply embedded sensual and emotional expression and allow the performer to transfer the representational and creative potential of energetic communication to an attentive as well as receptive audience. Fuller was alert to just this experience when watching Fanny Elssler "sparkling with life and joy."

Notes

1 Jeschke, Vettermann, and Haitzinger, *Interaktion und Rhythmus.*
2 Blasis, *Traité Elémentaire* and *The Code of Terpsichore.*
3 Bartholomay, *Tanzkunst,* 32; see Blasis, *Code of Terpsichore,* 439.
4 Bartholomay, *Tanzkunst,* 44; see Blasis, *Traité,* 42, fn. 1.
5 Bartholomay, *Tanzkunst,* 32–33.
6 Bartholomay, *Tanzkunst,* 36.
7 Bartholomay, *Tanzkunst,* 160.

Bibliography

Bartholomay, Paul Bruno. *Die Tanzkunst in Beziehung auf die Lehre und Bildung des wahren Anstandes und des gefälligen Äußeren.* Giessen: beim Autor, 1838.

Blasis, Carlo. *The Code of Terpsichore.* London: John Bullock, 1828.

Blasis, Carlo. *Traité Elémentaire, Théorique et Pratique de L'Art de la Danse.* Milan: Joseph Beati et Antoine Tenenti, 1820.

Jeschke, Claudia, Gabi Vettermann, and Nicole Haitzinger. *Interaktion und Rhythmus. Zur Modellierung von Fremdheit im Tanztheater des 19. Jahrhunderts.* München: epodium, 2010.

16

Hypnotic Dancing and the Science of Sleep and Dreams

The Controversial Case of Madeleine G.

CHANTAL FRANKENBACH

German theater dance was in the midst of a remarkable transformation at the turn to the twentieth century. Newspapers reported daily on the latest marvels in the new dance art that would overturn the banalities of nineteenth-century ballet for a novelty-hungry German public. Early in 1904, a rhymed satire in the popular humor magazine *Jugend* announced that dance-crazed Munich, having raved over la Tortajada, Guerrero, Saharet, Cléo de Mérode, la belle Otero, and Miss Duncan, was ready for a new dance idol. "And lo and behold," the dancer with "unconscious legs who only dances in dreams," Madeleine G., has arrived.[1] Known as *die Schlaftänzerin* (the "sleep dancer"), the latest dance phenomenon had her own peculiar innovations to capture public interest. Madeleine had created a sensation, wrote one journalist, that no other winter entertainment could compete with, for her dancing in a state of hypnosis captured "the two greatest powers of the mind—art and science."[2] A leading Berlin newspaper further characterized Madeleine as "the newest half-scientific, half-artistic wonder."[3]

Madeleine had indeed sparked a lively debate over the respective domains of art and science and their mutual fields of inquiry. Not just scientific experts, but those who saw Madeleine dance, were gripped by the questions she aroused. Was she really asleep? Was it a hoax? But most importantly, what did her spellbound movements reveal about the unconscious human mind? A leading doctor, commenting on the "passionate war of words" that followed one of Madeleine's performances, confirms that "everywhere you listened, this question was discussed almost continuously . . . during the breaks and after the concert."[4] Lampooning the publics' obsession with Madeleine, one satirist narrated his desperate journey by train across Germany to escape the incessant chatter about sleep dancing. Alas, as he reached the snowy peak of Germany's highest mountain, a lone alpinist greeted him with, "Tell me, have you seen Madeleine dance?"[5]

This essay examines Madeleine's sudden celebrity in Germany at a moment of intense interest in sleep and dreams as clues to the mystery of human creativity. Debates about the authenticity of Madeleine's sleep dancing in the popular press, medical journals, and scientific monographs reveal how interest in sleep and the subrational mind intersected with popular demonstrations of hypnosis such as Madeleine's, entangling theaters of medicine with theaters of art and entertainment in a search for the latent source of human creativity. Psychiatrists, hypnotists, neurologists, artists, satirists, philosophers: all scrutinized Madeleine in a discourse that brought fresh perspective to the developing science of the mind.

The deep interest in Madeleine's choreographic crossover from art to science was no accident. Her rise to fame was in fact engineered by an unusual combination of artists and scientists who brought her to Germany under the auspices of the Munich Psychological Society. Their chosen moniker for the trance dancer, "Madeleine G.," effectively invoked the pseudonyms given to celebrity psychiatric patients such as the famous Anna O. in Josef Breuer and Sigmund Freud's 1895 *Studies on Hysteria*.[6] Madeleine G. was born Emma Archinard in Tiflis, Georgia, in 1875. She married and had two children with a Parisian purveyor of building materials, Albert Guipet, and in April 1902 sought medical treatment from the celebrated French magnetist Émile Magnin.[7] Magnin was a therapist and teacher at the Paris École de Magnetisme when Madeleine came to him for incurable headaches "of a nervous origin."[8] In the course of Madeleine's hypnotic treatments, Magnin noticed her unusual sensitivity to music and a striking gestural expressivity that prompted him to seek evaluation by a professional musician.[9] What Magnin had taken for hysterical convulsions he now suspected were artistic revelations "that had sprouted fully formed from her unconscious mind."[10] Magnin arranged for Madeleine to be "displayed" twice at the Opéra-Comique in January 1904, and then in the Paris studios of sculptors Auguste Rodin and Albert Besnard and photographer Fred Boissonnas.[11]

From there, Madeleine's notoriety grew. After attending one of these exhibitions, the German psychiatrist Albert von Schrenck-Notzing arranged for Madeleine and Magnin to come to Germany, where her "haunting art," as critic Georg Fuchs wrote, "unleashed storms of delight and put thousands and thousands into a frenzy."[12] Along with the public, a spectrum of experts—professors, psychiatrists, pedagogues, philosophers, theologians, occultists, and aestheticists—jockeyed to see and to pass judgment on Madeleine's dream dancing: "so full of lightening-like inspirations, so full of true experience, so full of gripping natural poetry," wrote Leopold Weber, "as I have ever seen it manifested in the higher sphere of consciousness."[13]

The Science of Sleep, Dreams, and Hypnosis

Germans' keen interest in Madeleine's sleep dancing followed upon many decades of thought and experimentation in two key areas: nineteenth-century romantic nature philosophy, and scientific research on animal magnetism and hypnotism. Well before the surge of scientific interest in hypnosis and sleep, nineteenth-century German writers and philosophers had trolled the depths of the human psyche with a metaphysical view of the human subconscious informed by the awesome and infinitely mysterious power of nature. The nineteenth-century popularization of Darwinism in Germany aroused interest in human consciousness by challenging prior distinctions between the "civilized" and "instinctual" characteristics of human behavior. Darwin's *The Expression of the Emotions in Man and Animals* (1872) linked the human nervous system to ancient animal stimuli and behaviors, inviting research on dreams as a key to the regressive level of human impulse still coursing in the psyche of modern man.[14] For Romantic thinkers of the nineteenth century, this dual nervous system—one of the sleeping primordial past, one of the wakeful modern human—not only divided behavior into different realms of the self, but also, as art historian Marsha Morton explains, "established the unconscious as a state of being which unified man with other, more instinctual, forms of life."[15]

This instinctual life, buried in the body, was thought to be most accessible to the mind in the dream state. By 1814, German readers were introduced to Gotthilf Heinrich von Schubert's *Symbolism of Dreams* and Jean Paul's *Views into the World of Dreams,* which proposed that dreams were caused by somatic stimuli transmitted to the brain, and then transformed back to images taken from waking experience.[16] Arthur Schopenhauer's writing on the relationship of the body to the will and to consciousness presented dreams and somnambulism as "essential resources for the study of human behavior."[17] Another philosopher, Albert Scherner, posited sleep as a state of "assault . . . by the drives of the body," whose nocturnal "sovereignty" reigned unchecked by any "moral laws."[18] Attempts to relate the conscious to the unconscious mind incorporated the supernatural and the psychologically abnormal as prominent features of German Romanticism. Paranormal phenomena, séances, and hypnosis gained popularity via the sleepwalkers in stories and plays by Heinrich von Kleist, Arthur Schnitzler, Gottfried Keller, and especially E. T. A. Hoffmann. Morton concludes that in Hoffmann's thinking, "the trance approximated creative consciousness" and dreams became "artistic 'levers' to activate fictional transitions from the empirical realm to the fantastic."[19]

Turning back from the fantastic to the empirical, however, scientists sought clues to the hypnotic state in the long tradition of experimentation with electricity and magnetism.[20] Eighteenth-century searches for the principles of magnetic and electric phenomena cleared the way for nineteenth-century breakthroughs by Humphry Davy, André-Marie Ampère, and Michael Faraday, whose work in electromagnetic induction led him to conclude that "matter is a pattern of invisible energy."[21] This conflation of matter with energy found a home in the thought of German nature philosopher F. W. J. Schelling, who believed that nature itself operates beneath the rational as an "unconscious slumbering spirit."[22]

These scientific ideas found practical application throughout the nineteenth century in hypnosis. This practice grew most directly out of Franz Anton Mesmer's eighteenth-century theories of attraction and repulsion through the medium of animal magnetism, a fluid he believed was "universally widespread and pervasive."[23] Mesmer, who considered himself a physicist, studied medicine at the University of Vienna and wrote a dissertation on the effect of the sun and moon on the tides, including those in the bodily humors affecting the course of disease.[24] He established a clinic in Paris in 1778 and was flooded with patients seeking magnetic correction of their disrupted fluids with a treatment involving a wooden tub of "magnetized water" and iron filings that stored "animal magnetism."[25] Groups of the afflicted, joined by a cord, grasped movable iron rods protruding from the tub and applied them to their ailing body parts to improve "free circulation of the magnetic fluid."[26] Mesmer also believed he could improve magnetic flow by passing his hands over patients' magnetic "poles" to induce an often convulsive hypnotic trance. He realized the economic potential of these "mesmerisms" by performing them before curious onlookers in European salons, lecture halls, and carnivals.

Interest in magnetism grew throughout the nineteenth century. By 1817, the University of Berlin had appointed two professors of animal magnetism.[27] One of these, Karl Christian Wolfart, ran a large "magnetic polyclinic" visited by scientists, doctors, and philosophers from across Europe.[28] The term "magnetism" gave way to "hypnotism," and by 1894 the Berlin physician Jonas Grossmann had compiled twenty-nine expert reports showing that hypnotic suggestion was "a beneficial and low-risk form of treatment for a range of medical conditions" including epilepsy, hysteria, and the migraine headaches that afflicted Madeleine.[29] German psychologist Albert Moll's *Der Hypnotismus* (1899) brings us directly to the case of Madeleine, for in this work he carefully considered Schrenck-Notzing's advocacy of the dancer as a hypnotic subject. Two societies formed in Germany to encourage experimentation with hypnosis: Berlin's Society for Experimental Psychology and Munich's Psychological Society, which sponsored Madeleine's performances in Munich. Their experimentation tested the therapeutic efficacy of hypnosis, but also focused on the artistic potential of the dissociated mind, a question that informed much of the speculation about Madeleine.[30] Thus, two centuries of scientific inquiry into the nature of consciousness had set the stage for broad interest in Madeleine's "sleep dancing."

Madeleine's transference from Paris to Munich via Schrenck-Notzing's sponsorship paralleled a broader shift in the study of hypnosis and psychiatry from France to Germany.[31] The French rivalry in theories of hypnotic therapy between neurologist Jean-Martin Charcot at his Salpêtrière laboratory in Paris and Hippolyte Bernheim in Nancy was eclipsed in the 1880s by German-speaking neurologists and psychiatrists. Max Dessoir, Auguste Forel, Richard von Krafft-Ebing, Albert Moll, Adolf Weinhold, and Schrenck-Notzing all issued books on the therapeutic use of hypnotism between 1879 and 1896. With Freud's *Studies on Hysteria*, hypnosis entered medical discourse as a way to "create a vacancy of consciousness" and rid the patient of resistance to the subliminal self.[32] Schrenck-Notzing believed hypnosis afforded a "relative loosening of the psychological tissue," so that wakeful ideas "detach themselves," and are thus paralyzed or condemned to sleep. This "complete surrender" to suggested impressions was key to understanding Madeleine's choreographic outpourings, most evident in her "rapturous" responses to music.[33]

Particularly interesting to the case of Madeleine is the search for the source of human creativity in theories of ideomotor action, which held that the "idea" of a movement is "already the beginning of the movement itself" and that every state of consciousness "has a tendency . . . to transform itself into a movement or an act."[34] The ability to suppress automatic movement thus emerged as a key to understanding all human agency, hypnotism producing this transformation "in its purest form."[35] If volition could be understood as the mental power to stop movement, hypnosis became all the more attractive as a way to investigate the absence of willed agency over the autonomic responses of the body to the mind.[36] One report on the scientific basis of the Madeleine "phenomenon" comments on the "latent talents" of expressivity that are inversely "awakened" by hypnosis.[37] This happens because with the suppression of consciousness, "the susceptibility to suggestion . . . is almost without resistance."[38] Put another way, philosopher of art Theodore Lipps argued that any inhibiting factors disturbing the purity of Madeleine's dancing have "fallen asleep."[39] By dancing in a state of hypnosis, Madeleine provided a theatrical exhibition of ideomotor suggestion and its relation to the mysteries of human consciousness. As one observer noted, relationships between physical action and psychological perception have received "a new, particularly interesting and valuable confirmation from the Munich sleep dancer."[40] In Schrenck-Notzing's view, Madeleine's body became "an ideoplastic instrument in which every mental impulse finds its adequate expression."[41]

Madeleine's Hypnotic Dancing

Schrenck-Notzing considered Madeleine's sleep dancing an incomparable wealth of expressivity, making her case "equally valuable both psychologically and artistically."[42] He thus had to manage Madeleine's carefully staged program of musical and narrative "suggestion" as both physician and entrepreneur. Because public displays of hypnosis were illegal by 1903, he first presented Madeleine in a "séance" at his Munich residence.[43] More demonstrations followed in February 1904 for select circles of invited guests in the homes of a prominent banker and the Austrian Ambassador Count Zichy.[44] The *Berliner Morgenpost* reported on a performance in the salon of "an art-loving lady" in Munich where Madeleine "disarmed the most skeptical viewers."[45] In filmy blue robes, she succumbed to her magnetizer Magnin and "her flexible, elastic body became uncannily rigid." With her hypnosis complete, "a large, staring pair of eyes look directly into the electric spotlight directed at the figure. . . . Well-known doctors step to the podium and try to bend her outspread arms. . . . As sounds come from a piano . . . the eyes flash lightening and a vibration trembles through the still lifeless body of Madeleine. The expressions change abruptly, and . . . Madeleine follows the notes and struggles to do the strangest poses."[46]

Among the invited doctors, artists, writers, and journalists at these private soirées, some, like Alfred Keller, Alfred von Mensi-Klarbach, and Otto Julius Bierbaum, praised Madeleine's emotional immediacy.[47] Others suspected a hoax and shouted their objections. When dubious reports of these private showings hit the papers, Schrenck-Notzing hastened to defend his scientific reputation by booking Munich's intimate, experimental *Schauspielhaus* theater for six semi-public performances between March 9 and 22, where ticket-buyers paid an exorbitant twenty marks for entrance. Schrenck-Notzing's entrepreneurship, according to Corinna Treitel, "showed just how permeable the boundaries between science and art, research and performance could be in an age of emerging mass culture."[48] It also showed his defiance of regulations meant to protect the public from charlatans. According to one disgruntled source, Schrenck-Notzing observed certain pretenses "for the appearance of legality. . . . The Psychological Society circumvents the ban on hypnotic displays . . . by supplying an 'invitation card' for shops entrusted with the sale of the tickets to distribute to buyers."[49] One dubious critic signing as "Sleep writer" objected that Madeleine "will have made a small fortune" before the delicates of high society "wake from their own gullible doze."[50] Another scoffed that "when farce is passed off as science" for the overexcited Munich public, "everyone comes—everyone!"[51]

Madeleine's hypnotic dances followed a carefully controlled dramatic arc. The program for March 9, advertised as "a demonstration by the dream dancer Madeleine G. by the Psychological Society," contained three parts: an experimental prelude, consisting of Madeleine's hypnotism by Magnin; a musical demonstration, in which she danced to Chopin's *Funeral March,* Schubert's *Erlkönig,* and Wagner's *Lohengrin* performed by local professional musicians; and finally a dramatic pantomime to declamations by the actress Miss Lili Marberg.[52] Here Madeleine impersonated overwrought femmes fatales, what Don LaCoss calls "the immortal goddesses of hysteria"—among them Salomé, the Virgin Mary, Lucretia, and Helen.[53] Magnin's hypnosis employed a combination of eye fixation and hand passing. Schrenck-Notzing described the process: Madeleine sits or stands in front of the hypnotist, who takes her hands and fixes her eyes. After a few seconds, her gaze becomes rigid, she no longer blinks, and conjunctival reflex is weakened. Her facial expression becomes "mask-like," her arms fall to her sides and "a changed state of consciousness is quite unmistakable." Active somnambulism emerges slowly, as mild music takes hold of Madeleine's movements.[54]

The daily press reported Madeleine's performances in vivid detail. According to one account, the *Schauspielhaus* curtain went up on a bare stage with a lady in white sitting in a spacious armchair.[55] In tails, Magnin steps forward and performs "the familiar movements of the hypnotist." He steps back and the music begins. Respected musicians take turns at the grand piano, sometimes improvising. "The stately full figure of the hypnotized woman . . . rises from the armchair and makes a few tentative movements. . . . In the glaring glow of a blue lantern, the most wonderful and eerie dance begins. Every fiber of the woman's body, guided by no conscious effort, trembles in the onslaught of the sound waves like the leaf of a tree in the wind."[56]

When the music stops, Madeleine suddenly becomes rigid. Schrenck-Notzing calls on medical observers to investigate if they wish. "The doctors come, see, and are amazed."[57] Writing in the *Münchner Neueste Nachrichten,* Dr. Seif credited Madeleine's "hysterical disposition" for her extraordinary

Figure 16.1. "Arc de cercle" of a hysteria patient at the Salpêtrière. Paul Richer, *Études cliniques sur la grande Hystérie ou hystéro-epilepsie* (Paris, 1885), 69. Public domain. Accessed in HathiTrust Digital Library.

expressivity. "Her gestures and pantomimic expressions reveal sadness, bliss, rapture, rage—i.e., all the emotions—rather precisely, even according to pitch, volume, sound color, intervals, and rhythms."[58] The highly touted "hysterical" element of Madeleine's hypnotic condition was evident in her wild and violent expressions and movements. According to Weber, extremes of emotion poured from her body: "How she cringes to the core as she meets the roar of the funeral march, how she rears up, her whole face wracked with sorrow, to a violent level of pathos and pain, staggering rhythmically, shaken by massive grief."[59] Seif further described the "*arc de cercle*" (see figure 16.1 and figure 16.3) in which Madeleine's head bent back to touch her spine, and "the almost embarrassing groaning and screaming during Chopin's *Funeral March*, when she tries to dig her fingernails into the floor as the music slowly descends."[60]

Believers and Doubters

Reactions to the authenticity of Madeleine's hypnotic state crossed many of the carefully constructed boundaries separating lay hypnotists and medical

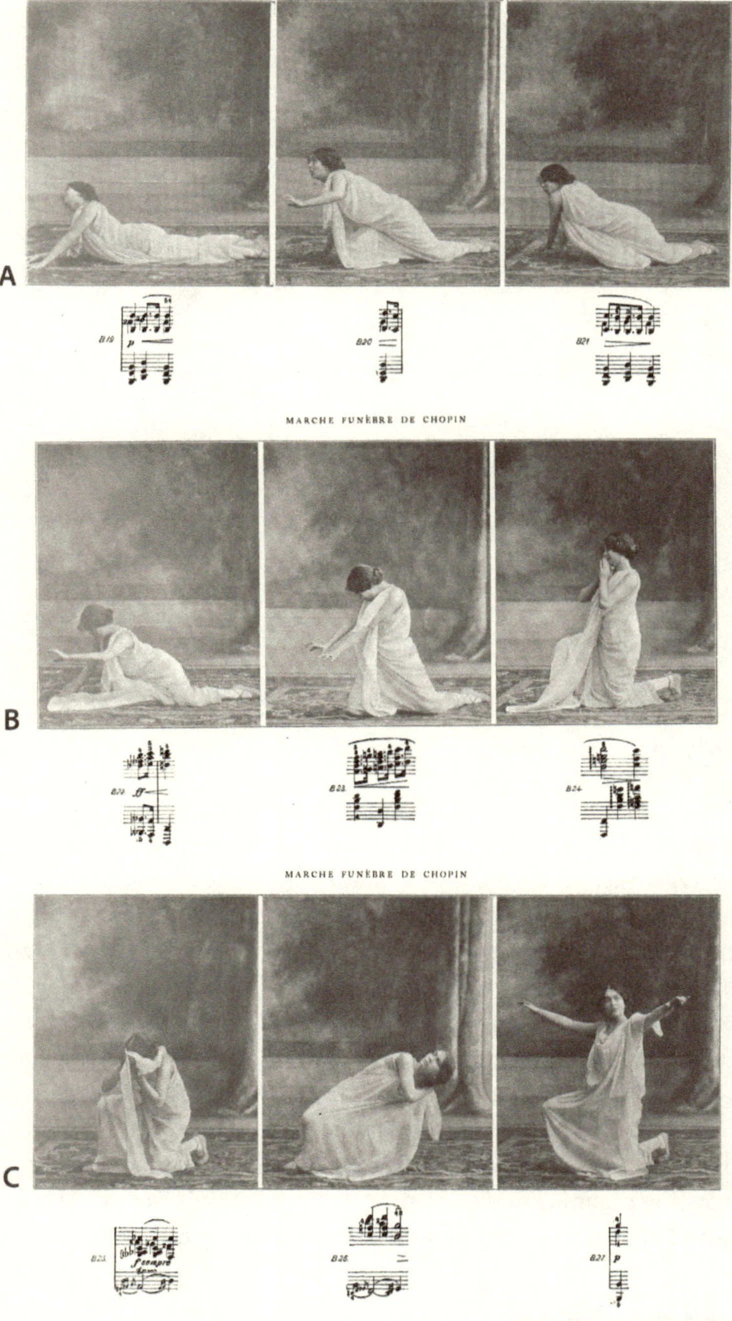

Figure 16.2a, 16.2b, 16.2c. Madeleine performing bars 19–27 of Chopin's *Funeral March*. Emile Magnin, *L'art et l'hypnose,* 2nd edition (Paris, 1907), 471, 473, 475. From a series of photographs by Fréderic Boissonnas. Public domain. Accessed in HathiTrust Digital Library.

practitioners. Critic and editor of the *Allgemeine Zeitung* Alfred von Mensi declared emphatically that "the cataleptic, inward-looking eye" maintained throughout Madeleine's hypnotic exhibition was not a fraud. "She danced and played for two hours, twice being awakened to rest and then put back to sleep. . . . All Schrenck-Notzing's colleagues found signs of an utterly cataleptic condition."[61] Neurologist Leopold Loewenfeld professed absolute confidence in Madeleine's hypnotic state; after participating in a special viewing for the Munich Medical Association, Loewenfeld attended Madeleine's performance at the *Schauspielhaus* on March 16 and conducted a backstage examination of the dancer. There, he induced catalepsy of both arms, and then a severe contraction of one leg. "The muscles of the leg felt like a board and the resistance . . . was so unyielding that, in my judgment, it could hardly have been overcome by a man of very significant muscle strength. With her rigidly stretched and raised legs, the bizarrely twisted arms . . . and the pulled back head, standing motionless on the uncomfortable cane chair, the woman offered a scene that dispelled any doubts about simulation."[62]

Yet Madeleine and Magnin did inspire doubts. These focused on Madeleine's professed lack of dance training and the possibility of simulation. In 1894, the Viennese neurologist Moritz Benedikt had published a scathing denunciation of hypnotic therapy, claiming that at least 90 percent of hypnotized patients had later confessed to pretending they were asleep; other doctors had followed suit.[63] Likewise, popular humor magazines and daily newspapers ridiculed Madeleine's "sleep dancing" as a sham. Several reviewers wondered why, if Madeleine was in a rapturous state of hypnosis, she never turned her back to the audience? How did she not collide with the stage props? And why did she adjust her hair and costume when they were out of place?[64] One skeptic notes that the appropriate movement "seems to have been triggered before the declamation or music has announced it."[65] Key to her critics' doubts was the claim that Madeleine had no dance training. *Jugend* lampooned this question in a series of comic drawings featuring one of Madeleine's supposed imitators, "Signora Schmuggolina." The spoof describes Schmuggolina's pantomime of Richard Wagner's "Ride of the Valkyries," wryly emphasizing that she has of course "never had a single riding lesson."[66] To counter such suspicions, Magnin assured audiences that Madeleine's "perfectly natural and involuntary expressions of instinct" were merely "the outflow of a natural talent, freed from all inhibitions by hypnosis and requiring no training."[67] Yet critic Detta Zilcken was not convinced: "I became certain," she cautions, "that this lady had frequent gymnastic exercise" and that a natural acting talent had been carefully developed. She had in fact "been preparing her role for a long time."[68]

Madeleine received intense medical scrutiny to further dispel such doubts. In addition to their observations during her performances, physicians performed tests on Madeleine in their laboratories and at the viewing for the Munich Medical Association. The *Münchner Neueste Nachrichten* reported that Schrenck-Notzing sought scientific verification of his observations from three neurologists—Dr. Seif, Dr. Hirt, and Dr. Feser—and from a specialist in the psychology of sound, Dr. Schultz. After an hours-long examination, all four scientific luminaries "are convinced Magdeleine's demonstrations take place in a state of hystero-hypnosis."[69] Further agreement came from Dr. Focke, director of the local insane asylum. Their tests focused on the sensory apparatus of the nervous system. While observing disturbances in Madeleine's eyes, the ophthalmologist Dr. Ancke noted "rigid pupils" and a complete absence of blinking for a duration of 11 ½ minutes.[70] In his book about Madeleine, Schrenck-Notzing provided the detailed findings of seventeen specialists who examined Madeleine during her five weeks in Munich, compiling data on her vision, hearing, reflexes, muscular strength, memory, intelligence, and motor control.[71] Never forgetting Madeleine's choreographic impact, Schrenck-Notzing was also attuned to the artistic potential of Madeleine's case. His stated aim, as coordinator of this research, was to document "the artistic and scientific sides of 'sleep dance' . . . in the dramatic-choreographic field."[72]

Yet as Schrenck-Notzing and many others stressed, Madeleine's impact was only "half-scientific." The complementary half of her importance was felt by artists. Munich's Psychological Society, since its formation, had in fact linked art with the study of the mind. The group's 1887 manifesto invited artists to consider the Psychological Society "a unique venue in which to apply new experimental psychology to the realistic treatment of psychological themes and topics in their work."[73] Society member Alfred Keller, who created upward of twenty paintings of Madeleine, acted in collaboration with Schrenck-Notzing as art director of her *Schauspielhaus* performances. Keller joined the Society in 1886, held more than fifty séances at his home, and made over a dozen paintings "involving scenes from psychical research."[74] Before becoming entranced with Madeleine, he was already using hypnotized models to access their emotional immediacy in depictions of nightmares and altered states of consciousness.[75] Many artists of the nineteenth century had similarly explored themes of the *somnambule.* Max Klinger's work from the 1870s and 1880s formed a provocative confrontation of the conscious and unconscious regions of the human mind in dreams, where, as Morton explains, "self-possession is involuntarily relinquished and figures are portrayed in hypnotic conditions whose dissociative character is conveyed by gaze, pose, and a sense of suspended animation."[76] Other artists belonging to the Psychological Society had also been experimenting with depictions of the insentient mind for over a decade. Their experimentation with occult phenomena put them, according to Treitel, in the vanguard of "an emerging modernist sensibility dedicated to exploring how the eruptions of unconscious drives, desires, and emotions played out on the surface of the human face and body."[77] Motivated by their interest in psychological realism, several of Munich's Decadent and Secessionist artists—Hugo von Habermann, Friedrich August von Kaulbach, and Franz von Stuck—made paintings of Madeleine.[78] In response to some of Madeleine's critics, Keller, von Stuck, and von Kaulbach signed a statement of support that appeared in the *Allgemeine Zeitung*: "After the declarations made in the newspapers by the scientific party which deals with the phenomenon of Mme. Magdeleine, we artists feel . . . we must publicly express our admiration and our gratitude for the extraordinarily artistic satisfaction we experienced in being able to study the force of her expressive power."[79]

Photography added another means to explore Madeleine's expressive range. Magnin's monograph on Madeleine included over 200 images taken by the Swiss photographer Fred Boissonnas. Their collaboration on *L'art et l'hypnose* combined artistic with scientific validation of hypnosis, presenting Madeleine as star witness (see figure 16.2). But Boissonnas also had precedents for his modernized "visual repertoire of hypnoses."[80] Two decades earlier, Charcot's assistant, Paul Richer, had meticulously photographed patients at the Salpêtrière neuropsychiatric hospital for women, revealing their "spectacle of pain" in a variety of cataleptic attitudes and poses, indexed in a table that depicts the four phases of a hysterical attack (see figure 16.3).[81] Richer's 1881 photographs and drawings form an uncanny prequel to Boissonnas's record of Madeleine, suggesting that he and Schrenck-Notzing, and possibly even Madeleine, had studied the cataleptic attitudes and gestures observed in Charcot's theater of hysteria.[82]

Theaters of Dance and Medicine

The documentary record of Madeleine's sleep dancing makes clear her appeal to both artists and scientists. Yet we must wonder why their typically different aims and approaches converged on a dancer and how this convergence ultimately broke down. I argue here that Madeleine's gestural enactment of the subconscious mind effectively cross-referenced the protocols of artistic and scientific inquiry. By transposing dance to fields of inquiry artists and scientists found equally legible, Madeleine also allowed each to borrow from the other's legitimacy.

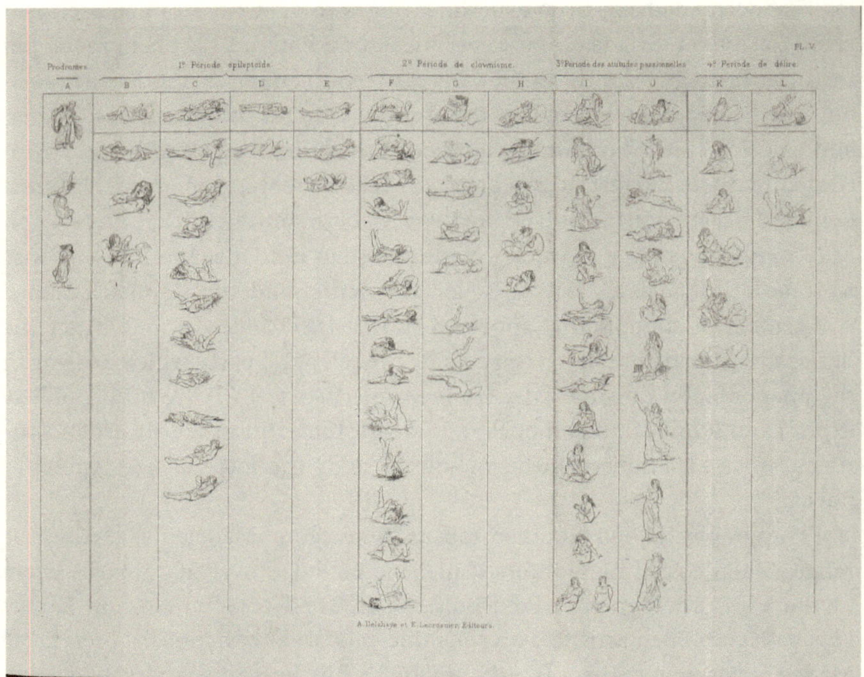

Figure 16.3. Synoptic table of the "complete and regular great hysterical attack," with its four periods and their variant poses. Richer, *Études cliniques sur la grande Hystérie ou hystéro-epilepsie.* (Paris, 1885), 168. Public domain. Accessed in HathiTrust Digital Library.

Medicine itself had a reputation for dramatically staged rituals as it sought to inspire respect and awe for the advances of modern surgery and psychotherapy. The nineteenth-century operating theater spawned its own genre of painting devoted to the drama of relations among celebrity surgeons, unconscious patient, and voyeuristic audience. Thomas Eakins's *The Agnew Clinic* (1889), in which the performance of a mastectomy absorbs the attention of an elite male audience, accesses many elements of traditional theater: dramatic arrangement of light and props, clearly delineated cast of characters, and rapt gaze of onlookers. Clinical advances in psychiatric treatment offered a similarly dramatic proscenium of discovery. Charcot performed demonstrations of the various stages of hysteria "in packed lecture halls, bathed in a spotlight before international medical and lay audiences."[83] The "theatrical bodies" of Charcot's patients exhibit, according to Didi-Huberman, "an extraordinary complicity between patients and doctors" that contributed not only to the spectacle of hysteria, but to its very invention, constructing a covert identification of hysteria with theater.[84] In *Une leçon clinique à la Salpêtrière* (1887), French artist André Brouillet depicts Charcot at one of his regular Tuesday lectures in a taut scene of suspense, a large audience of male students gazing intently on a female subject in a state of hysterical contracture. Didi-Huberman describes the dramaturgy of acts, scenes, and tableaux in the hysterical attacks of one of Charcot's star patients (Augustine), with intervening periods of "repose" and "entr'acte."[85] In this light, we see Charcot join Schrenck-Notzing as a theatrical "entrepreneur" of hypno-hysteria.[86]

Kélina Gotman further refines this entrepreneurship in terms of dance. Charcot's "convulsionary theater" displayed a science of neurology "born as a choreographic process,"[87] reproducing gestures in clinical settings where the patient's seizure "presented a dramaturgical arc, from onset to crisis and resolution."[88] The choreo-dramatic symptoms of Charcot's female hysterics do indeed emerge as an arc in the table constructed by Paul Richer depicting the phases of a hysterical attack (see figure 16.3). This arc seems to merge the horrors of female hysteria with Madeleine's very similar theatrical poses, exposing the darker, clinical symptoms latent in Madeleine's beautifully lucid dancing. As Gotman explains, the theatricality of psychiatric research occurred on a hierarchy of planes. In the first, for example, which might correspond to Richer's "Période épileptoïde," the female hysteric "performs her distress" by corporealizing "what is linguistically inexpressible." She is thus imagined to access "an ancient type of corporeality."[89] For Madeleine's observers, this ancient corporeality invoked the archaic and terrifying forms of animal instinct revealed by the precepts of Darwinian evolution, made both attractive and entertaining in the controlled conditions of theatrical hypnosis.

The cultural traffic between theaters of entertainment and medicine ran in both directions. Where medical experimentation sometimes appropriated elements of entertainment, public spectacles staged by lay hypnotists borrowed equal legitimacy from the rituals of scientific demonstration. Beginning in 1878, the most famous itinerant stage hypnotist, Carl Hansen, presented himself as a learned professor. His performances began with an introductory lecture on animal magnetism and proceeded to demonstrate the validity of his claims on "receptive" individuals from the audience (see figure 16.4). A report from 1880 described large audiences for a series of shows in Breslau where Hansen brought prominent citizens onto the stage. Hansen induced his magnetized subjects to eat raw potatoes, sing to the crowd, and perform "preposterous pantomimes."[90] According to eyewitness Stanley Hall, the civic authorities in Breslau invited the university professor and animal magnetist Professor Heidenhain to join in Hansen's exhibitions to offer "scientific" explanation.[91] At Hansen's shows in Chemnitz, physician and chemist Adolf F.

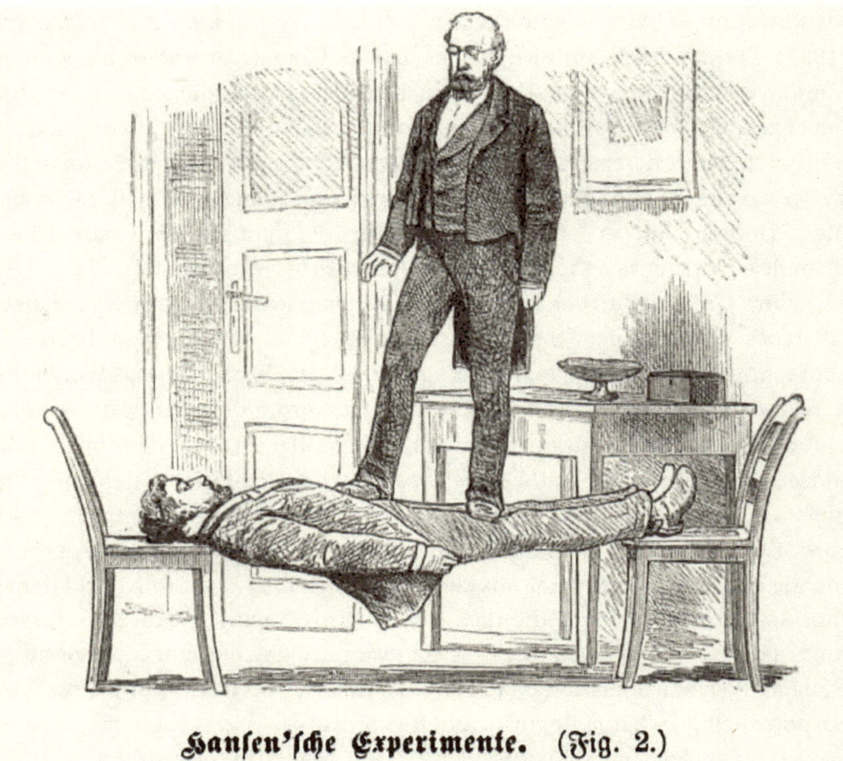

Figure 16.4. Carl Hansen's "human plank" tested credulity by placing a cataleptic subject between two chairs so Hansen could stand on the rigid body. Richard Rühlman, "Die Experimente mit dem sogenannten thierischen Magnetismus," *Die Gartenlaube* 28 (1880): 129. Public domain. Accessed in HathiTrust Digital Library.

Weinhold fulfilled the same function and also performed hypnotisms himself.[92] To the lay audiences that Hansen and a cadre of lesser-known traveling hypnotists entertained, these stage demonstrations differed little from the experiments being carried out in medical laboratories and clinical settings.

Among scientists, they sparked curiosity. Freud, Krafft-Ebing, and Charcot all attended Hansen's performances, which had in fact helped initiate the wave of scientific research on dreams, sleep, and hypnotism.[93] Gauld as well contends that German doctors' interest in hypnotism was "largely unleashed" by Hansen and his recruitment of medical men for his shows in cities and towns across Germany.[94] Thus, in spite of physicians' work to establish a medical monopoly over popular magnetizers like Hansen, the two classes of practitioner communicated and even collaborated in overlapping theaters of inquiry that eventually collided headlong in the case of Madeleine.

This collision reached maximum impact when members of the medical community publicly condemned Schrenck-Notzing's activities with Madeleine. As a leader of the campaign to separate medical hypnosis from lay mesmerism, Schrenck-Notzing had previously argued in several publications that hypnosis was a physical and psychological danger in the hands of lay practitioners.[95] Albert Moll, the esteemed neurologist who had worked with Schrenck-Notzing to outlaw the nonmedical use of hypnosis, bristled at the dual nature of Schrenck-Notzing's experiments: "sometimes of a scientific and sometimes of an artistic nature."[96] He considered Schrenck-Notzing's role in Madeleine's onstage antics a treacherous "insult to science," a duplicitous endorsement of quacksalvers and vaudevillians, and "a downright misdemeanor."[97]

Moll was not alone. Dr. Franz Roberts issued a twenty-four-page brochure attacking both Schrenck-Notzing and a gullible Munich public for falling prey to Madeleine's simulations.[98] Roberts asserts that the smart set of Munich was thoroughly duped by Schrenck-Notzing's skillful "launch" of the sleep dancer: "Very quietly, little notices appeared to prepare the public. Then came larger notices, then small articles, then raptures, then beatings of the tamtam and gong, and finally the trumpets of Jericho. The necessary authorities were recruited, the mass press persuaded, and 'all Munich' crept into the well-set trap."[99] In accordance with the motto, "Mundus vult decipi" (the world wants to be deceived), Roberts concludes with the idiom that Munich "has had a bear put on its back" ("Jemanden ein Bären aufbinden")—tricked, in other words, and left for fools. He asserts that Schrenck-Notzing is not just a mountebank, but a defiler of science for he has led Munich out of the light and into the "dark sciences" of the occult.

Further denouncing Schrenck-Notzing's crossing of this boundary with Madeleine and Magnin, another doctor recalls the early nineteenth-century practice of allowing the public into insane asylums for an entrance fee in order to "delight" in the inmates' abnormalities.[100] In his view, Madeleine's performances smack of "the barbarous practice of exhibiting insane patients to the gaping public."[101] Another critique came from a lay hypnotist who objected to a doctor becoming a "manager" of commercial exploitation, which could only be "repulsive and outrageous."[102] Viewing the "psychological agony" of temporary mental illness is no different than witnessing the trauma of a surgical incision, according to this critic, and no psychiatrist should allow spectators to observe the insane for their amusement. Madeleine's handlers do just this, and for a public "whose claim to this sight is based only on frivolous curiosity."[103] This moral theme motivated another critic who wished Madeleine would choose to be either an artist or a subject of scientific experimentation:

"If Madame G. really dances in a somnambulistic state, she . . . belongs not on the stage, but in a theater of abnormalities."[104]

Schrenck-Notzing issued a rebuttal, defending the Psychological Society's public showings of "this rare and interesting case."[105] Once the newspapers had revealed Madeleine's presence in Munich, he explained, dozens of organizations and all the major cities of Germany and Austria had eagerly requested private showings. The *Schauspielhaus* demonstrations—sponsored by a responsible organization in an artistically tasteful manner—were the lesser of many evils that might have befallen an entertainer in such demand. In this way, Madeleine could be compensated for her talents, sensation-seekers could be limited from entrance (thus the high ticket prices), and for three matinees the theater and orchestra could be devoted to specially invited groups of artists, students, and medical professionals. Faced with unforeseeable demand, the Psychological Society "fulfilled its duty as far as possible."[106]

Seeking the Source of Creativity

In order to understand Schrenck-Notzing's turn against decades of work on boundary-building between objective research and popular amusement, we must consider Madeleine's impact on his theories of hysteria and hypnosis. For Schrenck-Notzing, Madeleine's "mild hysteria" made her especially susceptible to hypnosis and its liberation of individual consciousness from "inhibiting and disruptive" psychological factors, thus laying bare the nature of artistic revelation.[107] He believed this relationship between hysteria and artistic sensitivity was in fact present in all great actresses—Sarah Bernhardt and Eleanore Duse in particular—and in every sensitive human as well.[108] What made Madeleine such an important case was the activation of her talent by hypnosis alone. "In the dream dancer we see a great born artist who did not develop in normal daily life, whose choreographic and mimic ability could only be brought to free development in a hypnotic state. What other artists achieved through laborious years of study, upbringing, and practice, Magdeleine received as a ready-made gift from nature."[109]

As a window into artistic inspiration, hypnosis struck some as a charade. The powerful critic Hanns von Gumppenberg found a "poverty" in Madeleine's performance that lay in the "immature passivity" of her condition. She was only a "puppet," lacking the "active, purposeful spirit" required of a true artist.[110] Yet some of the most astute critics of the time believed Madeleine's hystero-hypnosis revealed the source of artistic inspiration. Critic and keen cultural observer Georg Fuchs saw in Madeleine something akin to the mental state of brilliant actors. Once they are on stage, these artists "fall into a sort of trance, which allows them spiritual and physical abilities impossible under normal conditions."[111] To dance as she did, the otherwise unremarkable Madeleine "needed a changed state of consciousness and a dream-like veiling of her inner life."[112] Fuchs maintained that Madeleine's facial expressions, "equal to the high style of Duse," enabled her to evoke the extreme psychological states of Dionysian ecstasy in her depictions of Judith, Salomé, Klytemnestra, Helen, Antigone, and Cassandra.[113]

Many others saw in Madeleine's raw emotional impact a revelation of the hidden regions of the human mind and spirit. German novelist Otto Julius Bierbaum called it "a spectacle of almost uncanny violence and at the same time of absolute beauty. Never in my life, not even with the greatest actresses, have I seen a human body, a human face, with mental processes so strongly revealed."[114] Overwhelmed by Madeleine's expressive power, he accepted the dream state as the closest possible to being struck "with rare inspiration."[115] Fuchs also likened Madeleine's trances to divine inspiration, declaring that her dancing can only be considered "an artistic and cultural miracle."[116] With her consciousness redirected in the sleep state, she could represent more potently what occurs in all artists who "shape themselves under the pressure of an inner experience." All artists require a form of "auto-suggestion," akin to an ecstatic state of intoxication, allowing them to reveal their innermost being to an audience.[117] Fuchs believed hypnosis did this for Madeleine: "Like the angel who troubles the waters of Bethesda, the melodies stroke the surface of her soul: she trembles, she sways, she ruffles in gentle waves . . . she foams in chasing, howling storm surges . . . And what happens there in the dark regions . . . is brought out in the most perfect reflection before our delighted eye. 'What walks in the labyrinth of the heart through the night' comes through it to the light of earthly life: we can see it, we can experience it, pure and perfect as it happens."[118] Another critic declared that for many, Madeleine's "miraculous" expressivity revealed the possibility that creative forces lying in every human being "might be loosened by hypnosis."[119] Our artists, according to this observer, must consciously undertake "the same loosening of these fetters. . . . An artist is, as it were, one who is able to hypnotize himself." The urologist Dr. Felix Schlagintweit agreed. On the basis of Madeleine's demonstration for the Munich Medical Association, he concluded she was probably not hypnotized by Magnin, but was instead "auto-hypnotized" by the state of ecstasy all artists summon to unleash their creative powers.[120]

For some, Madeleine's expressive power even made the authenticity of her hypnosis unimportant. Theater critic Alfred Kerr declared her dancing "a fabulous phenomenon" in the art of acting.[121] Even while awake, "it would be great enough." Kerr extended this thinking to suggest that any creative act must induce a dream state to be fully expressive. Of writers, he asked, "Do we not lower our inner eyes to just one spot while paralyzing all our other faculties? Does a great, completely emotional actor do anything but sleep? Is Duse capable of effects which grip the very soul without the autosuggestion of sleep?" In Madeleine, the power of autosuggestion reached a higher level. "Is she a strongly hysterical performer . . . or is she a strong performer of hysteria?" Kerr concluded she is both, a "wildly great" example of the inhibitions "swept aside" and the "innermost turned outward."[122]

Back to *Jugend*

I began with *Jugend's* satire on Madeleine as the next in a long line of performers appearing in Munich's 1904 wave of dance novelties. In another rhymed mockery, *Jugend* satirized the Munich medical community's fight over Madeleine, concluding that the "arguing and ranting in the newspapers" is harmful to all parties: first, to Madeleine "whose hysteria is displayed for profit"; second to the public, "made stupid by the deception"; and especially to the medical profession, "which can never mask the fact that Magnin led it by the nose."[123] Yet the heated response to Madeleine's dances brought the science of sleep, hysteria, dreams, and hypnosis into direct dialogue with the nature of artistic creation. Madeleine G.'s impact, which one reviewer claimed made Isadora Duncan seem only a "weak dilettante," has been forgotten.[124] But as the science of neurology and psychiatry advanced apace with the aesthetics of dance, Madeleine's "unconscious legs" became a brief focal point of interest in the mysteries of the mind. Dancing in the bright light of the public theater, Madeleine brought science and art together in a common search for the dark secrets of the human psyche.

Notes

1 *Jugend,* Mar. 16, 1904, 258.
2 Zilcken, "Die Schlaftänzerin," 48.
3 *Berliner Morgenpost,* Feb. 24, 1904.
4 Maier, "Die 'Schlaftänzerin' Madeleine G.," 315.
5 H. St. "Die Schlaftänzerin. Drama in vielen Vorgängen," 1–2.
6 Eidenbenz, "Hypnosis at the Parthenon," 8.
7 Magnin, *L'Art et l'Hypnose,* 9.
8 Eidenbenz, "Hypnosis at the Parthenon," 1.
9 Börner, *So träumt man,* 10.
10 LaCoss, "Our Lady of Darkness," 55.
11 For the Opéra-comique program, see Magnin, *L'Art et l'Hypnose,* 397.

12 Fuchs, *Der Tanz,* 29–30.
13 Weber, *Der Kunstwart,* 90.
14 Morton, *Max Klinger and Wilhelmine Culture,* 258.
15 Morton, *Max Klinger and Wilhelmine Culture,* 271.
16 Morton, *Max Klinger and Wilhelmine Culture,* 259.
17 Morton, *Max Klinger and Wilhelmine Culture,* 259.
18 Scherner, *Das Leben des Traums,* 6. Cited in Morton, 260.
19 Morton, *Max Klinger and Wilhelmine Culture,* 275.
20 Wilson, "Matter and Spirit in the Age of Animal Magnetism," 331.
21 Faraday, cited in Wilson, "Matter and Spirit in the Age of Animal Magnetism."
22 Gauld, *A History of Hypnotism,* 142.
23 Wilson, "Matter and Spirit in the Age of Animal Magnetism," 331.
24 Gauld, *A History of Hypnotism,* 1–2.
25 Gauld, *A History of Hypnotism,* 5.
26 Gauld, *A History of Hypnotism,* 5.
27 Maehle, "A Dangerous Method," 198.
28 Gauld, *A History of Hypnotism,* 88–90.
29 Maehle, "A Dangerous Method," 198.
30 Wolffram, "'An Object of Vulgar Curiosity,'" 160.
31 Gauld, *A History of Hypnotism,* 419.
32 Hales, "Dancer in the Dark," 537.
33 Schrenk-Nozting, *Traumtänzerin,* 61–62.
34 Cowan, *Cult of the Will,* 90.
35 Cowan, *Cult of the Will,* 92.
36 Cowan, *Cult of the Will,* 90.
37 Henig, *Berliner Tagblatt,* Mar. 4, 1904, 6.
38 Henig, *Berliner Tagblatt,* Mar. 4, 1904, 6.
39 Lipps, "Zur Verständigung über die 'Schlaftänzerin,'" 3.
40 Henig, *Berliner Tagblatt,* Mar. 4, 1904, 6–7.
41 Schrenck-Notzing, *Traumtänzerin,* 121.
42 Schrenck-Notzing, *Traumtänzerin,* 9.
43 Treitel, *A Science for the Soul,* 116.
44 Zilcken, "Die Schlaftänzerin," 51.
45 *Berliner Morgenpost,* Feb. 24, 1904, 3.
46 *Berliner Morgenpost,* Feb. 24, 1904, 3.
47 Engels, cited in Maier, "Die 'Schlaftänzerin' Madeleine G.," 241.
48 Treitel, *A Science for the Soul,* 116.
49 Fürer, "Auch ein Beitrag," 471.
50 Kohl, "Das Märchen vom Schlaftanz," *Münchener Ratsch-Kathl,* Mar. 9, 1904, 2.
51 "Die Madeleine und ihr Impresario," *Münchener Post,* Mar. 12, 1904, 1.
52 Fürer, "Auch ein Beitrag," 471.
53 LaCoss, "Our Lady of Darkness," 65. For detailed description of these pantomimes see Zilcken, "Die Schlaftänzerin," 52.
54 Schrenck-Notzing, *Traumtänzerin,* 48.
55 Engels, cited in Maier, "Die 'Schlaftänzerin' Madeleine G.," 240–41.

56 Engels, cited in Maier, "Die 'Schlaftänzerin' Madeleine G.," 240–41.
57 Engels, cited in Maier, "Die 'Schlaftänzerin' Madeleine G.," 240–41.
58 Seif (1905), cited in Marx, "Madeleine: Two Reviews," 30.
59 Weber, *Der Kunstwart,* 90.
60 Seif (1904), cited in Schrenck-Notzing, *Traumtänzerin,* 41.
61 Von Mensi, cited in Maier, "Die 'Schlaftänzerin' Madeleine G.," 238–39.
62 Loewenfeld, "In Sachen der Schlaftänzerin," 570. The pulled back head likely indicates the *arc de cercle* pictured in Figure 1.
63 Maehle, "A Dangerous Method," 202.
64 Zilcken, "Die Schlaftänzerin," 52. See also *Münchener Post,* Mar. 12, 1904, 2.
65 Stümpke, *Bühne und Welt,* 476.
66 *Jugend,* Mar. 3, 1904, 218.
67 Zilcken, "Die Schlaftänzerin," 50.
68 Zilcken, "Die Schlaftänzerin," 55. On Madeleine's access to dance training see Börner, *So träumt man,* 10.
69 *Münchner Neueste Nachrichten,* Mar. 16, 1904. Cited in Magnin, *L'Art et l'Hypnose,* 374.
70 Schrenck-Notzing, *Traumtänzerin,* 35–36.
71 Schrenck-Notzing, *Traumtänzerin,* 23–26.
72 Schrenck-Notzing, *Traumtänzerin,* v.
73 Treitel, *A Science for the Soul,* 110.
74 Treitel, *A Science for the Soul,* 113.
75 Wolffram, *Stepchildren of Science,* 117 and Müller, *Albert von Keller,* 52.
76 Morton, *Max Klinger,* 252.
77 Treitel, *A Science for the Soul,* 114.
78 LaCoss, "Our Lady of Darkness," 58–59.
79 Cited in Magnin, *L'Art et l'Hypnose,* 375.
80 Eidenbenz, "Hypnosis at the Parthenon," 9.
81 Didi-Huberman, *Invention of Hysteria,* 1.
82 Richer, *Études cliniques.* See also Didi-Huberman, *Invention of Hysteria,* 8.
83 Lerner, "Hysterical Cures," 81.
84 Didi-Huberman, *Invention of Hysteria,* xi.
85 Didi-Huberman, *Invention of Hysteria,* 117.
86 Didi-Huberman, *Invention of Hysteria,* 115.
87 Gotman, *Choreomania,* 152–53.
88 Gotman, *Choreomania,* 156.
89 Gotman, *Choreomania,* 139–40.
90 Gauld, *A History of Hypnotism,* 303.
91 Hall, "Recent Researches," 98.
92 Mayer, *Sites of the Unconscious,* 114.
93 Mayer, *Sites of the Unconscious,* 268.
94 Gauld, *A History of Hypnotism,* 302.
95 Wolffram, "An Object," 152–53.
96 Moll, *Der Hypnotismus,* 475.
97 Moll, *Der Hypnotismus,* 534.

98 Roberts, *Die Schlaftänzerin Madeleine G.*, 5–6.
99 Roberts, *Die Schlaftänzerin Madeleine G.*, 6.
100 Fürer, "Auch ein Beitrag," 471.
101 *Journal of the American Medical Association* (April 16, 1904), 1033.
102 Grünwald, "Die Vorführung," 571.
103 Grünwald, "Die Vorführung," 571.
104 Verus, *Münchner Ratsch-Kathl*, Mar. 23, 1904.
105 Schrenck-Notzing, "Einige Bemerkungen," 667.
106 Schrenck-Notzing, "Einige Bemerkungen," 667.
107 Schrenck-Notzing, *Traumtänzerin*, 121.
108 Schrenck-Notzing, *Traumtänzerin*, 71.
109 Schrenck-Notzing, *Traumtänzerin*, 10.
110 Gumppenberg, *Der Kunstwart*, 698.
111 Fuchs, *Sturm und Drang*, 242.
112 Fuchs, *Sturm und Drang*, 242.
113 Fuchs, *Sturm und Drang*, 243.
114 Bierbaum, cited in Müller, *Albert von Keller*, 52.
115 Bierbaum, cited in LaCoss, "Our Lady of Darkness," 67.
116 Fuchs, *Der Tanz*, 21.
117 Fuchs, *Der Tanz*, 21.
118 Fuchs, *Der Tanz*, 25.
119 Engels, cited in Maier, "Die 'Schlaftänzerin' Madeleine G.," 241.
120 Schlagintweit, "Die Schlaftänzerin," 525.
121 Kerr (1905), cited in Marx, "Madeleine: Two Reviews," 31.
122 Kerr (1905), cited in Marx, "Madeleine: Two Reviews," 31.
123 *Jugend*, Mar. 23, 1904, 277.
124 "Magdeleine," *Der Kunstwart*, Mar. 1905, 852.

Bibliography

Journals and Newspapers

Berliner Morgenpost
Berliner Tagblatt
Deutsche Medizinische Wochenschrift
Die Gartenlaube
Journal of the American Medical Association
Jugend
Der Kunstwart
Mind
Münchener Medizinische Wochenschrift
Münchner Neueste Nachrichten
Münchener Post
Münchener Ratsch-Kathl

Psychische Studien
Die Schönheit
Der Tag

Books and Articles

Bierbaum, Otto Julius. *Münchner Neueste Nachrichten,* Feb. 19, 1904. Cited in Müller, Oskar A. *Albert von Keller, 1844 Gais/Schweiz - 1920 München.* Munich: Karl Thiemig, 1981.

Börner, Anna. "'So träumt man wohl zuweilen zu tanzen.' Zur Erscheinung des Schlaftanzes im ersten Drittel des 20. Jahrhunderts." PhD diss., Universität Lüneburg, 2014.

Cowan, Michael. *Cult of the Will: Nervousness and German Modernity.* University Park: Pennsylvania State University Press, 2008.

Darwin, Charles. *The Expression of the Emotions in Man and Animals.* London: John Murray, 1872.

Didi-Huberman, Georges. *Invention of Hysteria: Charcot and the Photographic Iconography of the Salpetriere.* Translated by Alisa Hartz. Cambridge, MA: MIT Press, 2003.

Eidenbenz, Céline. "Hypnosis at the Parthenon: Magdeleine G. Photographed by Fred Boissonas." Translated by John Tittensor. *Études photographiques* (Nov. 2011): 1–12.

Fuchs, Georg. *Der Tanz.* Stuttgart: Strecker and Schröder, 1906.

Fuchs, Georg. *Sturm und Drang in München um die Jahrhundertwende.* München: Verlag Georg D. W. Callwey, 1936.

Fürer, Dr. med. "Auch ein Beitrag zur Geschichte der Medizin." *Deutsche Medizinische Wochenschrift* 13 (Mar. 24, 1904): 471.

Gauld, Alan. *A History of Hypnotism.* Cambridge: Cambridge University Press, 1992.

Gotman, Kélina. *Choreomania: Dance and Disorder.* New York: Oxford University Press, 2018.

Grünwald, L. "Die Vorführung der Mad. Magdeleine in der Öffentlichkeit." *Münchener Medizinische Wochenschrift* (Mar. 29, 1904): 571.

Gumppenberg, Hanns von. "Vermischtes." *Der Kunstwart* 17 (Mar. 1904): 697–98.

Hales, Barbara. "Dancing in the Dark: Hypnosis, Trance-Dancing, and Weimar's Fear of the New Woman." *Monatshefte* 102, n. 4 (Winter 2010): 534–49.

Hall, G. Stanley. "Recent Researches on Hypnotism." *Mind* 6, n. 21 (Jan. 1881): 98–104.

Henig, Dr. R. "Die Schlaftänzerin: Eine Betrachtung zur wissenschaftlichen Erklärung des Phänomens." *Berliner Tagblatt,* Mar. 1, 1904, 1–2.

Jean Paul. "Blicke in die Traumwelt." In Part 1, Vol. 16 of *Jean Pauls Sämtliche Werke: Historisch-Kritische Ausgabe,* 139–66. Weimar: Hermann Böhlaus, 1938.

Kerr, Alfred. *Der Tag,* Feb. 8, 1905. Reprinted in Marx, Henry. "Madeleine: Two Reviews." *The Drama Review* 22, n. 2 (June 1978): 27–31.

Kohl, Amandus. "Das Märchen vom Schlaftanz." *Münchner Ratsch-Kathl,* Mar. 9, 1904, 2–3.

LaCoss, Don. "Our Lady of Darkness: Decadent Arts and the Magnetic Sleep of Magdeleine G." In *Neurology of Literature, 1860–1920.* Edited by Anne Stiles, 52–73. New York: Palgrave, 2007.

Lerner, Paul. "Hysterical Cures: Hypnosis, Gender and Performance in World War I and Weimar Germany." *History Workshop Journal* 45 (Spring 1998): 79–101.

Lipps, Theodore. "Zur Verständigung über die 'Schlaftänzerin.'" *Münchner Neueste Nachrichten*, Mar. 11, 1904, 3.

Loewenfeld, L. "In Sachen der Schlaftänzerin." *Münchener Medizinische Wochenschrift* (Mar. 29, 1904): 569–71.

Maehle, Andreas-Holger. "A Dangerous Method? The German Discourse on Hypnotic Suggestion Therapy around 1900." *Notes and Records of the Royal Society of London* 71, n. 2 (June 2017): 197–211.

Magnin, Emile. *L'Art et l'Hypnose: Interprétation plastique d'œuvres littéraires et musicales.* Paris: Félix Alcan, 1907.

Maier, Dr. Franz. "Die 'Schlaftänzerin' Madeleine G. in München." *Psychische Studien: monatliche Zeitsschrift vorzüglich der Untersuchung der wenig gekannten Phänomen des Seelenlebens* 31, n. 4 (Apr. 1904): 235–50.

Marx, Henry. "Madeleine: Two Reviews." *The Drama Review* 22, n. 2 (June 1978): 30.

Mayer, Andreas. *Sites of the Unconscious: Hypnosis and the Emergence of the Psychoanalytic Setting.* Chicago: University of Chicago Press, 2013.

Moll, Albert. *Der Hypnotismus.* Berlin: Kornfeld, 1889. Reprint, *Hypnotism: Including a Study of the Chief Points of Psycho-Therapeutics and Occultism.* Translated by Arthur F. Hopkirk. New York: Walter Scott, 1909.

Morton, Marsha. *Max Klinger and Wilhelmine Culture.* New York: Routledge, 2016.

Müller, Oskar A. *Albert von Keller, 1844 Gais/Schweiz – 1920 München.* Munich: Karl Thiemig, 1981.

Richer, Paul. *Études cliniques sur la grande Hystérie ou hystéro-epilepsie.* Paris: Adrien Delahaye et Émile Lecrosnier, 1885.

Roberts, Franz. *Die Schlaftänzerin Madeleine G.: Ein Protest gegen den Mißbrauch der Wissenschaft.* Munich: Birk, 1904.

Rühlman, Richard. "Die Experimente mit dem sogenannten thierischen Magnetismus." *Die Gartenlaube* 28 (1880): 127–30 and 9 (1880): 142–44.

Scherner, Albert. *Das Leben des Traums.* Berlin: Heinrich Schindler, 1861.

Schlagintweit, Dr. Felix. "Die Schlaftänzerin Mme. Magdeleine G. im Ärztlichen Verein zu München." *Münchener Medizinische Wochenschrift* 12 (Mar. 22, 1904): 524–25.

Schrenck-Notzing, Dr. Freiherrn von. *Die Traumtänzerin Magdeleine G. Eine psychologische Studie über Hypnose und dramatische Kunst.* Stuttgart: Ferdinand Enke, 1904.

Schrenck-Notzing, Dr. Freiherrn von. "Einige Bemerkungen über die Schlaftänzerin und ihr Auftreten in München." *Münchner Medizinische Wochenschrift* (Apr. 12, 1904): 667–68.

Schubert, Gotthilf Heinrich von. *Die Symbolik des Traumes.* Bamberg: Kunz, 1814.

Stümpke, Heinrich, *Bühne und Welt* 7, n. 1 (1904–1905): 476.

Treitel, Corinna. *A Science for the Soul: Occultism and the Genesis of the German Modern.* Baltimore: Johns Hopkins University Press, 2004.

Verus, S. E. "Miss Duncan and the 'M. Z.'" *Münchner Ratsch-Kathl.* Mar. 23, 1904.

Weber, Leopold. *Der Kunstwart* (Apr. 1904): 89–91.

Weinhold, Adolf Ferdinand. *Hypnotische Versuche: experimentelle Beiträge zur Kenntniss des sogenannten thierischen Magnetismus.* Chemnitz: Martin Bülz, 1879.

Wilson, Eric G. "Matter and Spirit in the Age of Animal Magnetism." *Philosophy and Literature* 30, n. 2 (Oct. 2006): 329–45.

Wolffram, Heather. "'An Object of Vulgar Curiosity': Legitimizing Medical Hypnosis in Imperial Germany." *Journal of the History of Medicine and Allied Sciences*, 67, n. 1 (Jan. 2021): 149–76.

Wolffram, Heather. *The Stepchildren of Science: Psychical Research and Parapsychology in Germany, c. 1870–1939*. Amsterdam: Rodopi, 2009.

Zilcken, Detta. "Die Schlaftänzerin Madeleine." *Die Schönheit* 2, n. 1 (1905): 48–55.

17

Ecstatic Fervors

Of Trance, Dance, and Self-Possession

Response to "Hypnotic Dancing and the Science of Sleep and Dreams: The Controversial Case of Madeleine G." by Chantal Frankenbach

Kélina Gotman

For Giorgio Agamben, "*Being-outside, and yet belonging*: this is the topological structure of the state of exception."[1] The sort of dance that takes place while under hypnosis involves fascination for viewers and medical practitioners alike: to dance while in an altered state of consciousness suggests the emergence of the body in an apparently primeval state. While "not conscious," one's body appears to move freely, sometimes more adeptly, than it does while one is apparently more conscious. I would like here to think critically and genealogically about ways corporeal states of noncontrol come to be equated with unfettered dancing specifically; and ways that theatricality comes to play a large part in conversations on the veracity of the dance as dance.

As Frankenbach suggests, Madeleine G. danced apparently brilliantly while under hypnosis (or, as it may be more apt to say, while "asleep"). Yet newspaper and medical debate—as well as popular conversation—queried whether she was "really" dancing while in this trance state or whether she might have been faking it. She rearranged her hair, kept her face to the audience, and so on. The problem this debate poses is that it offers a dichotomy between waking and sleeping states, and thus also a distinction between what it is to be conscious or unconscious. And yet the history of trance performance, as described by Michel Leiris, among others, suggests that one may be "acting" as one is also undergoing trance—that these modes are on a subtle continuum. To "act" is not merely to fake something, fully consciously, but rather to perform certain gestures, while in a comparatively disinhibited or reorganized cognitive state.[2]

The excitement Madeleine G.'s dancing provoked testifies to underlying desires to witness an apparently unfettered body, much as those bodies that Jean-Martin Charcot's audiences witnessed or appeared to witness in his clinic. That Madeleine offered stage performances hardly differs from the amphitheater performances that famous "hysteroepileptics" inhabited at the Salpêtrière. In those cases, suggestion took the place of outright "sleep," and clinicians were able subtly to direct the gestures that women (and some men) performed. "*Being-outside, and yet belonging,*" then, offers a figure for thinking about this way of being simultaneously "outside" the scene of dance, power, and control, and yet belonging to a theatrical situation within which one is the focus of attention—in many respects, the star. That these women might have experimented with a zone of indeterminacy between disinhibition, lack or loss of control, and control over the situation seems, I think, fairly unsurprising; to induce a trance state within oneself, even partially, might enable not only the loosening of self-control mechanisms, but also the passage into the ecstasy of another order of belonging. To read the history of "trance dance" then as one of *possession,* taken in the strongest sense of *self-possession,* suggests that the relation of voyeurism from audience to dancer might be overturned, or at least troubled. Taking possession of one's stage space while in an altered state of consciousness allows for just the sort of release and enjoyment that any celebration or *fête* might offer.

I have elsewhere called this genre of collective re-possession *ecstasy-belonging,* as a way to flip the terms of analysis away from the notion that ecstasy involves departure or dissolution of self (understood in the negative sense as loss), and rather to emphasize ways that *ecstasy-belonging* can highlight the possibility of collective banding through affective, cognitive, and gestural rearticulation, or reorganization.[3] To shift the terms of "rational" cognition and presence away from habituated models of self-control such as characterize a broadly performative and productivist Western modernity and toward something more open and expansive suggests arguably, or possibly, something of a cognitive decolonization, or in any event a cognitive reappraisal. To "dance" is at once to be present to the event of motion and, as trance performance and hypnotic or "hysteroepileptic" performances suggest, to dig into a different relation to presence and copresence. That viewers in the nineteenth century saw in this something of a return of the primitive or the primeval is perhaps hardly surprising; more important is that this wish for a sign of the continued existence of other entanglements between "mind" and "body" suggested and suggests still that something of the order of disinhibition and dream-states were lost in the over-colonization of the mind (and mind/body) through modern scientific approaches to questions such as the will.

Henry Maudsley suggests that the overcoming of the "ego" by means of cooperation was essential for survival in the primitive stages of human existence; that uniting the will and social energies enabled "social fusion," and that this capacity for cooperation determined who should live and who should not.[4] Questions of will—and free will—were hotly debated in the late nineteenth century. While conversations around the relationship between the "will" and physiological function often tipped into debates about what happens to the organic body in the event of death—relative to the continued life of the spirit, for example—as regards the movements of the dancer, this "preconscious" or primeval union of organic and spiritual function can be seen as dazzlingly visible, theatricalized, in the event of a dance performance. I am arguing, thus, that with dance, in the long nineteenth century, we find the dramatic evidence of an organic and spiritual fusion underpinning debates on the separation or union of body and mind. This is not quite the same thing as a debate on artistic genius or on the aesthetic merits of a particular set of gestures, but rather it situates the question of the choreographic—at once the consciously and unconsciously "danced"—at the heart of debates on what it means to be human and to be alive.

Notes

1 Agamben, *State of Exception,* 35 (italics in original).
2 See Leiris, "Acted Theater and Lived Theater in the Zar Cult," 113–17.
3 See Gotman, *Choreomania,* 196–223.
4 Maudsley, *Body and Will,* 163–84.

Bibliography

Agamben, Giorgio. *State of Exception.* Translated by Kevin Attell. Chicago: University of Chicago Press, 2005.

Gotman, Kélina. *Choreomania: Dance and Disorder.* New York: Oxford University Press, 2018.

Leiris, Michel. "Acted Theater and Lived Theater in the Zar Cult." Translated by James Clifford. *Sulfur* 15 (1986): 113–17.

Maudsley, Henry. *Body and Will, Being an Essay Concerning Will in its Metaphysical, Physiological and Pathological Aspects.* London: Kegan Paul, Trench & Co., 1883.

18

The Depths from the Surface

Interlaced Histories of Technologies and Dance in the Nineteenth Century

Janice Ross

Across the long *durée* of the nineteenth century and into the beginning of the twentieth, medical science, new technologies, and dance have had a deep and entwined history. More than just coincidental overlaps, these parallel quests shared a search for how to represent the body, its auditory, visual, and psychic interiors, with increasing external legibility. In the process, art and science have implicitly fueled one another's advancement, finding legitimization and resonance in one another's forward inquiries and also abetting their periodic retreats. This essay considers four scientific and technological innovations spanning this duration, each of which represents a deep quest for medical images from the interior body. As explored in this essay, each of these scientific innovations corresponded to the momentous shifting aesthetics of nineteenth-century dancing bodies. The scientific discoveries explored include the 1816 invention of the stethoscope by René Theophile Hyacinthe Laënnec, which made possible new clarity in the hearing of the heart and respiratory systems of patients; Thomas Edison's invention of the phonograph in 1877, preserving and making reproducible the human voice, and his embedding of a miniature phonograph in a doll in 1890; Wilhelm Röntgen's 1895 detection of X-rays, making possible the imaging of bones in living bodies; and, in 1921, Hermann Rorschach's creation of the inkblot perception test, eliciting and capturing unconscious psychological associations.

Each of these breakthroughs pierced the surface armor of the body conceptually, revealing sounds and sights of bodily interiors with unprecedented intimacy. Most profoundly for the arts, they unfolded against parallel aesthetic shifts in dance that gave emotional and psychological states new legibility through the dancing body, particularly in Western perceptions of concert dance. From the heroine of Romantic ballet, *Giselle,* whose fragile heart we sense through the anxious agitation and sudden arrest of her dancing, like a pulse made visible; to the mechanistic reproduction of the human in the dancing doll, *Coppelia;* to the celluloid censoring of Egyptian Ghawazi dancer Fatima's pelvic undulations and chest shimmies, as if visible through an X-ray in Thomas Edison's 1896 kinetoscope film of her; to Loie Fuller's rendering of the dancer's body as a whirling blur of fabric, a net on which to catch viewer's subjectivities—dance and science functioned in tandem across these decades to chart the pulse between discovery and invention, disclosure and concealment, of the living body.

This push toward externalizing bodily states and feelings for diagnostic and aesthetic ends accelerated across what Henri Bergson identified in 1889 as "the long durée" of the nineteenth century.[1] During this period, with the advance of new technologies, the relationship between a spectator's perceptions and an object's visual signs underwent recalibration. Bergson's theorization of time as continually unfolding, with the past, present and future interpenetrating, had particular relevance for dance. As a time-based art, dance has a distinctive way of manipulating temporality as medium, subject, and means: as dance deepened the body's capacities for expressiveness in tandem with the emergence of new technologies, it swung from anticipating to emulating trending changes at the nexus of popular culture and science.

William James and Indeterminism

Philosophers too pondered these shifts; the arts found a theoretical champion during this period in the writings of William James.[2] Reflecting on tensions between scientific determinism and individual autonomy, James lauded the sensation, experience, and *in*determinism of art over the traditional absolutism of science. He advocated indeterminism as a midpoint between the aesthetic idea of chance and the certainties of science—certainties shaped by the emergence of sharply defined academic and medical disciplines during the late eighteenth and early nineteenth centuries.[3]

James's thinking was exemplary in leading to an expanded view in which scientific utility was not the only metric of value; thus, among artists, philosophers, and audiences, a new appreciation developed for the role of the body in illuminating thought-in-action responses to live events.

The recognition that emotions and intellect were just two parts of what was considered the triad of human experience allowed a new valuing of the third part—the physical reality of the body.[4] Art historian Jonathan Crary, writing from the perspective of a discipline that de facto privileges the visual, identifies this recognition as a profound shift in the early nineteenth century, when visual experience—the act of seeing—was transformed from an objective into a deeply embodied experience: "What begins in the 1820s and 1830s is a repositioning of the observer, outside of the fixed relations of interior/exterior presupposed by the camera obscura and into an undemarcated terrain on which the distinction between internal sensation and external signs is irrevocably blurred."[5] Dance—initially, for James and his contemporaries, this meant Romantic ballet—was regarded as an ideal art form for rendering external signs of internal sensations in real time, since it spoke through the moving body. Across the nineteenth century, science would increasingly probe and medicalize the body, refining what was regarded as "anatomical understanding" through several major inventions of diagnostic medical devices.[6]

In this essay, I ask that we take an embodied approach to science retrospectively. Can we posit that in the nineteenth century, as science was making significant strides in visualizing the body in medicine, psychiatry, and technology, systemic shifts and a new logic emerged in the performing arts, and in particular in dance, with regard to how corporeal meanings were read as embedded in the body?

Romantic Ballet and the Stethoscope

In the early decades of the nineteenth century, living bodies were, largely, opaque mysteries. Thus, sound served as a particularly valued diagnostic tool for assessing physical health in medical contexts while also serving as a symbolic medium through which the arts represented emotional and psychological well-being. This was true for dance, as demonstrated in the signature Romantic ballet *Giselle.* With a plot line that moves forward through mimed and danced references to listening and to hearing hidden or distant sounds, *Giselle* offered glimpses of the interior body metaphorically. In *Giselle,* choreographer Jules Perrot choreographed sound. Depicting dancers miming the actions of listening, knocking, and hearing, supported by leitmotifs in Adolph Adam's musical score, Perrot made ballet a portal to the body's interior feelings, fears, and passions in the period when medical technology was devising new means to actually hear the inner workings of the body.

In the opening section of this 1841 ballet, Prince Albrecht, disguised as a peasant, summons the young peasant girl Giselle after first putting his ear to the door of her cottage and listening for sounds of life within, like a doctor listening to the chest of a patient. Hearing human activity, he then mimes softly rapping his knuckles on the door of the cottage where Giselle lives with her mother. Adam's score mimics the sound of knocking with a passage of fast, lightly plucked chords from the strings. These sounds animate Giselle, who bursts out of the door expectantly. Failing to see Albrecht, who is hiding, she mimes his knuckles rapping on the door and her hearing this, as if to reassure herself that this was indeed an external sound, not something in her imagination. Listening, as the counterpart to the making of sounds, becomes her theme, and animates Giselle to venture outside as she is summoned by her would-be lover's tapping on the door to her private domestic space.

The importance of attending to sounds is a cautionary theme that reverberates throughout the ballet. The most profound of these moments is that of heeding the beating of Giselle's own weak heart and the mortal risk she runs of stressing it fatally with too much activity (dancing) and too much excitement (an unfaithful lover). A beating heart is the core of this ballet—its pulse reflects the heart and also the fragile emotions of a young woman new to love.

Building the tension in the ballet around Giselle's own weak heart brought a new dimension of corporeality to the ballerina as well as an enhanced reflection of the realities of dance and its toll on the body. These layered ballet narratives—particularly narratives told through dance about dancing bodies—required modifications in what the body could do. Thus, ballet class became, according to sociologist Sophie Merit Müller, a "body workshop."[7] The phrase illuminates parallels between scientific and aesthetic ways of modifying bodies and, in particular, implies how dance training, analogous to surgical practices, imposes changes upon the body. There are, of course, differences: as Müller notes, "Ballet bodies are not shaped with a scalpel, but with exercise, and the work relies on the embodied person's vigilance, rather than that person's (anesthetic) evacuation. Also, while surgery (and medicine in general) aims at making broken bodies whole again, ballet is interested in constantly disturbing bodies in the interest of recrafting them in step with anatomical knowledge" and with aesthetic objectives.[8]

Indeed, the era of Romantic ballet stepped up the demands and interventions on the ballerina technically with intensified regimens of training in the quest for greater virtuosity onstage.[9] Pointe work became more sustained, jumps lighter and higher, and transitions from one pose to the next swifter and more fluid. To support these new technical demands, dancing masters developed approaches to training the body, and to understanding it as a moldable technological device. Such rigorous studio instruction Müller sees as the conceptual and material reconfiguring of classical-ballet bodies as instruments, "opening the black box 'body' for technical improvement" by analytically honing the body to perform.[10] She notes that this "technologization of the body" is closely tied to growing scientific and anatomical knowledge about the body that intertwines with pedagogic practices in ballet training. It also intertwines in *Giselle* with choreography, as in the case of the listening theme throughout the ballet.

As described above, *Giselle* opens with Albrecht listening at the portal of Giselle's cottage by first putting his ear to the door as if he were a doctor monitoring the lungs of a consumptive patient. Giselle mimes having heard the knock at her door when she first bursts out onto the stage, but does not see Albrecht, who hides himself; later, a hunting horn sounds, summoning the hunting party that triggers the revelation of Albrecht's deception and leads to Giselle's despair and death; and finally the tolling of the morning bells signals Albrecht's redemption after a night of being sustained by Giselle despite Myrtha's decree that he dance himself to exhaustion and death.

Just as Michel Foucault noted that regimens for disciplining bodies spatially and kinesthetically evolved in tandem with discoveries about human anatomy, so too the symbolic weight of the beating of hearts and the sustaining of breath in *Giselle* echoes an important invention for monitoring acoustic readings of the body's circulatory and respiratory systems.[11] The Romantic period of ballet unfolded in the wake of a pivotal scientific invention: the stethoscope. A few years prior to the premiere of *Giselle* the stethoscope was invented by French doctor René Laënnec, who used sound as a diagnostic aid by listening to patients' breath and heartbeats. Laënnec, also an accomplished flautist, reported that the following incident sparked his invention, out of necessity and anxieties about women's sexualized bodies and moral sensitivities:

> In 1816, I was consulted by a young woman laboring under general symptoms of a diseased heart, and in whose case percussion and the application of the hand were of little avail on account of the great degree of fatness. The other method just mentioned [direct auscultation—listening with the ear against the chest] being rendered inadmissible by the age and sex of the patient, I happened to recollect a simple and well-known fact in acoustics . . . the great distinctness with which we hear the scratch of a pin at one end of a piece of wood on applying our ear to the other. Immediately, on this suggestion, I rolled a quire of paper into a kind of cylinder and applied one end of it to the region of the heart and the other to my ear, and was not a little surprised and pleased to find that I could thereby perceive the action of the heart in a manner much more clear and distinct than I had ever been able to do by the immediate application of my ear.[12]

Laënnec improved on his improvised solution to hearing the beating of a live heart by building a hollow wooden cylinder that he called the stethoscope (from the Greek *stethos* for chest and *scopos* for examination). Considering *Giselle* in the context of the stethoscope allows a more complex reading of the ballerina's corporeality and a reflection on the realities of nineteenth-century dance, dance training, and its toll on the female body. The result is that rather than being just pathologized with madness as a physical identity, Giselle gains a pulse, breath, and a body on which the traces of physical exertion and emotional stress are evident. The librettist for *Giselle,* Théophile Gautier, championed what dance historian Joellen Meglin called "the translucent female body," alluding to the preponderance of images in the ballet focusing on "the whiteness, paleness, phosphorescence, bareness, iciness, transparency, and general see-through quality of the female body."[13] Working in the shadow of the invention of the stethoscope, a device created to allow deeper, and yet decorous, monitoring of the female body, the creators of *Giselle* offered a more medicalized portrait of this newly translucent dancing woman than generally acknowledged.

Coppelia and Edison's Mechanical Talking Doll

In counterpart to the early Romantic ballet's reflection of nineteenth-century scientific techniques for understanding the anatomical body, the Romantic era in ballet history closed with art and science correspondences of a different tenor. *Coppelia,* a ballet considered by some the capstone of the Romantic era,[14] premiered in 1870 during a period of Victorian audiences' fascination with mechanical wonders displayed at public exhibitions. The same year that the choreographer of *Coppelia,* Arthur Saint-Léon, premiered this ballet to Charles-Louis-Étienne Nuitter's libretto in Paris, inventors in France and the United States were marketing mechanical dolls that mimicked human motion. These dolls ranged in form from crawling infants to mechanized mothers pushing prams, designed to suggest a deception: the miracle of life within inanimate objects. The mechanical crawling baby, wound up with a key, was actually an elongated body rolling along on two hidden wheels that turned a spring-driven brass mechanism, causing the toy's arms and legs to slide along the ground, in imitation of crawling.

In *Coppelia* the plot centers around a living woman impersonating a mechanical doll in an effort to win back her boyfriend, who has fallen for a seductive automaton conjured by a lascivious inventor, Dr. Coppelius. Both the spirited living woman, Swanilda, and the android doll, Coppelia, are danced by ballerinas skilled at performing layered identities; Swanilda must represent a live woman trying to appear mechanical and Coppelia is charged with doing the reverse, as a living woman impersonating an automaton that tries to appear human. Beyond the virtuosity driving both expectations is a fascination with what constitutes human physical action, what of it is transferable to the nonhuman through imitative action, and which of these figures holds the greater interest and appeal. *Coppelia*'s play with mechanical dolls focused on imitating the physical, and in 1877 inventor Thomas Edison put forward his own mechanical doll that complicated these kinetic explorations: with the addition of sound, his invention could suggest imitation of a mind as well as of a body. In parallel, William James was exploring the philosophical parameters of mind/body dualisms and the boundaries between physical and mental states.

Edison, a prolific inventor, had created the phonograph in 1877—the first machine to record the human voice. He soon began brainstorming uses for the phonograph; one that held promise of being marketable was a talking doll. Over a period of six weeks in 1890 his laboratory produced several hundred dolls with ceramic heads, long hair, metal bodies, and wooden limbs. Inside each torso was a tiny phonograph with a wax record of one of twelve different recordings of a young woman reciting a children's nursery rhyme. Romantic ballerinas of course were mute on stage, which made the shaping of their expression through movement all the more condensed and focused and allowed for an easier impersonation by a mechanical facsimile. Edison's talking doll was a complicated endeavor for mechanical reproduction. The addition of a human voice, hinting at a mind inside the artifice, turned out to be a major challenge. Voices lack the neutrality of mechanistic actions, as recordings of them carry not just audible words but also the emotional affect of the speaker. The women employees from Edison's factory who were assigned to read the nursery rhymes had to shout their recitations in order to have their declaration of "Hickory Dickery Dock" and "Jack and Jill" register loudly enough to be captured on the wax records.[15]

The results of these recordings were tremulous voices that sounded both frightened and frightening. Children found the dolls scary, the phonograph mechanisms broke, the dolls were heavy, and the quality of the recordings was poor. The dolls did not sell well and, within a matter of weeks, Edison deemed them a mistake, halted their production, and destroyed many of the dolls. While the dolls failed as toys, they marked the first entertainment recordings, the first children's records, and the first recordings of women's voices ever made. They also illuminate hidden aspects of the *Coppelia* ballet, bringing into perspective the liabilities that can arise when men attempt to replicate female bodies and female subjectivities through mechanistic objects of men's invention. The organic female body triumphs in the ballet *Coppelia* as it did in Edison's ill-fated experiment of a talking doll. In the ballet, Swanilda destroys the facsimile woman. With Edison's talking doll, the real women's voices betrayed their hired status in service to a man's projection, disrupting the intended fantasy. Unintentionally, they encapsulated anxiety and perhaps even a snapshot of assembly-line stresses on working women.

Edison's talking-doll machine, like the woman/doll Coppelia, represented nineteenth-century mechanized attempts at mechanical embodiment, attempts that backfired. The result was that these mechanical female objects became abhorrent, more specters than cuddly companions. Edison's doll carried the untheatricalized and unmediated vocal register of tremulous unease rather than of maternal warmth. The ballet *Coppelia* similarly performed the comically inept failure of the mechanical doll as a perverse and eroticized fantasy of an old man. It thus offered an implicit reassurance that the mechanical is finally a broken, illusory artifact lacking the elements of cognition and will, qualities that ultimately set the human apart from too-perfect artificial beauty. As Bettyann Kevles has noted, the last decades of the nineteenth century saw an unacknowledged competition between the domestication of sound and that of the visual. "With the invention and almost instant commercialization of the telephone in 1876, sound seemed to take the lead," she writes. "For the first time in history, the human voice could be transmitted miles from the mouth of the person speaking."[16]

Laënnec's stethoscope, which was such a novelty in the early nineteenth century, had by the final decades become a device all doctors carried for listening to the chest as a routine diagnostic method for hearing disease. When Edison introduced the possibility of recording on a phonograph, sound continued as the leading sense in a technological explosion, but within a few years sight would take the lead and Edison would again be in the forefront of an invention shaping the reception of dance.

X-Rays and Thomas Edison's First Kinetoscopes of Dancers

The project of seeing into the body acquired a new depth, literally, with the discovery of X-rays in 1895 by physicist Wilhelm Röntgen. This innovation marked a radical shift toward inventions that sought to visually reveal, not ventriloquize, the body. Rather than merely addressing the body as a mechanical system, Röntgen's X-rays allowed deeper anatomical knowledge about the body as a work of highly complex physiology, facilitating fact-based diagnoses of the body's condition and opening the way toward visualizing treatment. The discovery of the X-ray shattered the boundaries of human experience about the body: for the first time, science had penetrated the flesh of the live body, without surgery, to reveal its visual secrets.

Distinct anxieties arose about female bodies as the targets of this new diagnostic: as Kevles has noted, "In making the interior of the body visible, X-rays shifted an assumption that had guided human behavior throughout recorded history. Visual penetration of one living person by another had been blocked by the skin. That barrier had dissolved. The X-ray unilaterally altered ages of societal accord by making public what was once private."[17] These anxieties intensified regarding women's bodies as the gaze, rather than touch, was reinforced as a cultural indicator of the eroticized body. With the X-ray, suddenly what had been private could now be public, as visuality threatened to penetrate the deepest secrets of the body. Surface seemed to have disappeared and the public now had access to things previously only seen by a surgeon. At the center of anxieties were concerns about protecting the modesty of the female body. A London clothing company advertised "X-ray proof clothing for modest women," while anxieties about the possibilities of X-ray vision—people literally peeking at others' bones—was a subject of widespread speculation.[18] The surface of the skin had gone from opaque to transparent, revealing the living skeleton.

The first X-ray image Röntgen made was of his wife's left hand, showing the bones of her wrist and fingers clearly, with the band of her wedding ring hovering over her ring finger. Thus, the image simultaneously disclosed the interior of her body and signaled her status through marriage to Röntgen, thus legitimating his sharing of what was, until then, the most intimate image ever generated of a living woman's interior body. "I have seen my own death," Frau Röntgen presciently commented after sitting for the fifteen-minute exposure to radiation that this first X-ray required. Up until this moment, not only the living human body, but the spaces it occupied, had been opaque, shielded from gazes into their interiority. Just as domestic spaces were protected from strangers' eyes by dense drapes and window coverings, individual attire for proper men and women concealed their skin under layers of clothing, even when swimming.[19] "It was a world where full-length mirrors were a luxury, and few people ever saw, much less examined, their own naked bodies. Skeletons, which could be seen only after death, quite reasonably symbolized death," Kevles writes, calling it "an historical irony that the discovery of rays that could penetrate clothing and skin and leave an image of living bones appeared in the most inhibited period in Western history," with the result that "the discovery of x-rays was one of the nails in the coffin of Victorian prudery."[20]

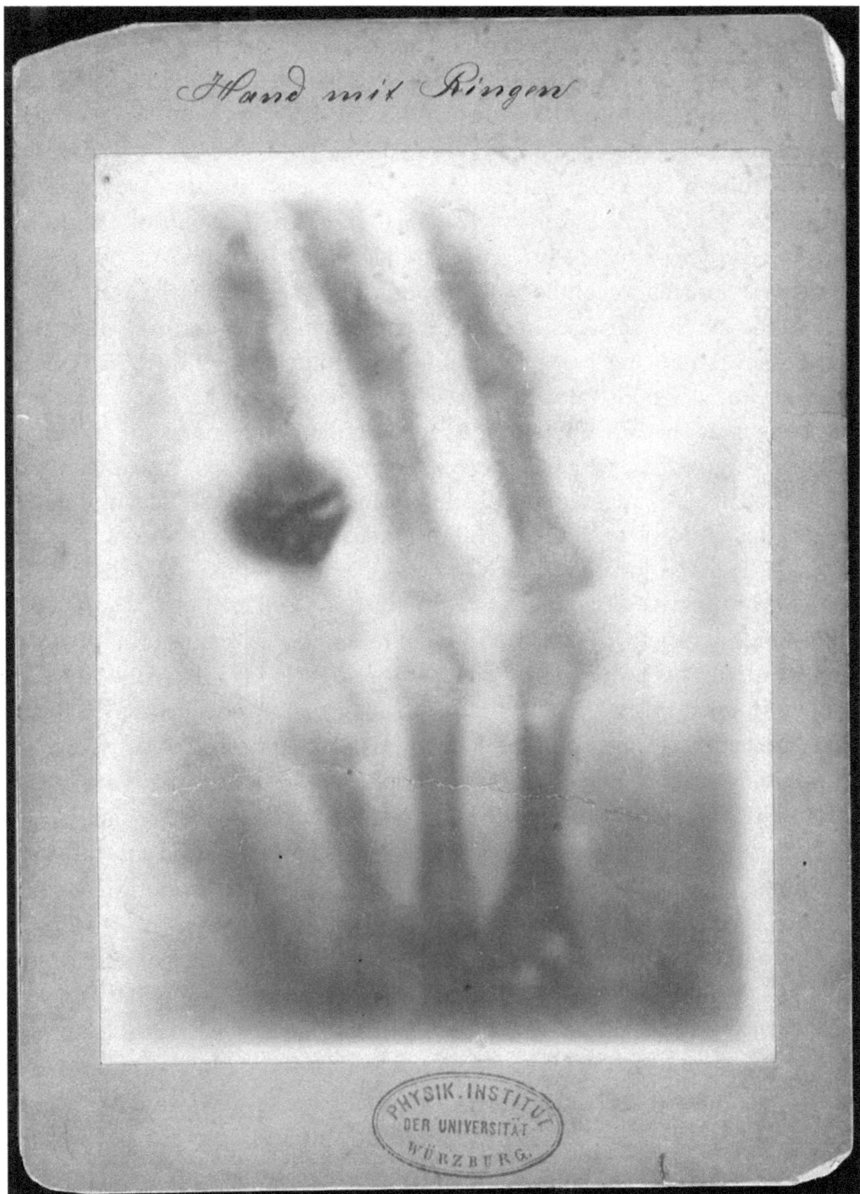

Figure 18.1. The bones of a hand with a ring on one finger, viewed through X-ray. Photoprint from radiograph by W. K. Röntgen, 1895. Reference: 32971i. One of the first X-rays by Wilhelm Röntgen (1845–1923) was of the left hand of his wife Anna Bertha Ludwig. It was presented to Professor Ludwig Zehnder of the Physik Institut, University of Freiburg, on January 1, 1896. Wellcome Collection, licensed through Creative Commons, https://creativecommons.org/licenses/by-nc/4.0/.

Dance intersected with the advent of the X-ray in a number of ways. For some, like the dancer Isadora Duncan, the X-ray was a reality and a metaphor that she used to assert the healthful and natural dimensions of her own dancing while gesturing toward a reality of the skeletally deformed body of fashion, deep inside the corset. Supported by images of both healthy and twisted bones, Duncan railed against ballet as shaping an unnatural body, one she likened to "a deformed skeleton dancing." Asserting her feminist authority over her own anatomy, she referenced the skeleton as a diagnostic subject and X-rays as confirmation of the ways fashionable corsets disfigured and crushed girls' and women's ribcages, stunting their growth and deforming their bodies. Duncan's dancing presented a reimagined image of the woman's body, with a healthy, unconstrained skeleton as a starting point. Her statement may have begun as rhetoric but with the rapid circulation of images of real bones inside living bodies, which Röntgen's X-ray set in motion, a woman's fully developed, healthy skeleton now had the validation of medical technology.

As Duncan was leveraging the interior diagnostics of X-rays to support her sexually liberated, white woman's body as a model in the public sphere, this same technology was being used to circumscribe and exoticize Middle Eastern women dancers. Thomas Edison had followed Röntgen's development of the X-ray with his own experiments at imaging the body internally. However, Edison stopped this work after his leading assistant on the project fell ill and died of cancer as a result of excessive radiation exposure. Edison persisted in pursuing another invention that he hoped would "do for the eye what the phonograph did for the ear."[21] The result was the "kinetoscope," a machine, invented in 1894, that had a peephole through which one viewed short, soundless films. Dancers were a favorite subject for Edison: he made thirty-seven films of dancers in the first two years of developing the kinetoscope.

One of the earliest of these was his 1896 film of the Egyptian dancer Fatima Djemille, a global star whom Edison may have seen performing her *danse du ventre* (belly dance), titled the *Coochee Couchee Dance,* at Chicago's Columbia World Exposition in 1893. The one-and-a-half-minute film Edison made of Djemille shows her dressed in a two-layered full skirt and blouse adorned with chains of sparkling coins as she shimmies her shoulders and does traveling movements with pelvic circles and a slow front pelvic lock.[22] Her expression is matter-of-fact, while her body vibrates with fast, rhythmic pulses—as much a demonstration of virtuosity as a tease of sexuality. Reformers who attempted to regulate kinetoscope peepshow viewings for the paying public, however, apparently found the dance alarmingly erotic.[23] Responses to the spectacle of a film of a woman's rhythmically pulsing body viewed through the intimacy of a one-person kinetoscope, made this private spectatorship seem as objectionably erotic as an X-ray film. As the accessibility of films like this exploded in 1905–1907, larger audiences attracted more aggressive responses, perhaps motivating film censors in 1907 to add the grid of white bands, forty seconds into the short film, painted over the areas of celluloid where Fatima's chest and hips appeared.[24] This shielded her shimmying breasts and pelvis from view just as one might add a leaden covering to hide body parts from X-ray penetration.[25] The mere presence of X-rays as a means for depicting bodily interiors triggered broad anxieties about bodies suddenly becoming available to those eager for prurient and covert peeping; the kinetoscope commodified this new appetite.

Rorschach's Inkblot and Loie Fuller

By the turn of the twentieth century, diagnostic technology had a new target within the body—the emotional functioning of the mind. In the late nineteenth century, physicians and psychologists in Germany and France experimented with inkblots as the basis for games and creativity tests, for the purpose of studying associative consciousness and imagination. In 1921, the Swiss psychiatrist and psychoanalyst Hermann Rorschach published *Psychodiagnostik*, a series of ten cards with inkblot images honed for their diagnostic value as psychometric tools to gauge an individual's perceptions. Rorschach created these deliberately ambiguous yet circumscribed inkblot images as a means of testing markers for personality characteristics and emotional functioning through his patients' associations. During the same period, fin de siècle dance innovators were engaged in corresponding projects of capturing associations from spectators through abstract images created by the dancers' bodies.

American dancer Loie Fuller, a leading innovator of imagistic dances, used lengths of what was known as "Chinese silk," a strong, yet sheer, glossy fabric for lining quilts and clothing, which she manipulated around her body by holding the material aloft on long sticks.[26] Fuller's serpentine dances, which she lit with colored lights placed under glass-covered openings in the stage, effectively metamorphosed her body into a blank screen onto which viewers could project their own narrative meanings. While appearing free-form and organic, Fuller's dances, like Rorschach's inkblot cards, employed visual prompts, subtle triggers that seemed unpremeditated yet were essentially choreographed to arouse a circumscribed range of interpretive possibilities in the perceiver. As Ann Cooper Albright has noted, Fuller thought of the stage as a laboratory where she manipulated light and movement, changing from darkness to dim lighting to reveal small motions of her hands and of the silks, the movements gradually growing in scale to suggest the metamorphosis of insects, waves, and clouds in a crescendo of light and motion that subsided back into darkness and quiet.[27] Similarly, Rorschach's inkblots might initially appear as anonymous smears of paint, but they were carefully selected and tested with his patients to conform to his criteria of presenting incomplete and competing visual stimuli.[28]

The impact of what I call Fuller's "projective dances" was to rewrite the job description of the spectator from observer to constructor. This act of perception in watching Fuller dance, like the movement memories animated by Rorschach's inkblot test, positioned those who interpreted the dancer or the inkblot as projectors of meanings onto ambiguous forms and bodies. These bodies were live in the case of the dance and a spectral presence in the case of the inkblots. Such interpretive processes allowed for insights about perception while also serving as pressure-releases for the subjective and subconscious life of the beholder. Fuller herself also played with what Albright calls "corporeal writing," a game in which she had friends sign their names in liquid ink along the fold of a piece of paper that she then doubled over, producing a mirror-like image of a signature—"little shadow skeletons," according to Albright.[29]

Bodily motion perceived, imagined, or remembered became an important sensory channel that was activated when people viewed Rorschach's inkblot cards.[30] "The sensation of movement was experienced by proxy," Naamah Akavia has argued, explaining how in the Rorschach test the subject's dynamic impulses were projected onto the static inkblot. "The result was that the Rorschach blot came to embody the seemingly paradoxical condition of physical restraint giving rise to associational freedom," as the seated test-taker reflexively used movement memory in assigning meaning to the ink patterns.[31] At the turn of the century, intellectuals in the West had a particular obsession with "unmasking," with freeing and unfettering the body, impulses that played out among artists as well as scientists.[32] Both realms evidenced fascination with reading the depths from the surface, with objectifying interiors into an external image.

Rorschach inkblots, similar to Fuller's seemingly ambiguous dances, might look at first sight like nothing more than watery spills across paper, muddy bleeds that stop short of dripping off the page. In itself, Fuller's dancing was unremarkable compared to the customary image of a classically trained dancer: she was a full-figured woman with plain features, no real dance training, and a functional way of navigating the stage. Off the stage, she might have been as unremarkable as an ordinary inkblot, but when Fuller stepped onstage and began moving, her form suddenly became a pulsing canvas of images. Her body was blurred by the silks swirling around her and illuminated by banks of hidden colored lights, obscuring the form of her body. Instead of specificity, Fuller gave her audiences mystical uncertainty—the female dancer's body as a transformational blank canvas. The audience projected their own visual metaphors, passions, emotions, and kinesthetic sensations onto the billowing silk screens of her costumes.[33]

Albright, who re-created some of Fuller's dances, writes about what she calls "Fuller's eventual disappearance into her costume," by way of describing her self-transformation from a skirt dancer to a manifestation of abstract visual imagery. "The vivid contrast of light and dark allows her torso to blend into the light background and her hair to merge with the dark butterfly figures, leaving only her face and arms to stand out. Eventually, even these parts of her body would become obscured by her increasingly voluminous costume."[34] To the audience, she was no longer a woman, but "A Giant Violet," a "Butterfly," a "Slithering Snake," or "A White Ocean Wave," the titles of her dances for her debut performance at the Folies Bergère in Paris in 1892.

Fuller's performances have been described as a place for the "re-direction of physical force," an invitation to "displace erotic kineticism—onto her costumes, her technology, the air, and ultimately onto the members of Fuller's public."[35] Indeed, Fuller's performances made space for her audience to project their own desires as she became "a tabula rasa, a living projection screen." La Loïe, as poet Stéphane Mallarmé dubbed her, "invented an art form balanced between the organic and the inorganic, one in which the literal drama of theatrical transformation played out on stage."[36] Fuller did not play roles or imitate things, she subsumed her physical self into her subject, presenting her body as a landscape that the viewer was invited to narrate.

The mind—particularly the subconscious—as a landscape ripe for discovery, was part of the new territory Rorschach was exploring. While it is not known if Rorschach frequented dance performances, the cities through which he circulated—Zurich, Berne, and Moscow, among others—were all rich with viewing opportunities of the new form of dance abstraction Fuller helped pioneer. European high culture embraced Fuller and used her often as an object of aesthetic contemplation. Mallarmé and William Butler Yeats wrote about her, René Lalique and Louis Comfort Tiffany fashioned her image in glass and crystal objects; Jules Massenet and Claude Debussy composed music for her; and her close friend Auguste Rodin made bronze casts of her hands. Fuller also fascinated the world of academic science, gaining the admiration and friendship of Marie and Pierre Curie as well as of astronomer Camille Flammarion, whose laboratories she regularly visited.[37] These scientists, who were drawn to Fuller, understood that the world is not grasped objectively but imaginatively, in part through metaphors generated by embodied experience.

Fuller achieved her stardom by presenting to her audience images balanced delicately between the organic and the inorganic, playing out onstage the literal drama of theatrical transformation.[38] Her work exaggerated a deep aspect of performance—the unique doubleness where the dancer is a living individual and also the image she is performing. This approach allows bodily experience to be shaped into a visual image and to be projected metaphorically. Contemporary reviews document Fuller's power as coming from her thrilling enactments of precisely this metamorphosis: she seemed to retain her body and yet at the same time to merge with the realm of the nonhuman. Her doubling, folding, and symmetrical forms were among the images that astounded her audiences. Garelick describes her performance in language that recalls Rorschach's inkblots and the process of personal disclosure they prompted: "She astounded audiences by allowing them to watch her as she split and doubled herself over and over again, mutating into a series of dramatic and ever-changing shapes."[39] Meanings spilled out. This kind of mirror-like image, of course, was also an essential visual feature of Rorschach's inkblots, with their doubling on the folded paper.

Transformation and doubling had been popular themes in the Romantic ballets of the mid-nineteenth century (*Robert le Diable, Giselle, La Sylphide*), but what made Fuller's work so modernist and captivating for audiences was how she displaced these themes from narrative plot to dance's blank page—the volumetric space of the stage. She was making emotions and images that had remained below the surface in ballet, hidden in the messy depths of the psyche, suddenly visible as the spectator's own projection. Fuller was also an enactment of the newly blurred boundaries of the fin de siècle—specifically, the boundary between the human and the machine, and even that between the artist and the scientist. Fuller's performance offered a vigorous alternative to the nineteenth-century Freudian hysteric through her presence as a woman who consciously controls the illusion that her movement is out of control.

These qualities of displacement and transformation are important elements reflecting how the subjects in Rorschach's experiments "projected their dynamic impulses onto the static inkblot."[40] The phrase "dynamic impulse" is key here, for it is the dynamism of the memory of movements, which mediated subjects' responses to the inkblots. "Rorschach insisted that the movement 'seen' in the blot be immediately and viscerally 'felt' by the subject as an empathetic movement in the subject's own body," Akavia wrote.[41] Movement remembered thus shaped movement perceived, a phenomenon Rorschach called "kinesthetic engrams." The term "kinesthesia," denoting the sensation of movement, entered scientific discourse in the late nineteenth century.[42]

Both kinesthesia and a term sometimes used interchangeably with it, "proprioception," point toward the centrality of bodily experience and memory in reading the inkblots and decoding the arrested movement and organic images embedded in them. Fuller's inclusion in this conversation shows how pervasive the fascination of the period was in relying on images to penetrate the mind directly. Complex and ambiguous ideas were transformed into visual clarity through "embodied perception"—the term Rorschach used for this kind of memory.[43] Because the ink blots deliberately made the subject work at perceiving and describing them, Rorschach's experiment became about interpretation.

Rorschach also had a particular interest in responses that interpreted his blots through form and color, as well as kinesthetic perceptions. Subjects, "whose responses were marked by a predominance of movement responses, were characterized by greater creative ability and by more inner life," Akavia explains about interpretations of responses to Rorschach's diagnoses. This argument for memory as a form of embodied perception, which can also embrace the idea of kinesthesia, helps explain why a subject's response to a Rorschach blot reflected the presence of encoded movement. Rorschach's experiments were important steps toward uncovering how the mind held and manipulated physical memories as muscular traces. This connection between kinesthetic and subjective worlds similarly shaped the reception of Fuller's dancing.

* * *

This retrospective approach to the embodied historical nature of scientific investigation and tools across the later decades of the nineteenth century and into the beginning of the twentieth reveals concert dance of the era as engaged in remarkably similar empirical research. As referenced in the four twinned examples discussed, as scientific technologies sought to externalize the interior of the body and mind for diagnostic purposes, dancers modeled new ways of excavating subjective experience from the depths to the performable surface. Choreography as a technology of discovery and disclosure that performs its revelations through and on the living body shifts the cultural climate in ways remarkably similar to those of science. These correspondences in the developments of scientific technology and dance through their shared anchorage in the living body are being renewed today in emerging artificial-intelligence and machine-learning technologies.

With machine learning developing increasingly complex technological surrogates for functions of the live human body, dance is re-emerging as a platform for cyborgs to demonstrate their verisimilitude. Entertaining yet menacing, dancing robots simultaneously display how clumsy they are at trying to appear human and yet also how startlingly close they are to approaching human functionality. An example is the Boston Dynamics's *Spot's on It*, a virtuoso kickline of seven identical yellow dog-like Spot robots programmed to move in a sleek, smooth ensemble.[44] They perform the impossibly tight geometry of a Busby Berkeley routine with the mechanistic choppiness of inanimate objects programmed to look like humans trying to look inanimate. The more modernist and percussive the dance genre is, the easier it is for choreorobotics to closely parse the distinction between the robot and the live person. The gestural vocabulary of the android has in fact been in circulation for years, drawing on the now nearly ubiquitous lexicon, shaping stage and screen dancing, of virtuoso urban dance styles like robotic popping, locking, and bone breaking. Smooth and fluid actions without a visible pulse seem among the most elusive dance qualities in programmed robots, exposing the point of slippage between the technological and the human.

What began as a reciprocal evolution between dance and science in the early nineteenth century by means of the body has evolved in the twenty-first century into ever-deepening penetrations of science into the living and digitally constructed body. Here, technology is seeking new ways to bring embodiment, and with it the effect of mobile and sentient bodies, into the construction of surrogate machine bodies. In the twenty-first century, motion, and embodied presence—"morphological intelligence," as it has been dubbed—are frontiers of discovery for the constructed body in science the way sound and visuality were in the nineteenth century.[45] We are in the midst of a virtual—and literal—race between science and art to penetrate further into the body and psyche and to claim them for medical and aesthetic contemplation. Thus, these overlaps between science, technology, and dance initiated in the early nineteenth century have a legacy well into the twenty-first century. The frontiers of machine learning suggest that these connections will continue to grow and deepen, challenging both domains and forcing a new reckoning between the boundaries of the aesthetic, the diagnostic, the dancer, the machine, and the medicalized body.

Notes

1 Bergson, *Time and Free Will*, 104.

2 For example, see James, "What Is an Emotion?"

3 On the emergence of new disciplines in the late eighteenth and early nineteenth centuries, see Katz, "The Role of American Colleges in the Nineteenth Century."

4 Crary, *Techniques of the Observer*, 20–21.

5 Crary, *Techniques of the Observer*, 24.

6 Vedar, *The Living Line,* 2.
7 Müller, "Distributed Corporeality," 871.
8 Müller, "Distributed Corporeality," 871–72.
9 See, for example, dance treatises by Blasis, *Traité élémentaire, théorique et pratique de l'art de la* danse (1820) and Saint-Léon, *La Sténochorégraphie ou L'art d'écrire promptement la danse* (1852).
10 Müller, "Distributed Corporeality," 869.
11 See Foucault, *The Birth of the Clinic.*
12 Barton, "The Story of Rene Laennec," 1.
13 Meglin, "Behind the Veil of Translucence, Part III," 118.
14 See for example Banes, *Dancing Women,* 35.
15 "Edison Talking Doll Phonograph and Cylinder," The Henry Ford Museum of American Innovation.
16 Kevles, *Naked to the Bone,* 7.
17 Kevles, *Naked to the Bone,* 3.
18 Badash, "Radium, Radioactivity," 145.
19 Kevles, *Naked to the Bone,* 14.
20 Kevles, *Naked to the Bone,* 14.
21 Edison, *The Diary,* 64.
22 Fatima Djemille was one of two dancers who performed as "Fatima," and the first of three dancers who performed the belly dance under the name "Little Egypt." An annotated version of Edison's film of Djemille can be viewed on YouTube.
23 Francesca Royster notes that this film belongs to a genre of Edison films of this period treading the line between commercial and scientific interests, revealing ways that racialized bodies were seen as signifying evolutionary and racial difference by means of gesture and movement signs (*Becoming Cleopatra,* 67).
24 Gunning, "Making Sense of Film."
25 "The First Censored Movie, Fatima's Coochee Coochee Dance," YouTube.
26 Beginning in the late nineteenth century, in response to the devastation of silkworm disease, the manufacturing of "Chinese silk" had shifted to new and cheaper factories in the United States, so Fuller's use of it can be seen as adding a layer of imperialist trade histories to her appropriating and amplifying traditional Chinese silk scarf dancing. See Ma, "The Modern Silk Road," 335.
27 Albright, *Traces of Light,* 8–9.
28 Meyer and Viglione "An Introduction to Rorschach Assessment," 281–336.
29 Albright, *Traces of Light,* 42.
30 Akavia, *Subjectivity in Motion,* 126.
31 Akavia, *Subjectivity in Motion,* 128.
32 Franko, *Dancing Modernism/Performing Politics,* 5.
33 Albright, *Traces of Light,* 45.
34 Albright, *Traces of Light,* 21.
35 Albright, *Traces of Light,* 170.
36 Albright, *Traces of Light,* 5.
37 Flammarion even arranged for Fuller to become a member of the French Astronomi-

cal Society, in recognition of her investigations into the physical properties of light in her performances.

38 Garelick, *Electric Salome,* 5.

39 Garelick, *Electric Salome,* 154.

40 Akavia, *Subjectivity in Motion,* 54.

41 Akavia, *Subjectivity in Motion,* 37.

42 Akavia, *Subjectivity in Motion,* 269.

43 Akavia, *Subjectivity in Motion,* 24.

44 Boston Dynamics, "Spot's on It," YouTube.

45 Keyan, "Morphological Intelligence," 1.

Bibliography

Akavia, Naamah. *Subjectivity in Motion: Life, Art and Movement in the Work of Hermann Rorschach.* New York: Routledge Press, 2012.

Albright, Ann Cooper. *Traces of Light, Absence and Presence in the Work of Loie Fuller.* Middletown, CT: Wesleyan University Press, 2007.

Badash, Lawrence. "Radium, Radioactivity, and the Popularity of Scientific Discovery." *Proceedings of the American Philosophical Society* 122, n. 3 (June 9, 1978): 145–54.

Banes, Sally. *Dancing Women: Female Bodies on Stage.* London: Routledge, 1998.

Barton, Marc. "The Story of Rene Laennec and the First Stethoscope." *Past Medical History,* Sept. 20, 2016. https://www.pastmedicalhistory.co.uk/the-story-of-rene-laennec-and-the-first-stethoscope/.

Bergson, Henri. *Time and Free Will: An Essay on the Immediate Data of Consciousness,* trans. F. L. Pogson. Mineola, NY: Dover, 2001.

Blasis, Carlo. *Traité élémentaire, théorique et pratique de l'art de la danse, contenant les développements et les démonstrations des principes généraux et particuliers qui doivent guider le danseur.* Milan: Joseph Beati et Antoine Benenti, 1820.

Boston Dynamics. "Spot's on It." https://www.youtube.com/watch?v=7atZfX85nd4. Accessed Dec. 26, 2023.

Crary, Jonathan. *Techniques of the Observer on Vision and Modernity in the Nineteenth Century.* Cambridge, MA: MIT Press, 1990.

"Edison Talking Doll Phonograph and Cylinder Recording of 'Twinkle, Twinkle, Little Star,' 1890." The Henry Ford Museum of American Innovation. https://www.thehenryford.org/collections-and-research/digital-collections/artifact/334590. Accessed Dec. 26, 2022.

Edison, Thomas A. *The Diary and Sundry Observations of Thomas Alva Edison.* Edited by Dagobert D. Runes. New York: Philosophical Library, 1948.

Edison, Thomas A. "Fatima's Coochee-Coochee Dance." Film produced by James A. White and shot by William Heisey for the Edison Manufacturing Company, 1896. https://www.youtube.com/watch?v=AxLJJK_ZQyM. Accessed Dec. 26, 2022.

Edison, Thomas A. "The First Censored Movie, Fatima's Coochee Coochee Dance." https://www.youtube.com/watch?v=9Y2Di2h8XBI. Accessed Dec. 26, 2022.

Foucault, Michel. *The Birth of the Clinic: An Archaeology of Medical Perception.* London: Tavistock Publications, 1973.

Franko, Mark. *Dancing Modernism/Performing Politics.* Bloomington: Indiana University Press, 1995.

Garelick, Rhonda K. *Electric Salome: Loie Fuller's Performance of Modernism.* Princeton, NJ: Princeton University Press, 2007.

Gunning, Tom. "Making Sense of Film." From *Making Sense of Evidence* series on *History Matters: The U.S. Survey on the Web.* https://historymatters.gmu.edu/mse/film/. Accessed Dec. 26, 2022.

James, William. "What Is an Emotion?" *Mind* 9, n. 34 (Apr. 1884): 188–205.

Katz, Michael. "The Role of American Colleges in the Nineteenth Century." *History of Education Quarterly* 23, n. 2 (Summer 1983): 215–23.

Kevles, Bettyann Holtzmann. *Naked to the Bone: Medical Imaging in the Twentieth Century.* New Brunswick, NJ: Rutgers University Press, 1997.

Keyan, Ghazi-Zahedi. *Morphological Intelligence: Measuring the Body's Contribution to Intelligence.* New York: Springer, 2019.

Ma, Debin. "The Modern Silk Road: The Global Raw-Silk Market, 1850–1930." *Journal of Economic History* 56, n. 2 (1996): 330–55.

Meglin, Joellen A. "Behind the Veil of Translucence: An Intertextual Reading of the Ballet Fantastique in France, 1831–1841. Part III: Resurrection, Sensuality, and the Palpable Presence of the Past in Théophile Gautier's Fantastic." *Dance Chronicle* 28, n. 1 (2005): 67–142.

Meyer, Gregory, and Donald Viglione. "An Introduction to Rorschach Assessment." In *Personality Assessment,* edited by R. P. Archer and S. R. Smith, 281–336. London: Routledge, Taylor & Francis, 2008.

Müller, Sophie Merit. "Distributed Corporeality: Anatomy, Knowledge, and the Technological Reconfiguration of Bodies in Ballet." *Social Studies of Science* 48, n. 6 (Dec. 2018): 869–90.

Royster, Francesca T. *Becoming Cleopatra: The Shifting Image of an Icon.* New York: Palgrave Macmillan, 2003.

Saint-Léon, Arthur. *La Sténochorégraphie ou L'art d'écrire promptement la danse.* Paris: Brandus, 1852.

Vedar, Robin. *The Living Line: Modern Art and the Economy of Energy.* Hanover, NH: Dartmouth University Press, 2015.

19

Movement-Machines

Reflecting New Technologies in Doing and Scoring Dancing

Response to "The Depths from the Surface: Interlaced Histories of Technologies and Dance in the Nineteenth Century" by Janice Ross

CLAUDIA JESCHKE

Janice Ross provides provocative trips into topoi drawn from the "deep and entwined history" of "medical science, new technologies, and dance" in their "search for how to represent the body, its auditory, visual, and psychic interiors, with increasing external legibility." These technologies crossed borders from science into dance. Supplementing that cultural discourse, my approach offers a different methodological and technical debate: it refers to the many praxeological writings on doing dancing and their legibility inside the nineteenth-century dance scene itself. The scoring of motion phenomena—of dance notations written by practitioners (dancers, choreographers, pedagogues)—constitutes a body of first-hand materials reflecting discursive aspects of deep intra-dance knowledge. Thus, I explore dance notations of the period as relevant responses to the boundary-crossing of (new) technology—and vice versa.

Among the many technological innovations dealing with the phenomenon of motion (ranging from automata to cinematics), the concepts of doing and scoring dancing met the scientific challenge by using pioneering figurations that transferred analytical dance knowledge into dance practice and art. Markers of these figurations can be detected in the multiple zones between the perceptions of doing, writing, and creating dance. They negotiate moving bodies as complex systems articulating (e)motion in the liminal space between the mechanical and the imaginative.

The "Body-Machine" and Choreo(-)graphy

The concept of the "body-machine," dating back to Jean-Georges Noverre (1727–1810), Carlo Blasis (1797–1878), and other theoreticians of the time committed to physical and aesthetic codification, initiated scientifically tinged changes for the dance scene of the first third of the nineteenth century.[1] As a method for the improvement of dance technique, the concept of the "body-machine" became physiological and mechanistic by multiplying itself as *corps de ballet*[2] and by equally promoting the virtuosity of individual ballet artists. This development of staging as well as performative maneuvers expanded the workings of choreo(-)graphy as compositions on stage as well as written notations.

Notations—especially—captured the opaque zone between biomechanics and aesthetics by systematizing, articulating, visualizing, and mediating dance action. They transferred the complexity and diversity of dance concepts and practices into scientifically tinged, stable strategies by applying specific devices (drawings, verbal descriptions, signs for movements).

The assumption of this essay is that notational strategies in nineteenth-century dance share (absorb as well as generate) the timely trend of new technologies, explored in Ross's chapter with a focus on moving subjects and objects. Thus, notations can be understood as a para-scientific methodology objectifying physical premises like anatomy, mechanics, and motricity into the external media of notation and, reversely, subjectifying physicality as idealized imaginings of human movement. Two examples of notation from the middle of the century can be understood as methodological and technical blueprints of physical as well as testable activities: the drawings of Giovanni Léopold Adice in his unpublished manuscript *Note sur l'arabesque*, and of Arthur Saint-Léon in his book, *Sténochorégraphie*.[3] In transferring the construction of the "body-machine" to the (bio)mechanics of moving, the figurative designs of the two examples display segmentation and differentiation. Both invent fresh technologies for writing dance, making the action of notating more effective and reproducible by using templates (Adice) and notation fonts (Saint-Léon).

"Movement-Machines": Exploring the Para-Science of Moving

Using a fine-pointed pen, Adice dynamically describes complex models of arabesques in detailed poses/stills, covering roughly 322 pages of text and images. He both employs and dissects the terms of the ballet vocabulary. In his drawings, Adice zooms in on isolated body parts like arms, hands, head, legs, feet; the drawings of these parts also show the pasting of fixed stencils/templates. Dotted lines connect the respective moving parts, with verbal comments clarifying the locus and kind of physical activities. Adice's writing technique nearly always depicts each single arabesque with two figures, which he draws in mirror-image according to the three body directions (up-down/to the sides; up-down/forward-backward, forward-backward/to the sides). The movements appear as a three-dimensional cross in motion, the intersection point usually being along the spine, at the waist. The three-dimensionality of movement thus communicates—that is, struggles with—the two-dimensionality of the paper. On the one hand, the depicted body looks like a marionette, with mechanical movements; on the other hand, due to the displaceability of the axis crossing the three body planes, the figures on the papers and the papers with figures allow the movements to appear as if floating, transparent, suggesting slow motion—a vision between movement and stillness. Adice's descriptions, as instructions for pupils, can serve the fundamentals of arabesques and their virtuoso execution. Virtuosic or pedagogically uniform—Adice's concept of the arabesque can be read or performed in both ways; thus, it integrates open-ended solutions beyond function and aesthetics.

In contrast, Saint-Léon's stick figures use pictographs, a stenographical procedure of continuous movement activities perceivable in a quasi-cinematic view as a horizontal sequence seen from the front. He provides a scheme of five lines for the activities of the legs, which he dissects and highlights, and one additional line for the épaulement with its movements of arms, shoulders, and head. In so doing, Saint-Léon arranges body positions, frame by frame, in distinct zones; he is also able to differentiate between front and back views of the body, although his writing technique appears to be less three-dimensional than Adice's drawings. Adice deals with the three-dimensionality of the positions and the transitions between them, while Saint-Léon generates a sense of space by depicting the sequentiality of changing positions. The underlying concept is, for both Adice and Saint-Léon, segmentation and structured re-figuration through imagination.

As evident in both cases, segmentation operates when notations analyze and depict body parts as singular energetic elements, departing from the aesthetically justified and codified look of a position or a movement, while always leading back to it. As in the then-popular constructions of automata, to which Ross refers in the case of Edison's dolls, the ability of re-figuration is produced as well as perceived and notated as mechanical, and/or pre-cinematic motricity.

In nineteenth-century notation, the analysis of the body and its parts in space and time gives preference to particular relations between form and energy, between mechanics and dynamics. Coordination, synchronization, and sequencing of movements, thus, became the issue of the quasi-scientific research in and of doing dancing. The negotiations of automata and the cinematic during the nineteenth century feature artificial-mechanical movements, as do the dance notations in exploring the dance-specific (e)motional range of human physicality and biomechanics. It might be appropriate to consider notations of the period as models of doing dancing that "admit degrees of affective and physiological elasticity into the biotechnical category of the automatic"[4] and, as might be added, of the cinematic. The notations, then, equally admit the biotechnical category of the automatic (cinematic), the affective, and the physiological elasticity of doing dancing. The "articulate body," as represented in the examples of Adice's and Saint-Léon's "movement-machines," reflects and responds to the technologized qualities of automata and of film, presenting dance notations in their notable capacity as precursors of filmic capture and the inventors of these notations as "directors."

Notes

1 Jarasse, "Corps-machines à L'Opéra," paragraph 3.
2 Jeschke, "Dancing Figures /Figuring Dance," 30–41.
3 Adice, *Note sur l'arabesque;* Saint-Léon, *La sténochorégraphie.*
4 Austin, "Elaborations of the Machine," 66.

Bibliography

Adice, G. Léopold. *Note sur l'arabesque,* ms. Bibliothèque nationale de France, vol. III, B-61 (3). https://catalogue.bnf.fr/ark:/12148/cb40947788q. Accessed May 23, 2022.

Austin, Linda M., "Elaborations of the Machine: The Automata Ballets." *Modernism / modernity* 23, n. 1 (2006): 65–87.

Jarasse, Bénédicte. "Corps-machines à L'Opéra: l'imaginaire scientifique dans les écrits sur la danse (1830–1860)." Des corps dans la ville: norme et écart au XIXe siècle. *Arts et Savoirs* 16 (2021). https://journals.openedition.org/aes/4075.

Jeschke, Claudia. "Dancing Figures /Figuring Dance. Searchlights on Nineteenth-Century Composition Manuals." *Zwischenzonen: Bewegungskünste im 19. Jahrhundert – Tanz, Oper, Zirkus, Varietè. Tanz&Archiv. Forschungsreisen* 8 (2020): 30–41.

Saint-Léon, Arthur. *Sténochorégraphie, ou Art d'écrire pomptement la danse.* Paris: Chez l'auteur, 1852.

IV

"Outro"

20

Observing the Observers

Emily Coates

The cosmic information I had imagined was not so much missing, as nonexistent. I pictured finding extensive data on the moon, planets, and stars transiting across the heavens. Instead, I found weather records:

Cloudy. Fair.
Fair.
Cloudy. Rainy.
Cloudy.
Rainy.
Fair.
Fair.
Fair.
Fair.
Fair.
Fair.
Cloudy. Rainy.
Rainy. Fair.
Cloudy.
Fair. Cloudy. Fair.
Fair.
Fair.
Fair.

I was searching through the archives of Shattuck Observatory, built in 1854 at Dartmouth College, while in residence at the college's arts center to create a performance about dance and astronomy.[1] My guiding light had little to do with the time period: I had begun to collect references to cosmic dances over the past several millennia that traced the patterns of the stars. Twentieth-century dance history is littered with choreographers who mined material gathered from art history, anthropology, or archaeology to make dances—

many of them fraught, some of them heroic. My idea was to put a spin on the research-into-performance method by creating a self-reflexive sort of cosmic dance from the observatory's material remains. Reanimating not just an archive, but a *scientific* one. Confronted with a blank canvas and a trove of details, I did what post-postmodern dance artists do: scavenged for usable gestures, choreographic structures, and bodily forms.

The lineup of adjectives came from the oldest papers in the observatory's holdings, which predated its construction. The cover of the notebook read:

> Thermometrical
> Register *(this script looks pre-printed)*
> Nov 1, 1827 *(this looks handwritten)*

And after a space: Ebenezer Adams Jr.[2] I learned later that Ebenezer was the son of Dartmouth's first professor of mathematics and natural philosophy. This notebook marked the first scientific research done at the college and started one of the longest, continuous meteorological data collections in the United States.[3] Here was a person creating a discipline.

The 1827–1828 diary included hand-drawn columns for temperature measured morning, noon, and night, and narrowly codified adjectives, one per day. March 1828 did not go out like a lamb:

> Fair.
> Fair.
> Fair.
> Fair.
> Cloudy.
> Fair.
> Fair.
> Cloudy.
> Fair.
> Fair.
> Cloudy.
> Cloudy.
> Variable.
> Cloudy.

I was captivated by Ebenezer's weather records. In their daily ritual and devotion to the weather, they reminded me of art making. Adams Jr. wrote with such impeccably tiny penmanship the words appeared to be mechanically produced. Some individuality did get through. I discovered he had ended a column on one page with a curiously jagged edge. In the margins on an-

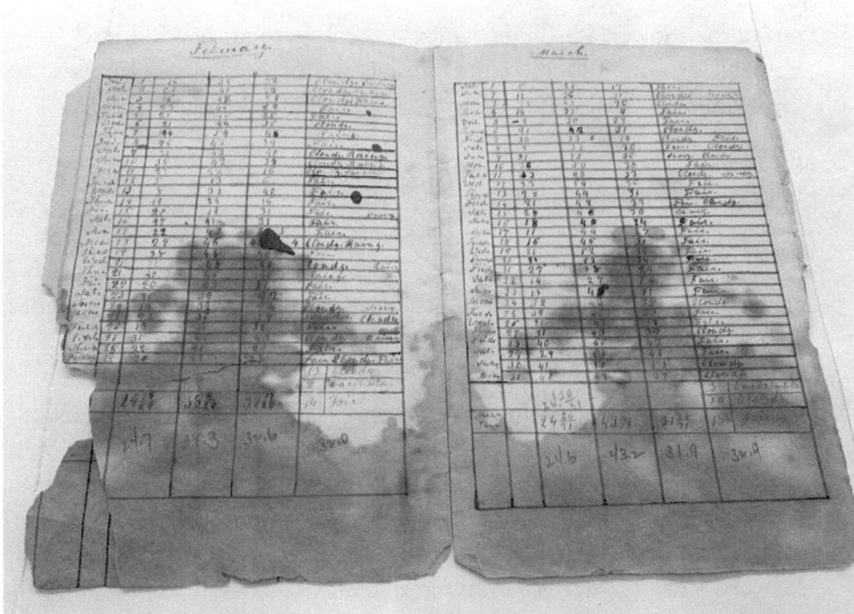

Figure 20.1. Page from "Thermometrical Register, Ebenezer Adams, Jr. / Nov 1827–Aug 1828," Ebenezer Adams Jr & Ira Young Registers, DA-9, Box 2704, Dartmouth Library Archives and Manuscripts, Rauner Special Collections, George C. Shattuck Observatory Records. Fair use; photograph by Emily Coates.

other page, he had doodled a tiny house, with a small person inside (a cry for help?).

I learned about nineteenth-century meteorology as I went. Illegibility—a lack of familiarity with the system, discipline, or historical period—taught me how to *notice*. To close-read the details and scrutinize weird things at the periphery. One value of not knowing, upon entering an archive, is that it allows you to come into understanding through more phenomenological than discursive means. I searched for traces of sensation and stories.

Sometime around the 1920s, someone thought it would be wise to transcribe the mid-nineteenth-century weather registers into sturdier, leather-bound volumes prefabricated for meteorological use. An archive copying itself has a strange self-reflexivity, caught in a loop of observing its own observations. From the Meteorological Observation Registers 1850–1859, carefully rewritten over half a century later by one Gladys Moore:

> June 1851
> Caterpillars made their first appearance on the currant bushes.

August 1851
During the shower, a brilliant rainbow, the upper part of a hill, was divided, or rather doubled, thus [draws arc] probably some mirage, the phenomenon passed, but a few moments and then disappeared.

July 1852
First ripe red raspberries. Aur. light at 10 P.

Peas in blossom.

Yellow corn silked out. Picked Raspberries, Strawberries, and Blueberries,

Picked first green cucumbers from our vines, also green peas.[4]

These vivid descriptions continued for several years until tapering off near the end of 1854 to become almost exclusively numbers, in columns for Barometer, Thermometer, Hygrometer, Clouds, Wind, and Precipitation. Quantitative measurement took over, one language consuming another. Gladys Moore copied it all.

The word "cosmos" derives from the ancient Greek word *kosmos,* meaning *order,* which connected the Greeks' conceptualization of societal organization with the organization of the universe. The scholar Barbara Rappengluck further distills the links between cosmic dances and stellar motion into the three elements: body, space, and time, "the basic parameters of dance," which "also describe important characteristics of celestial movements."[5] This seemed helpful, but also rather obvious. So I began to create my own cosmic dance canon—examples of the form that you might not think of as such. I identified several features:

Cosmic dances attempt to break human scales of time and space, both in terms of human lifespan, and in the delimitations of the human body.

Cosmic dances create networks of interrelationship: between humans, between human and nonhuman, between animate and inanimate, between the present and the past.

Cosmic dances inescapably use the materials and means of their historical moment, even as they attempt to escape (and sometimes transcend) the confines of the present.

Cosmic dances are archives of observation.

If I stretched my definition of choreography, Adams Jr.'s weather records became to me a cosmic dance, moving from the particular—the immediate temperature taking, barometer readings, and atmospheric observations in Hanover, New Hampshire—into pooling data later in the nineteenth century with a national and ultimately international meteorological community. Weather recorders like Adams Jr. attempted nothing less than to touch the cosmos. As I researched in Shattuck Observatory's archives, I began to draft notes for other stories and references that I might fold into my piece.

notes for a performance

Human
Ebenezer Adams Jr. (1827)
Lincoln Kirstein (1934)
Ancient Egyptian astronomer-priests
Plato (timeless)
Nyota Inyoka (1928)

Observation

Technologies
Shattuck Observatory—> 19th century Meteorology
Emily Dickinson's astronomical textbook—checked out of my university library!
Meteorology by Charles Wilkes, U.S.N. (1851)—weather collecting as migrating imperialism (a heavy volume that occupied a corner of my desk for a year—imperialist even in the afterlife)
Abenaki astronomy
Cosmic dances

I had arrived at cosmic dance through Lincoln Kirstein, who in his 1935 book *Dance: A Brief History of Theatrical Dance* included a description of ancient Egyptian astronomer-priests and their Dance of the Stars. Kirstein might have gotten this idea from Plato—who, he says, "seems to have had close contact with its observance."[6] (Close contact in what sense? His wording is ambiguous.) Or perhaps Kirstein learned of the dance first from Francis de Miomandre, a contemporary and a French poet and critic, who in his book *Danse,* also published in 1935, described the exact same choreography: novitiate priests inside their temples, rotating around their altars, mimicking the signs of the zodiac.[7] Kirstein cites four or five Egyptological texts of the late nineteenth and early twentieth century as his sources, yet none of them mention such a dance. Miomandre cites the dancer and choreographer Nyota Inyoka, who was famous at the time of his writing yet subsequently got lost to dance history, until recent efforts to resuscitate her work and life.

Dance scholars have discovered that Nyota Inyoka invented her name, her place of birth, and her cultural origins. She liked to tell the press she was a Princess of Egypt or India. In fact, she was born in Paris in 1896, to a mother who was a French domestic worker and a father who was either Egyptian or Indian. She claimed a sweeping synthesis of all ancient Eastern civilizations in her choreography: Indian, Balinese, Cambodian, Egyptian.[8] She did her research so thoroughly in the libraries and museums of Paris and through her travels that in her dancing, Miomandre thought he saw the ways that ancient Egyptians three thousand years ago must have seen the skies. Through their carefully calculated gyrations, Nyota Inyoka's movements made him feel he was in the presence of the spinning planets and their mysterious lives. The ancient Egyptians must have been noble and harmonious, Miomandre gushed, with such a magical and natural initiation.[9]

There is no reason to suspect the caterpillars crawling on the currant bushes that June day in 1851 did not exist (the hand-copying decades later could have introduced errors, though Gladys Moore's precise handwriting seems unimpeachably concerned with accuracy). In contrast, Miomandre observing Nyota Inyoka in performance, who herself was embodying an ancient Egyptian archive through dance, might get dismissed as artful fiction—at the risk of overlooking the transmissions and conjuring of knowledge generated through observation in motion. We don't yet know which dance Miomandre was looking at, or her sources for the choreography. Nyota Inyoka's 1920s Egyptian dances included *Prière aux dieux solaires (d'après papyrus hiéro-gliphiques), Disciplina,* and *Danse Égyptienne Bédouine.*[10] Who is to say that some actual relic of the Dance of the Stars, circa 3500 BCE, was *not* reconstituted through Nyota Inyoka's research and imagination, in one of these dances?

I am situating dance as a mutual partner to scientific disciplines such as meteorology and astronomy in the human effort to accumulate knowledge through observation. Recognizing that Nyota Inyoka's choreography *observes*—transmitted through archives, fully reembodying and reanimating epistemological histories in motion—expands the definition of observation and upends the misguided dichotomy between observer and observed. As observation became more instrumentalized and narrower in focus in Western science, the human data collector was set at a greater remove from their object of study. Scientific instruments exaggerate the human body, compensating for its limitations, while also concealing just how much *body* nonetheless remains in the records that they create. As more information is excluded from that which gets documented (such as quantitative measurements displacing qualitative description), important dimensions of that which is sought—the object of study, the natural world—recede farther from view. To show the embodied, the sensorial, the kinesthetic, the indigenous knowledge that has coexisted all along in the positivist archive demands a resuscitation effort.

Writing in the beginning of the twenty-first century, the poet Susan Howe calls out this seething life in archives: "—quick thought—reached through the fire of art—/ To reach is to touch."[11] Howe's poetry exhumes bodies in texts. Discovering an eighteenth-century diarist who lived and wrote not far from her own home in Connecticut, Howe senses the same atmospheric conditions across three hundred years, transmitted through language: "through an occult invocation of verbal links and forces, the qualities peculiar to our changing light and color."[12] The force of *poesis* is one of the ways that art observes.

What new sort of *poesis* do I unleash when situating fragments from a nineteenth-century astronomical observatory's archives inside of twenty-first-century choreography? What is to be gained by juxtaposing two very different historical subjects—Ebenezer Adams Jr., and Nyota Inyoka—separated by one hundred years and ever increasingly disastrous effects of violent colonial expansion? Is there a way of thinking through their *poesis* comparatively—one registering experiential measurements in writing on the page, and the other enlivening an ancient archive of astronomical observation in choreographic form? What kind of new cosmic dance am I choreographing, patching together these stories that stretch the very definition of both cosmic and dance? I am thinking this through as I write.

I respond to the rhythms and the rituals I discover in the archive. But soon enough I need to know something about the historical context. Not least to dispel the illusion that the scientific records I am looking at are hardened facts when the reality is more complex. Inquisitive data gatherers of the nineteenth century understood that immediate atmospheric conditions offered a slender view of some larger cosmic picture, hence meteorology and astronomy were interrelated: the part stood in for the whole. But they also filled in for lack of understanding with fantastical reasoning. Those 1851 caterpillars—real or imaginary—were recorded during an era when the most preeminent astronomer believed that barometric pressure created atmospheric waves, like the rippling waters of an ocean.[13] Shattuck Observatory oversaw the meteorological measurements but mostly did not gather star charts, for the basic reason that its telescope was used predominantly not for research, but instead for teaching: to educate young male Dartmouth students in the natural patterns created by their Christian god.[14] Observations by humans and their instruments are always caught within an anthropocentric view of the universe, never quite able to get the *humanness*—or human imagination—out of the way.

Cosmic dances may reflect more empirical realities, and nineteenth-century natural philosophers more fantastical fictions, than either one tends to get credit for.[15]

Humans do try to get around themselves. In their fastidious focus on daily atmospheric conditions, the meteorological registers ignore political history. I do not know the extent to which Adams knew that the European colonial violence that had begun several centuries earlier had enabled the existence of Dartmouth College and his presence there. He wrote a mere sixty years after the first students of the college spent part of their lessons cutting down 270-foot-high white pine trees to claim the land, displacing the Abenaki people who then lived there and eventually moved across the river, unwilling to fight.[16] There is tragic irony in attempting to destroy an ecosystem and indigenous knowledge only to turn around and measure, record, categorize, and otherwise scramble to rediscover that same knowledge in a different form.[17] Forty years later the records march through the Civil War, seemingly oblivious to their times.

Such massive amounts of meteorological data had piled up in the first half of the nineteenth century that by the second half experts cried "stop!" The 1860 International Statistical Congress devoted a keynote to stemming the flow: "Loading the mind, as well as the shelves, with overwhelming accumulations of facts, only causes distaste, if not oppression, even among the most zealous; and then the progress of science suffers from its misdirected, though earnest, votaries," the speaker complained.[18] Those attempting to standardize meteorological observation across international boundaries clearly did not know what to do with the awe-struck language of Samuel Duncan, Dartmouth class of 1858, when he wrote:

magnificent rainbow[19]

The qualitative descriptions in Shattuck Observatory's weather records have never been used.

Type "nineteenth century meteorology and astronomy" into my university library's search engine and primary sources pop up. And for whatever reason, the nineteenth century is still in circulation. I have carried with me on errands a scientific journal issue from 1883 that contained an article by Charles Young, the second director of Shattuck Observatory, who later left for Princeton and founded astrophysics. I read Young's 1880 booklet "Spectroscopic Notes" on Metro North. *Meteorology* by Charles Wilkes, U.S.N. (1851) weighs a few pounds. The tome holds weather records recorded throughout the American naval officer's extensive travels, a voracious coupling of colonialist impulses with scientific data collection. These books seem indestructible.

Indigenous knowledge filters through, pressured and partially obscured. In Eugene Vetromile's *The Abnakis and their History*, published in 1866 (and which, I later discovered, enjoys a digital afterlife on Kindle), the Italian philologist and missionary includes a chapter titled "Astronomy and the Division of Time." The Abenaki, the Algonquin-speaking tribe of northeastern North America, held in-depth knowledge of the land now called Hanover, New Hampshire, which must have included the view from the hill upon which Shattuck Observatory rests. The writer notes the Abenaki's sharply different methods: "True, they have no astronomical instruments, and whether they ever had any is a question at present now involved in darkness. Yet nature seems to have endowed them with very acute senses, and they use them with much skill and accuracy." Vetromile credits the Abenaki with observational powers, their ability to perceive "many small things" which "pass unobserved by the whites."[20] The Abenaki must have developed sophisticated physical techniques and practiced them repeatedly to achieve this level of perceptual power. Vetromile recognizes this way of knowing, without fully realizing that through their physical tuning practices they transformed the human body itself into an instrument.

On stellar and planetary scales, the Abenaki knew Orion and Sirius, the Pleiades, Venus, the seven stars of Ursa Minor, and the Milky Way. While the tribe also had superior knowledge of the sun (the writer credits the ancients with Copernican understanding of earth's rotation around the sun, pre-Copernicus), their rituals depended most upon the moon.[21] The Abenaki organized their daily and monthly rhythms according to the lunar cycle. February was *Taquask'nikizoos*, "moon in which there is crust on the snow." March was *Pnhodamwikizoos*, "moon in which the hens lay." April was *Amusswikizoos*, "moon in which we catch fish." July was *Atchittaikizoos*, "moon in which berries are ripe."[22] Shattuck Observatory's 1850s weather records show the influence of this indigenous cosmology in documenting the growth of vegetables and wild berries alongside temperature and atmosphere. Well before nineteenth-century natural philosophers headed out with their chronometers and achromatic telescopes, the Abenaki had connected the earth and the sky.

These layered histories vibrated as I left the Dartmouth Library last spring and wound my way up between several buildings to reach the grounds of Shattuck Observatory, which sits on a hilltop, just above a slightly younger red-brick chemistry building, with woods behind. The central observatory structure forms a bright white T: two modestly sized rooms extend outward on either side, and the third protrudes out the back, which is the transit room. The windows are small and thickly paned, in the quaint colonial style. The signature gray dome soars above, modest by twenty-first-century standards, but sizable enough to look from afar like a massive eyeball peering at the sky.

The summer before, my musician collaborator and I learned how to open the telescope dome by climbing a ladder inside and turning a large metal wheel. On the sloping lawn in front of the building, we presented some of the material we had developed for the performance, for an audience at sunset. I told stories about cosmic observers, while moving through choreographies of accumulation and decomposition. Ebenezer Adams Jr., Lincoln Kirstein, Plato, Francis de Miomandre, Nyota Inyoka, and the mysterious ancient Egyptian astronomer-priests. The musician orbited the premises, improvising on a rigged cornet. Another collaborator read Ebenezer Adams Jr.'s peculiar shorthand: "C'dy, c'dy, c'dy." An astronomer and a theoretical physicist joined me in conversation.[23] Darkness gradually fell as we cranked opened the dome. Afterward we had to chase a squirrel off the tracks before shutting the eaves.

Transforming archival research into performance is strangely cyclical. I used the first draft of this essay to create a script, which my collaborators and I then whittled into choreographic form. Now many revisions later, this essay has reabsorbed the text for the final performance, the rhythms and content of which guide my edits. In a flash, the performance reconstituted from historical materials has turned back into writing and photographs. Animated liveness can hold out only so long before succumbing to the archive's gravitational pull. And yet the impact of the live performance we created now hums in the words on this page.

The spring that I returned to Shattuck Observatory, I tried to hold the voices of the nineteenth-century Northeastern college and the sensations of my live performance from the previous summer in my head. But everything slipped away. So instead, I noticed the temperature. For late April, the air felt crisp. A brisk wind brushed against my cheeks—probably a 3 on the wind scale. Had I been tasked with describing the weather, I would have written: "Cloudy." The light settled into a grayish late afternoon haze. I circled behind the observatory to a clearing with a view of the hills across the river and observed a trembling weathervane affixed to the tip of a metal tower. Sensitive to touch, the small arrow wobbled first to the right, then to the left, reflecting the whimsical inconsistencies of the breeze. Paying close attention to such a rudimentary instrument made me suddenly pay attention to the modulations of the wind in force and vector. The arrow sensitized my body, as if I had become a weathervane, too. I rested amid one among countless days that represent larger patterns of the cosmos. The hours passed uninscribed.[24]

Notes

1 With gratitude to Mary Lou Aleskie and the Hopkins Center for the Arts at Dartmouth for commissioning the performance project and supporting its development.
2 "Thermometrical Register, Ebenezer Adams, Jr. / Nov 1827-Aug 1828," George C. Shattuck Observatory Records, Box 2704, Dartmouth Library Archives and Manuscripts (hereafter DLAM). Ebenezer Adams Jr. was the son of Ebenezer Adams, who served as professor of mathematics and natural philosophy 1810–1833 ("Dartmouth College Department of Physics and Astronomy, Biography").
3 "Description," George C. Shattuck Observatory Records, DLAM.
4 "Meteorological Observations Registers 1850–1859," George C. Shattuck Observatory Records, transcribed by Gladys Moore, undated, DLAM, DA—9, Box 2709.
5 Rappengluck, "Cosmic Dance," 308.
6 Kirstein, *Dance,* 13.
7 Miomandre, *Danse,* 5. "The altar, placed in the center of the temple, represented the Sun, and the officiants, sometimes representing the signs of the zodiac, sometimes the seven planets or the constellations, revolved around it in the direction of the evolution of the celestial bodies. There was a kind of lesson there." Translation by Lynn M. Brooks.
8 Jahn, "Cutting Into History," 188–89; López Arnaiz, "Living Archive."
9 Miomandre, in *Danse,* writes, "How noble and harmonious a society it must have been whose members received this naturalist and magical initiation," 5. Translation by the author.
10 Email correspondence with Dr. Iréne López Arnaiz, Feb. 10, 2023. I am grateful to Dr. Arnaiz for searching through Nyota Inyoka's 1920s programs to seek this information for me.
11 Howe, *Spontaneous Particulars,* 60.
12 Howe, *Spontaneous Particulars,* 59.
13 Gibbons, "Grasping the Elusive," 85.
14 Conversation with Richard Kremer, Dartmouth College professor of history of science emeritus, Apr. 23, 2021, and Apr. 30, 2021.
15 A similar point is made by Nancy Rose Marshall: "Science does not remain objective, timeless, universal, and stateless while art gets to be imaginative, subjective, and culturally constituted; science is not the straight man to art's funky and unpredictable comedian." Introduction to *Victorian Science and Imagery,* 15.
16 Conversation with Richard Kremer.
17 Amitav Ghosh aptly indicts the close ties between scientific practices and violent colonial expansion in his recent book *The Nutmeg's Curse.*
18 "Meteorology," 4.
19 "Meteorological Records, Ira Young, Samuel Duncan (class of 1858) /Jan 1850-Aug 1859," George C. Shattuck Observatory Records, DA-9, 2:7, DLAM.
20 Vetromile, *The Abnakis and their History,* 73.
21 Vetromile, *The Abnakis and their History,* 77–78.
22 Vetromile, *The Abnakis and their History,* 80.
23 My collaborators for that event were musician Taylor Ho Bynum (a role later played by

Charles Burnham), actor Derek Lucci as the reader, and director and dramaturg Ain Gordon. The scientists were Elizabeth Newton and Devin Walker. The material is part of the piece *We,* a shared program I co-created with my fellow dancer-choreographer Emmanuèle Phuon.

24 With thanks to Lynn Matluck Brooks, Garth Grimball, and Amanda Reid for their astute comments on different drafts. This essay is part of a larger work-in-progress book manuscript titled *Science Dances: Choreographies on the Edge of Knowledge.*

Bibliography

"Dartmouth College Department of Physics and Astronomy." Biography. Dartmouth Library Archives and Manuscripts website. https://archives-manuscripts.dartmouth.edu/agents/corporate_entities/636. Accessed Feb. 20, 2022.

Dartmouth Library Archives and Manuscripts, Rauner Special Collections Library Repository, George C. Shattuck Observatory Records.https://archives-manuscripts.dartmouth.edu/repositories/2/resources/525. Accessed May 18, 2022.

Gibbons, Carey. "Grasping the Elusive: Victorian Weather Forecasting and Arthur Hughes's Illustrations for George Macdonald's at the Back of the North Wind." In *Victorian Science and Imagery: Representation and Knowledge in Nineteenth Century Visual Culture,* edited by Nancy Rose Marshall, 79–110. Pittsburgh: University of Pittsburgh Press, 2021.

Ghosh, Amitav. *The Nutmeg's Curse: Parables for a Planet in Crisis.* Chicago: University of Chicago Press, 2021.

Jahn, Tessa. "Cutting Into History: The 'Hindu Dancer': Nyota Inyoka's Photomontages." In *Exploring Alterity in a Globalized World,* edited by Christoph Wulf, 187–96. New York: Routledge, 2016.

Howe, Susan. *Spontaneous Particulars: The Telepathy of Archives.* New York: New Directions Books, 2014.

Kirstein, Lincoln. *Dance: A Short History of Theatrical Dancing.* New York: G. P. Putnam's Sons, 1935.

López Arnaiz, Iréne. "Living Archive. Nyota Inyoka's Archive: Traces of the Ephemeral and Ancient Dances Reenactment." *Anales de historia del arte* 32 (2022): 327–50.

Marshall, Nancy Rose. "Introduction." In *Victorian Science and Imagery: Representation and Knowledge in Nineteenth Century Visual Culture,* edited by Nancy Rose Marshall, 3–27. Pittsburgh: University of Pittsburgh Press, 2021.

"Meteorology. Suggestions Intended to Promote Correspondence Between Meteorological Observers." Fourth International Statistical Congress. London, England, 1860. 6th Section. London: H. M. Stationery office, 1860.

Miomandre, Francis de. *Danse.* Paris: Flammarion Press, 1935.

Rappengluck, Barbara. "Cosmic Dance: Correlations Between Dance and Cosmos-Related Ideas Across Ancient Cultures." *Mediterranean Archaeology and Archaeometry* 14, n. 3, 307–17.

Vetromile, Eugene. *The Abnakis and their History.* New York: J. B. Kirker, 1866.

Wilkes, Charles. *Meteorology.* Philadelphia: C. Sherman, 1851.

21

Querying the Cosmos

Response to "Observing the Observers" by Emily Coates

Christian DuComb

The interpretours of Plato do thinke that the wonderfull and incomprehensible ordre of the celestial bodies, I mean sterres and planettes, and their motions harmonicall, [give] to them . . . in the sondrye diuersities of nombre and tyme, a fourme of imitation of a semblable motion, whiche they called daunsinge.

—Thomas Elyot, *The Boke named the Governour,* 1531[1]

In "Observing the Observers," Emily Coates follows the connection between dance and the stars from Plato and the astronomer-priests of ancient Egypt to her own historical and choreographic research at Dartmouth College's Shattuck Observatory, touching on several points in between. As her range of examples suggests, the impulse to connect the movement of human bodies to the movement of celestial bodies is widespread, across cultures and across time. Indeed, the scope of her argument could be expanded to include Renaissance scholars like Thomas Elyot (quoted above), the early modern court dances of France and Spain, and indigenous understandings of the cosmos from both North and South America. But what is the nature of this pervasive connection between dance and the stars? Is dance a metaphor for astronomical motion, as in Plato's *Timaeus?* Is dance an allegory of royal power, as in the court ballets performed by and for Louis XIV, the "Sun King"? Or is there a material relationship between dance and the cosmos, as in some indigenous epistemologies?

In one oft-told story among the Haudenosaunee nations of the eastern Great Lakes region, the Pleiades (or Seven Sisters) came down to earth to bring people joy by teaching them to dance.[2] In mid-November, as the Pleiades rise in the east, the Barasana people of the northwest Amazon basin "dance to the Milky Way in order to urge the stars along their course."[3] These examples move beyond metaphor and instead reflect a concrete belief in a physical relationship between the cosmos and the dancing body. Plato's *Timaeus,* on the other hand, compares the movement of the planets and stars to *choreia,* the choral dances of ancient Athenian theatrical performance.[4] Although metaphorical rather than material, this "Platonic vision of the choral cosmos," as James L. Miller calls it, left a deep imprint on Western literature and philosophy.[5] When Plato laments the inadequacy of language to describe the movement of the stars, dance serves as a proxy for the incomprehensible patterns traced by the celestial bodies above.

As France and Spain expanded their colonial empires in the Americas in the seventeenth century, the dances performed at court inverted Plato's metaphor. Rather than expressing the limits of human ability to observe and describe the cosmos, dance instead served symbolically to bring the power of the stars down to earth and into the political realm. In 1653, King Louis XIV of France famously appeared as the Rising Sun in the *Ballet de la nuit,* a performance which helped to consolidate his power.[6] In Spain, King Philip IV "was praised by poets as *el rey planeta,*" and "many courtly and theatrical works allegorized him in this light" throughout the seventeenth century.[7] As early modern Europe witnessed colonial expansion abroad and the rise of absolute monarchies at home, Western astronomy was meanwhile transformed into a modern science by Hans Lippershey's invention of the telescope in 1608, which enabled the precise, empirical study of the stars.[8] As Emily Coates illustrates in vivid detail, dance can be conceptualized as "a mutual partner to scientific disciplines such as meteorology and astronomy," insofar as both dance and science accumulate knowledge through embodied practices of observation. Broadly speaking, science represents knowledge through data and analysis, whereas dance represents knowledge through movement and mimesis. By juxtaposing the history of dance and the history of science, Coates suggests that the metaphorical comparison of dance and the cosmos has material implications for the scientific observation of the sky. In other words, we come to understand the movement of the stars and planets at least in part through the movement of our own bodies, through our physical capacity to observe, represent, and imitate the order of the cosmos.

Coates not only explores the imbrication of dance and science; she also challenges "the misguided dichotomy between observer and observed." This vein of her argument troubles the distinction with which I begin this essay—namely, the distinction between metaphorical and material conceptions of the relationship between dance and the stars. As linguists George Lakoff and Mark Johnson argue, "human *thought processes* are largely metaphorical," and "the human conceptual system is metaphorically structured and defined."[9] In this light, Plato's metaphorical vision of the choral cosmos is not so different from the material relationship to the movement of the Pleiades in the storytelling traditions of the Haudenosaunee and the Barasana. For Plato, as for some indigenous peoples, comprehension of the stars flows through both language and the body, blending metaphorical and material ways of knowing. To call dance a condition of possibility for understanding the cosmos would be to overstate the case. But it would be impossible to query the cosmos without putting our own bodies into motion.

Notes

1 Elyot, *The Boke named The Governour,* quoted in Major, "Moralization of the Dance," 33.
2 Mann, *Iroquoian Women,* 105–106.
3 Aveni, *Star Stories,* 66.
4 Plato, *Timaeus,* 40c–d.
5 Miller, *Measures of Wisdom,* 15.
6 Cowart, *Triumph of Pleasure,* 19.
7 Brooks, *Art of Dancing,* 30.
8 King, *History of the Telescope,* 30–32.
9 Lakoff and Johnson, *Metaphors We Live By,* 6 (emphasis in original).

Bibliography

Aveni, Anthony. *Star Stories: Constellations and People.* New Haven, CT: Yale University Press, 2019.

Brooks, Lynn Matluck. *The Art of Dancing in Seventeenth-Century Spain: Juan de Esquivel Navarro and His World.* Lewisburg, PA: Bucknell University Press, 2003.

Cowart, Georgia. *The Triumph of Pleasure: Louis XIV and the Politics of Spectacle.* Chicago: University of Chicago Press, 2008.

King, Henry C. *The History of the Telescope.* Mineola, NY: Dover Publications, [1955] 2003.

Lakoff, George, and Mark Johnson. *Metaphors We Live By.* Chicago: University of Chicago Press, [1980] 2003.

Major, John M. "The Moralization of the Dance in Elyot's *Governour.*" *Studies in the Renaissance* 5 (1958): 27–36.

Mann, Barbara Alice. *Iroquoian Women: The Gantowisas.* New York: Peter Lang, 2004.

Miller, James L. *Measures of Wisdom: The Cosmic Dance in Classical and Christian Antiquity.* Toronto: University of Toronto Press, 1986.

Plato. *Timaeus and Critias.* Translated and annotated by Desmond Lee and T. K. Johansen. London: Penguin Books, 2008.

CONTRIBUTORS

Lynn Matluck Brooks is Arthur and Katherine Shadek Humanities Professor Emerita at Franklin & Marshall College, where she founded the Dance Program in 1984. At F&M, she was awarded the Bradley R. Dewey Award for Outstanding Scholarship and the Christian and Mary Lindback Award for Teaching. Brooks holds degrees from the University of Wisconsin and Temple University (master's and doctorate) and is a Certified Movement Analyst. Her dance history research earned grants from the Fulbright/Hayes Commission, the Pennsylvania Council on the Arts, the National Endowment for the Humanities, and the Andrew W. Mellon Foundation. Brooks has authored books and scholarly articles on dance history and movement analysis and has served as performance reviewer for *Dance Magazine*, editor of *Dance Research Journal* and *Dance Chronicle*, and writer and editor-in-chief for thINKingDance in Philadelphia. Her current research focuses on dance discourses in antebellum Philadelphia/US and on interconnections of dance, science, and cultural discourse.

Pallabi Chakravorty is Stephen Lang Professor of Performing Arts in the Dance Department at Swarthmore College. She is a visual anthropologist, a Kathak dancer, and a choreographer. Her interdisciplinary research on embodied practices includes long-term and multisited ethnographic engagements in India. She is the artistic director and founder of Courtyard Dancers, a nonprofit based in Philadelphia and Kolkata. Her recent publications are *This is How We Dance Now: Performance in the Age of Bollywood and Reality Shows*, and, as coeditor, *Dance Matters Too: Markets, Memories, Identities.* She is currently writing a book titled *Critical Postures: Towards a New Somatics of Yoga, Performance, and Healing.*

Elizabeth Claire is associate professor of history (CNRS, CRH-EHESS) in Paris, France, where she directs the Cultural History of Dance Seminar (https://ahc-danse.hypotheses.org/), and publishes on dance, performance, medicine, imagination, and gender. She edited *Clio: Women, Gender, History: Dancing; Chorégraphier l'égalité culturelle. Vers une histoire décentrée de la danse*; and, with Alessandro Arcangeli, *Ludica: Performance, Politics and Play*. Current research projects include a monograph, *Imagination Embodied. Dancing Eloquence, Enthusiasm & Contagion (1754–1815)*, on philosophies of imagination and dance in

Enlightenment Europe; a bilingual book series *Techniques, Imagination, and the Body*; an interdisciplinary project, "DisOrienting Bodies," with Mariem Guellouz (Tunisia/France) and Felicia McCarren (USA/Morocco); and collaborations with the *Dance in African History* network organized by Cécile Bushidi, Zoe Groves, and 'Funmi Adewole. Claire performed professionally (1994–2015) with PearsonWidrig DanceTheater, David Dorfman Dance, Jess Curtis, Dogtroep (Netherlands), Yvonne Rainer, Nancy Stark Smith, Au Cul du Loup (FR) and Richard Schechner.

Emily Coates has performed internationally with New York City Ballet, Mikhail Baryshnikov, Twyla Tharp, and Yvonne Rainer, and developed a body of her own work at the intersection of art and science. Her choreographic projects have been presented by the Baryshnikov Arts Center, Guggenheim Works & Process, Quick Center for the Arts, Hopkins Center for the Arts, Danspace Project (NYT Critic's Pick 2017, Fall Dance to Watch 2018), Performa (NYT Best Dance 2019, with Yvonne Rainer), and in *Hard Return* at Neuberger Museum. She is professor in the practice and director of dance studies at Yale University, where she created the dance studies curriculum. She coauthored *Physics and Dance* with physicist Sarah Demers and coedited *Remembering a Dance: Parts of Some Sextets, 1965/2019* with Yvonne Rainer. She holds a BA in English and a PhD in American studies from Yale.

Susan C. Cook retired from the faculty of the University of Wisconsin-Madison in 2024 where she had taught as professor of musicology since 1991. She served in key administrative positions, including a decade-long term as director of the Mead Witter School of Music. She has also taught as a Fulbright Senior Distinguished Professor in the Netherlands. Her research continues to focus on American musical practices and social dance, demonstrating her commitment to feminist methodologies and interdisciplinary cultural criticism. She is coeditor of the award-winning *Cecilia Reclaimed: Feminist Perspectives on Gender and Music* and, with Sherril Dodds, of *Bodies of Sound: Studies Across Popular Music and Dance.* She has published essays in *The Cambridge History of Twentieth-Century Music,* the *Garland Encyclopedia of World Music,* and *The Arts of the Prima Donna.* Her essay "Watching Our Step: Embodying Research, Telling Stories," on the gendered and racialized meanings of ragtime social dance, won the Lippincott Prize from the Society for Dance History Scholars. She is currently working on a monograph, "Watching Your Step: Ragtime Culture 1896–1916," exploring local and transatlantic practices of social dance and popular music.

Jane Desmond is professor of anthropology, gender/women's studies, and dance at the University of Illinois at Urbana-Champaign, where she serves as executive director of the International Forum for U.S. Studies. Formerly a professional modern dancer and choreographer, she holds a PhD in American studies from Yale University and has served on the faculties of Cornell, Duke, the University of Iowa, and Etvos Lorand University Budapest, Hungary. Her scholarship addresses issues of embodiment and social identity, grounded in dance studies, tourism, cultural studies, performance, visual studies, and animal studies. Her solo-authored books include *Staging Tourism: Bodies on Display from Waikiki to Sea World* and *Displaying Death and Animating Life: Human-Animal Relations in Art, Science, and Everyday Life,* as well as two edited volumes in Dance Studies, Meaning in Motion: *New Cultural Studies of Dance* and *Dancing Desires: Choreographing Sexuality on and off the Stage.* Desmond was a 2021 Fulbright Professor in Germany, a 2023 British Academy Research grantee, and a 2023–2024 fellow in bioethics at the Harvard Medical School.

Christian DuComb is associate dean of the faculty for faculty recruitment and development and associate professor of theater at Colgate University. He has previously taught at Haverford College and Brown University, where he received his PhD in theater and performance studies in 2012. His first book, *Haunted City: Three Centuries of Racial Impersonation in Philadelphia,* traces the deep roots of Philadelphia's annual Mummers Parade through the city's history of blackface minstrelsy and other forms of racial impersonation. His essays and reviews on a variety of topics have appeared in *Theater* magazine, *Modern Drama, Performance Research, Theatre Journal, TDR: The Drama Review,* and the *Washington Post,* as well as several edited collections.

Chantal Frankenbach is professor of historical musicology in the School of Music at California State University, Sacramento. Her research on music and dance in the context of German nationalism appears in the *Journal of Musicology,* the *Journal of the Society for American Music, Dance Chronicle,* and various book chapters. Her current book project is a cultural history of modern dance pioneer Isadora Duncan's early career in Germany, examining public and critical reaction to Duncan's dance reforms within the complex interaction of politics, art, and public persuasion in prewar Germany.

Sariel Golomb is lecturer in the Princeton Writing Program. She holds a BA in dance and English from Columbia University and a PhD in theater and performance studies from Stanford University. Her research explores the body in twentieth- and twenty-first-century theatrical dance alongside race/coloniality, medicine, biopower, and the politics of representation. Her writing appears in *Theatre Journal, TDR: The Drama Review, Dance Research Journal, The Brooklyn Rail,* and *ODC.Dance.Stories.* She is the recipient of a 2023–2024 Mellon Dissertation Fellowship, two Carl Weber Memorial Fellowships (2021–2023), and a 2022 Selma Jeanne Cohen Award for excellence in graduate dance scholarship.

Kélina Gotman is professor of performance and the humanities in the English Department at King's College London. She is author of, among others, *Choreomania: Dance and Disorder* and *Essays on Theatre and Change: Towards a Poetics Of;* editor of *Theories of Performance: Critical and Primary Sources;* and coeditor of *Foucault's Theatres* and *Performance and Translation in a Global Age.* She writes widely on dance, the history and philosophy of ideas, science, medicine, disciplines, and institutions. She has served as guest or visiting professor at the University of Toulouse, the Goethe University in Frankfurt, Cornell University, the School of the Art Institute of Chicago, the New School, and Bard College; she was also a Fellow at New York Academy of Medicine. In 2022–2023 she was awarded a Leverhulme Research Fellowship for her book project on early twentieth-century work practices. She trained professionally in theatre and dance and has collaborated internationally across the arts and museum sectors.

Garth Grimball is a writer based in Oakland, CA. He is the editor of *ODC.dance.stories.*

Steven Ha is assistant professor in the School of Dance at the University of Oklahoma. His research on twentieth-century ballet has been published with the Deutsche Shakespeare-Gesellschaft (German Shakespeare Society) and in *Dance Chronicle.*

Andrea Harris is the author of *Making Ballet American: Modernism Before and Beyond Balanchine,* named a 2018 CHOICE Outstanding Academic Title. She is also the editor of *Before, Between, Beyond: Three Decades of Dance Writing,* the last collection of the writings of dance historian Sally Banes. Harris's essays on American dance history appear in *Dance Research Journal, Dance Chronicle, Discourses in Dance,* the *Journal of American Drama and Theatre,* and *Performing Arts Resources,* as well as chapters in edited volumes. Her current book project explores the influence of nineteenth-century physiology and physiological psychology on the therapeutic impetus in early modern dance. Harris is a professor in the Dance Department at the University of Wisconsin-Madison where she is the Buff Brennan Faculty Fellow in Dance and a senior fellow at the UW-Madison Institute for Research in the Humanities.

Claudia Jeschke is professor emerita at Salzburg University, dance historian, reconstructor, choreographer, and curator. Along with her studies in Munich and a doctoral dissertation on the history of dance notation systems, Jeschke trained professionally in various dance forms. Her scholarly and practical expertise allows her to approach dance heritages *in actu* on stage (e.g., the restoring of Nijinsky's *Afternoon of a Faune* with Ann Hutchinson Guest), to curate exhibitions, to create television programs, and to teach and write in academic contexts. Her international teaching career has included positions at the universities of Munich, Leipzig, Cologne, Salzburg, and Linz, among others worldwide. Her extensive publication corpus focuses on historical and theoretical dance issues; on movement analyses, notations, and scores; and on discursive and praxeological transfers between these fields of research (recently, the construction of "otherness" in nineteenth-century dance and editorial work on Nijinsky's Notation Notebooks).

Whitney Laemmli is a historian of science and technology and assistant professor in the Department of Social Science and Cultural Studies at the Pratt Institute. Her research focuses on the scientific and technological construction of the human body in the modern United States and Europe, and her work has appeared in journals including *Technology and Culture, History of the Human Sciences, Osiris, Grey Room,* and *Information and Culture.* She is currently finishing a book project, *Measured Movements,* that explores how bodily movement became a central object of scientific, political, and popular concern over the course of the twentieth century. Laemmli received her PhD from the University of Pennsylvania, and prior to arriving at Pratt, was a member of the Department of History at Carnegie Mellon University and a fellow at the Columbia University Society of Fellows in the Humanities. Her research has been supported by the SSRC, the ACLS/Mellon Foundation, the Max Planck Institute for the History of Science, the NYU Center for Ballet and the Arts, and the Institute for Advanced Study. Laemmli received the 2018 Abbott Payson Usher Prize from the Society for the History of Technology.

Tiziana Leucci, dance historian and anthropologist, is associate professor of research and senior fellow at the Centre d'Etudes Sud-Asiatiques et Himalayennes of the French National Center for Scientific Research (CNRS, Paris). Leucci's master's degree (Bologna University, Italy) and PhD thesis in social anthropology (École des Hautes Études en Sciences Sociales, Paris) dealt with South Indian dancers' and courtesans' culture. After studying ballet and modern dance at the National Academy of Dance in Rome, she learned Bharatanatyam and Odissi dances first in Italy with Kama Dev and for twelve years in India at Kalakshetra and with hereditary artists V. S. Muthuswamy Pillai, M. Selvam Pillai, Kadur Venkatalakshamma, and Kelucharan Mohapatra. She authored several books and articles on the history and representation of the Indian dancer (*bayadère*) in European travelers' accounts, literature, and theatrical works from the thirteenth century onward, including those relevant to Marius Petipa's *La Bayadère* in the opera houses of Milan, Rome, Paris, and St. Petersburg. She also researches the history of circulations and interactions among Indian, European, and American artists. Leucci teaches Bharatanatyam dance at the Conservatoire de musique et danse 'Gabriel Fauré,' Les Lilas, France.

Dick McCaw has edited and introduced three books: *With an Eye for Movement*, on Warren Lamb's development of Rudolph Laban's movement theories, *The Laban Sourcebook*, and *The Art of Movement: Rudolf Laban's Unpublished Writings*. He has written *Bakhtin and Theatre, The Actor's Body—A Guide*, and *Rethinking the Actor's Body—Dialogues with Neuroscience*. He has authored articles and chapters on actor training, movement training, and neuroscience and performer training. McCaw qualified as a Feldenkrais practitioner in 2007 and became an instructor of Wu Family Tai Chi Chuan in 2016.

Johanna Pitetti-Heil is senior lecturer (Akademische Rätin) for gender and diversity studies in the English Department I of the University of Cologne, Germany. She is coeditor of *gender forum* and a board member of SLSAeu. Her current book project, *Becoming-Body: Practices of Freedom and Technologies of the Self in American Modern Dance*, explores American modern dance techniques in the context of the history of materialist philosophy and understandings of subjectivity. She is author of *Walking the Möbius Strip: An Inquiry into Knowing in Richard Powers's Fiction*. Her work in American literary and cultural studies and dance studies has appeared in *Dance Chronicle, Hypatia, Amerikastudien /American Studies, WiN*, the forum of *J19*, and in edited collections.

Janice Ross is professor emerita, Department of Theater and Performance Studies, Stanford University. She has a BA from UC Berkeley and MA and PhD degrees from Stanford. Her books include *Like a Bomb Going Off: Leonid Yakobson and Ballet as Resistance in Soviet Russia; Anna Halprin: Experience as Dance*, winner of a de la Torre Bueno Award 2008 Special Citation; *San Francisco Ballet at 75;* and *Moving Lessons: The Beginning of Dance in American Education*. Ross is coeditor, with Susan Manning and Rebecca Schneider, of *Futures of Dance Studies*. Her newest book, *The Choreography of Environments: How the Anna and Lawrence Halprin Home Transformed Contemporary Dance and Urban Design*, is forthcoming in 2025. Her awards include Guggenheim and Fulbright Fellowships, two Stanford Humanities Center Fellowships, a 2022 Bogliasco Foundation Fellowship, an NYU Center for Ballet and the Arts 2018–2019 Fellowship, Dean's Award for Distinguished Teaching at Stanford (2021–2022), and Honorary Fellow of the Jerusalem Academy of Music and Dance, Israel (2022).

Olivia Sabee is associate professor of dance and a member of committees on comparative literature and interpretation theory at Swarthmore College. A scholar of French and Italian dance and literature, she is particularly interested in how the publishing industry shaped the dissemination and reception of early modern dance texts, and subsequently of early modern dance theory, the subject of her book *Theories of Ballet in the Age of the Encyclopédie.* She is currently working on two projects: one focused on mannerism, imitation and copying, and the belle danse; and a second, funded by a Mellon New Directions Fellowship, on ballet between France, Saint-Domingue, and United States in the last decade of the eighteenth century.

Alexander H. Schwan is a dance scholar and theologian with a research focus on spirituality, religion, and ethics. He is the author of the book *Schrift im Raum. Korrelationen von Tanzen und Schreiben bei Jan Fabre, Trisha Brown und William Forsythe.* His current book project, *Theologies of Modern Dance,* researches theological implications in the works of modernist choreographers in Europe, Israel, and the US. He has been a visiting lecturer at University of California-Santa Barbara and has held visiting fellowships at UC Berkeley, Princeton University, and Harvard. His article "Queering Jewish Dance: Baruch Agadati" (*Dance Research Journal* 2022) was awarded an Honorable Mention for the Dance Studies Association's Gertrude Lippincott Award for the best English-language article in dance studies.

Carrie Streeter is lecturer of US history at Appalachian State University. A specialist in the cultural history of gender, race, and health, her current research examines the role of elocution and dance curricula in redefining citizenship during the Reconstruction Era.

INDEX

Page numbers in *italics* refer to illustrations.

www.ingramcontent.com/pod-product-compliance
Lightning Source LLC
LaVergne TN
LVHW100922110826
845155LV00036B/45
* 9 7 8 0 8 1 3 0 8 1 8 1 6 *